Springer Water

The book series Springer Water comprises a broad portfolio of multi- and interdisciplinary scientific books, aiming at researchers, students, and everyone interested in water-related science. The series includes peer-reviewed monographs, edited volumes, textbooks, and conference proceedings. Its volumes combine all kinds of water-related research areas, such as: the movement, distribution and quality of freshwater; water resources; the quality and pollution of water and its influence on health; the water industry including drinking water, wastewater, and desalination services and technologies; water history; as well as water management and the governmental, political, developmental, and ethical aspects of water.

More information about this series at http://www.springer.com/series/13419

Lu Chen · Shenglian Guo

Copulas and Its Application in Hydrology and Water Resources

Springer

Lu Chen
School of Hydropower & Information Engineering
Huazhong University of Science and Technology
Wuhan, Hubei
China

Shenglian Guo
State Key Laboratory of Water Resources and Hydropower Engineering Science
Wuhan University
Wuhan, Hubei
China

ISSN 2364-6934 ISSN 2364-8198 (electronic)
Springer Water
ISBN 978-981-13-0573-3 ISBN 978-981-13-0574-0 (eBook)
https://doi.org/10.1007/978-981-13-0574-0

Library of Congress Control Number: 2018942911

Printed on acid-free paper

This Springer imprint is published by the registered company Springer Nature Singapore Pte Ltd.
The registered company address is: 152 Beach Road, #21-01/04 Gateway East, Singapore 189721, Singapore

Contents

Chapter 1
Introduction

1.1 Univariate Hydrological Frequency Analysis

Univariate hydrological frequency plays a vital role in estimating the recurrence of floods or rainfall, which is used for designing structures such as dams, bridges, culverts, levees, highways, sewage disposal plants, waterworks and industrial buildings. By using the univariate hydrological frequency analysis method, the probability for a given event can be estimated, and the value of a T-year design rainfall or flood also can be calculated. The main objective of univariate hydrological frequency analysis therefore is to establish a relationship between flood or rainfall magnitude and recurrence interval or return period.

Two kinds of samples are usually utilized for univariate hydrological frequency analysis. One is the annual maximum (AM) sampling method, and the other is peaks-over-threshold (POT) (or partial duration series (PDS)) sampling method. The AM series includes the maximum peaks or volumes for every year in the observational period, and the other includes all and only the peaks for events that exceed a given threshold (Ben-Zvi 2009). The univariate return periods are traditionally estimated by fitting a probability distribution function to the historical observations, such as AM and POT hydrological extreme series (Li and Zheng 2016).

The procedures for hydrological frequency analysis mainly include two steps: selection of an appropriate parent distribution and estimates of the parameters for the selected distribution. The distribution for univariate hydrological analysis has been discussed and investigated by many research. Numerous probability distribution models have been used in hydrological frequency studies, including two-parameter distributions, i.e. Gumbel, Weibull, Gamma and Lognormal (Du et al. 2015; Giraldo and García 2012; Jiang et al. 2015; Villarini et al. 2009), and three-parameter distributions, i.e. general extreme value (GEV) (Cannon 2010; El Adlouni et al. 2007) and Pearson type III (Chen et al. 2010, 2015). The commonly used distributions in hydrology are summarized in Table 1.1. To determine whether

L. Chen and S. Guo, *Copulas and Its Application in Hydrology and Water Resources*, Springer Water, https://doi.org/10.1007/978-981-13-0574-0_1

Table 1.1 Commonly used distributions in hydrology

Distributions	CDF	PDF	Parameters
Normal	$F(x) = \int_{-\infty}^{x} f(x)\mathrm{d}x$	$f(x) = \frac{1}{\sqrt{2\pi\sigma}}\exp(-\frac{(x-\mu)^2}{2\sigma^2}) \quad x \in (-\infty, +\infty)$	σ-scale parameter, > 0; μ-location parameter
Gamma	$F(x) = \int_{0}^{x} f(x)\mathrm{d}x$	$f(x) = \frac{1}{\beta^\alpha \Gamma(\alpha)} x^{\alpha-1}\exp(-\frac{x}{\beta}) \quad x \in [0, +\infty)$	α-shape parameter, > 0; β-scale parameter, > 0
Gumbel	$F(x) = \exp(-\exp(-\frac{x-\mu}{\sigma}))$	$f(x) = \frac{1}{\sigma}\exp(-\frac{x-\mu}{\sigma} - \exp(-\frac{x-\mu}{\sigma}))\, x \in (-\infty, +\infty)$	σ-scale parameter, > 0; μ-location parameter
Pearson type III (P-III)	$F(x) = \int_{\gamma}^{x} f(x)\mathrm{d}x$	$f(x) = \frac{1}{\beta^\alpha \Gamma(\alpha)}(x-\gamma)^{\alpha-1}\exp(-\frac{x-\gamma}{\beta}) x \in [\gamma, +\infty)$	α-shape parameter, > 0; β-scale parameter, > 0 γ-location parameter
Log normal	$F(x) = \int_{\gamma}^{x} f(x)\mathrm{d}x$	$f(x) = \frac{1}{(x-\gamma)\sigma\sqrt{2\pi}}\exp(-\frac{(\ln(x-\gamma)-\mu)^2}{2\sigma^2})$ $x \in (\gamma, +\infty)$	σ-scale parameter, > 0; μ-scale parameter; γ-location parameter
Generalized normal	$F(x) = \Phi(y)$ $y = -\frac{1}{k}\ln(1 - \frac{k(x-\mu)}{\alpha})$	$f(x) = \frac{1}{\sqrt{2\pi}(\alpha - k(x-\mu))}\exp(-\frac{(\ln(1-k(x-\mu)/\alpha))^2}{2k^2})$ $\begin{cases} x \in (-\infty,\ \mu+\alpha/k) & \text{for } k > 0 \\ x \in (\mu+\alpha/k,\ +\infty) & \text{for k} < 0 \end{cases}$	α-scale parameter, > 0; k-shape parameter, $\neq 0$ μ-location parameter Φ-the standard normal CDF
Weibull	$F(x) = 1 - \exp(-(\frac{x-\gamma}{\beta})^\alpha)$	$f(x) = \frac{\alpha}{\beta}(\frac{x-\gamma}{\beta})^{\alpha-1}\exp(-(\frac{x-\gamma}{\beta})^\alpha)\, x \in [\gamma, +\infty)$	α-shape parameter, > 0; β-scale parameter, > 0

(continued)

Table 1.1 (continued)

Distributions	CDF	PDF	Parameters
Generalized Pareto (GP)	$F(x) = 1 - (1 + k\frac{x-\mu}{\sigma})^{-1/k}$	$f(x) = \frac{1}{\sigma}(1 + k\frac{x-\mu}{\sigma})^{-1-1/k}$ $\begin{cases} x \in [\mu, +\infty) & \text{for } k > 0 \\ x \in [\mu, \mu - \sigma/k] & \text{for } k < 0 \end{cases}$	k-shape parameter, $\neq 0$; σ-scale parameter, > 0; μ-location parameter
Exponential (Ex)	$F(x) = 1 - \exp(-\frac{(x-\gamma)}{\alpha})$	$f(x) = \frac{1}{\alpha}\exp(-\frac{(x-\gamma)}{\alpha})\ x \in [\gamma, +\infty)$	α-scale parameter, > 0; γ-location parameter
Generalized extreme value (GEV)	$F(x) = \exp(-(1 + k\frac{x-\mu}{\sigma})^{-1/k})$	$f(x) = \frac{1}{\sigma}\exp(-(1 + k\frac{x-\mu}{\sigma})^{-1/k})(1 + k\frac{x-\mu}{\sigma})^{-1-1/k}$ $\begin{cases} x \in (\mu - \sigma/k, +\infty) & \text{for } k > 0 \\ x \in (-\infty, \mu - \sigma/k) & \text{for } k < 0 \end{cases}$	k-shape parameter, $\neq 0$; σ-scale parameter, > 0; μ-location parameter
Generalized logistic (GL)	$F(x) = \frac{1}{1-(1+k\frac{x-\mu}{\sigma})^{-1/k}}$	$f(x) = \frac{(1 + k\frac{x-\mu}{\sigma})^{-1-1/k}}{\sigma(1 + (1 + k\frac{x-\mu}{\sigma})^{-1/k})^2}$ $\begin{cases} x \in (\mu - \sigma/k, +\infty) & \text{for } k > 0 \\ x \in (-\infty, \mu - \sigma/k) & \text{for } k < 0 \end{cases}$	k-shape parameter, $\neq 0$; σ-scale parameter, > 0; μ-location parameter
Wakeby	$F(x) = \varsigma + \frac{\alpha}{\beta}(1 - (1-x)^{\beta}) - \frac{\gamma}{\delta}(1 - (1-x)^{-\delta})$	$f(x) = \frac{(1 - F(x))^{\delta+1}}{\alpha((1 - F(x))^{\beta+\delta}) + \gamma}$ $\begin{cases} x \in [\varsigma, +\infty) & \text{if } \delta \geq 0 \text{ and } \gamma > 0 \\ x \in [\varsigma, \varsigma + \frac{\alpha}{\beta} - \frac{\gamma}{\delta}] & \text{if } \delta < 0 \text{ or } \gamma = 0 \end{cases}$	β, γ and δ-shape parameters; ς, α-location parameters
Kappa	$F(x) = \{1 - h[1 - \frac{k(x-\varsigma)}{\alpha}]^{1/k}\}^{1/h}$	$f(x) = \frac{1}{\alpha}[1 - \frac{k(x-\varsigma)}{\alpha}]^{1/k-1}[F(x)]^{1-h}$	α-scale parameter; k, h-shape parameter; ς-location parameter

the distribution model can fit the data series properly, certain goodness-of-fit test, such as Kolmogorov-Smirnov, Anderson-Darling, and Chi-squared tests, can be used.

The parameter estimation methods, which are widely used for univariate hydrological frequency analysis, are the method of moments (MOM), the maximum likelihood (ML) method, and the L-moments method. The limitation of the MOM is that the moments of data series are equally influenced by small values, and the higher moments (e.g., coefficient of variation and skewness) are much affected by extremes in the data series (Haddad and Rahman 2011). An alternative method to MOM is ML for estimating the parameters of a distribution. Haddad and Rahman (2011) indicated that the ML is a robust method in most cases and will provide estimators with good statistical properties. Another method, L-moments are highly recommended by many researches. This method is less affected by extremes in the data series (Hosking 1990). Also, the L-moments provide more weights to the larger values in the hydrological series and hence are expected to provide better fits to the upper tail of the distribution (Wang 1997).

The method mentioned above mainly focus on the univariate hydrological analysis. A complex phenomenon is often characterized by multiple aspects. Several hydrological phenomena are described by two or more correlated characteristics. For example, for a flood event, which can be characterized by flood peak, flood magnitude and duration, a univariate probability distribution analysis is apparently not enough, since these three random variables are not mutually independent due to the multivariate nature of the phenomenon. For a system with two or more variables, the return period is not equal to the forcing return period of a variate (Hawkes et al. 2002). In the case of flood frequency estimation, merely analyzing the flood peak or flood volume frequency will lead to an underestimation or overestimation of risk (De Michele et al.2005; Yue and Rasmussen 2002). Therefore, a multivariate statistical analysis is required for a more complicated hydrological phenomenon with more variables (Grimaldi and Serinaldi 2006a, b).

1.2 Multivariate Hydrological Frequency Analysis

As hydrological phenomena are usually described by two or more correlated variables, a multivariate statistical analysis and dependence analysis are required. The most significant issue of multivariate probability analysis is the construction of dependence structure for the involved correlated random variables (Li and Zheng 2016). Multivariate distribution functions have been widely used in the literature for modeling two or more dependent hydrological variables and their dependence structure (Salvadori and De Michele 2007). The multivariate hydrological analysis mainly includes the following three elements: (1) showing the importance and explaining the usefulness of the multivariate framework, (2) fitting the appropriate multivariate distribution in order to model hydrological phenomenon, and estimating the corresponding parameters, and (3) studying multivariate return periods or other related hydrological analysis and simulation (Chebana and Ouarda 2011).

In the past years, some multivariate approaches have been introduced in hydrological and environmental applications. The most widely used joint cumulative distribution function (CDF) is the Gaussian one, but it has the limitation that the marginal distributions must be normal. Then bivariate distributions with non-normal margins have been proposed, such as bivariate exponential (Favre et al. 2002), bivariate gamma (Yue et al. 2001), and bivariate extreme value distributions (Adamson et al. 1999). Favre et al. (2004) summarized the drawbacks of these types of distributions are that (1) the same family is needed for each marginal distribution, (2) extensions to more than the bivariate case are not clear, and (3) parameters of the marginal distributions are also used to model the dependence between the random variables. To overcome these shortcomings, copula functions that represent the most recent and promising mathematical tool for investigating multivariate problems have been applied in the hydrological analysis (Xiao et al. 2008). The advantages in using copulas to model joint distributions are manifold: (1) flexibility in choosing arbitrary margins and structure of dependence, (2) extensions to more than two variables, and (3) split of marginal and dependence structure analysis (Salvadori et al. 2007; Serinaldi et al. 2009).

In the past decade, copulas have been used for multivariate hydrological analyses. Favre et al. (2004) used 2-copula to describe the dependence between flow peak and volume. Shiau et al. (2006) analyzed the bivariate frequency of flood peak and volume. Zhang and Singh (2006) exploited Archimedean copulas to build bivariate distributions of flood peak and volume, flood peak and duration, and flood volume and duration. Grimaldi and Serinaldi (2006a, b) built a trivariate joint distribution of flood event variables using the fully nested or asymmetric Archimedean copula functions and performed extensive simulations to highlight differences with the well-known symmetric Archimedean copulas. Salvatore and De Michele (2007) presented some advances in hydrological modeling that exploit copulas, such as the calculation of conditional probabilities and return periods of bivariate events. Zhang and Singh (2007a) used the Gumbel–Hougaard copula to derive trivariate distributions of flood peak, volume and duration. Kao and Govindaraju (2008) examined a non-Archimedean copula from the Plackett family and applied it to the study of the temporal distribution of extreme rainfall events. Serinaldi et al. (2009) applied copulas to the probabilistic analysis of drought characteristics. Until now, the utilization of copulas in hydrology and water resources can be summarized as: rainfall frequency analysis (Michele and Salvadori 2003; Grimaldi and Serinaldi 2006a; Kao and Govindaraju 2007; Zhang and Singh 2007a; Kuhn et al. 2007; and Keef et al. 2009), flood frequency analysis (Favre et al. 2004; Shiau et al. 2006; Zhang and Singh 2006, 2007b; Renard and Lang 2007; Xiao et al. 2009), drought frequency analysis (Shiau 2006; Kao and Govindaraju 2010; Song and Singh 2010), sea storm analysis (Michele et al. 2007), streamflow simulation (Chen et al. 2015), and some other theoretical analyses of multivariate extreme problems (Salvadori et al. 2007; Salvadori and Michele 2010). Therefore, copula function has been proved to be very useful and effective tools for multivariate hydrological analysis and simulation.

1.3 Dependence Analysis

There are several well-known methods that describe the stochastic dependence. One is a linear relation which mainly exists in regression models measured by covariance and correlation coefficient, and it is based on the multivariate normal distribution (Xu 2005; Zhao and Lin 2011). Calsaverini and Vicente (2009) clarified the danger of using linear correlation as a measure of dependence for, e.g., portfolio optimization or time series analysis, as this measure is bound to underestimate the dependence that would be better captured by easily estimated marginal invariant measures. Zhao and Lin (2011) stated that the method of linear relationship ignores some fluctuations, such as high peak and fat tail relative to kurtosis and skewness, which have frequently been reported in data analyses. The drawbacks of the linear correlation method are summarized as: (1) it only applies to a linear correlation, (2) it tends to focus on the degree of dependence, and ignore the structure of the dependence, and (3) it is a dimensionless quantity, and is difficult to compare with three or more sets of variables (Zhao and Lin 2011).

Two important measures of dependence (concordance) known as Kendall's tau and Spearman's rho, provide perhaps the best alternatives to the linear correlation coefficient as a measure of dependence for non-Gaussian distributions, for which the linear correlation coefficient is inappropriate and often misleading. The Spearman rank correlation coefficient is its analog when the data is regarding ranks. One can therefore also call it correlation coefficient between the ranks. Kendall's tau is equivalent to Spearman's rho, with regards to the underlying assumptions, but Spearman's rho and Kendall's tau are not identical in magnitude, since their underlying logic and computational formulae are quite different. The main advantage of using Kendall's tau over Spearman's rho is that one can interpret its value as a direct measure of the probabilities of observing concordant and discordant pairs. The disadvantage of the rank-based correlation coefficient is that there is a loss of information when the data are converted to ranks; if the data are normally distributed, it is less powerful than the Pearson correlation coefficient (Gauthier 2001). Furthermore, they cannot be used to detect the dependence when more than two variables are involved.

Kendall's tau and Spearman's rho correlation coefficients have a relationship with the copula function, and are usually used to estimate the parameters of bivariate copulas. Thus, the copula function also can measure the dependence relationship, which has been applied to investigate nonlinear dependence and has received a lot of attention in recent years (Zhao and Lin 2011). The copula function is capable of exhibiting the type of the dependence between two or more random variables, and has recently emerged as a practical and efficient method for modeling the general dependence in multivariate data (e.g., Joe 1997; Nelsen 2006). The advantages in using copulas to model joint distributions have been described above.

Another method is based on entropy theory. A comprehensive review of the use of information theory in hydrology and water resources can be found in Singh (1997, 2011). In the entropy theory, mutual information (MI) has been successfully employed as a nonlinear measure of inference among variables by many researchers (e.g., Khan et al. 2006; Molini et al. 2006; Ng et al. 2007; Hejazi et al. 2008). Mutual information, defined as the difference between marginal and conditional entropy, is a measure of the amount of information that one random variable contains or explains about another random variable. It can be used to indicate the dependence or independence between variables. If the two variables are independent, the mutual information between them is zero. If the two are strongly dependent, e.g., one is a function of another; the mutual information between them is large (Li 1990). The use of mutual information has become popular in several fields of science to measure the dependence between variables (Alfonso et al. 2010). For example, using the MI method, Harmancioglu and Yevjevich (1987) analyzed three types of information transferring among river points. The mutual information was also used for the network design (Alfonso et al. 2010). Some of the advantages of this method have been reported widely (e.g., Li 1990; Singh 2000; Steuer et al. 2002). The advantages of MI are summarized as follows: (1) it is a non-linear measure of statistical dependence based on information theory (Steuer 2006), (2) it is a non-parametric method and makes no assumptions about the functional form (Gaussian or non-Gaussian) of the statistical distribution that produce the data, and (3) it can be extended to higher dimensions.

Therefore, copulas and entropy theory have been taken as effective tools for measuring the non-linear dependences of multi-variables in hydrology and water resources. This book will propose a new method based on these two theories for the multi-variate dependence analysis.

1.4 Scope and Organization of the Book

This book presents an overview of the copula theory and its applications in hydrology and water resources. The specific applications include the studies of flood frequency analysis, drought frequency analysis, flood coincidence risk analysis, stochastic simulation using copulas and dependence analysis. In this book, we also extend the traditional bivariate copula model to a trivariate or multivariate model. This book provides valuable knowledge, useful methods and practical applications with respect to multivariate hydrological analysis using copulas. Researchers, scientists and engineers in the fields of hydrology and water resources can benefit from this book. This book is also useful for graduate or doctoral students with basic knowledge of copula functions who want to learn about the latest research developments in this field.

Chapter 1 reviews the univariate hydrological frequency analysis, multivariate hydrological frequency analysis and dependence analysis. Copula and copula entropy methods are introduced in hydrology and water resources.

Chapter 2 gives the detailed information of copula theory. The contents involve the definition of copulas, introduction of two kinds of copulas widely used in hydrology, parameter estimation methods, goodness of fit methods, and copula entropy theory.

Chapter 3 proposes an annual maximum flood frequency analysis method considering flood occurrence dates and magnitudes as well as a bivariate flood frequency analysis method with historical information considering flood peaks and volumes.

Chapter 4 proposes a new seasonal design flood method that considers the flood occurrence dates and magnitudes of the peaks (runoff) based on copula functions.

Chapter 5 presents a method for estimating the return periods of drought events based on copulas, in which four drought characteristics, namely drought duration, severity, time interval and the minimum SPI values, are considered.

Chapter 6 analyzes the coincidence of flood flows of the mainstream and its tributaries by considering flood magnitudes and time (dates) of occurrence based on the four-dimensional copula functions.

Chapter 7 introduces a new copula-based method for generating long-term multisite monthly and daily streamflow data.

Chapter 8 introduces a hydrologic uncertainty processor (HUP) based on a copula function, in which a Bayesian copula processor associated with the Bayesian model averaging (CBP-BMA) method is presented with ensemble lumped hydrological models.

Chapter 9 proposes a copula-based uncertainty evolution (CUE) model to describe the evolution of streamflow forecast uncertainty.

Chapter 10 proposes a new method based on the copula entropy (CE) theory to identify the inputs of ANN-based flood forecasting models.

Chapter 11 measures the total correlation between the mainstream and its upper tributaries by using the copula entropy method.

References

Adamson PT, Metcalfe AV, Parmentier B (1999) Bivariate extreme value distributions: an application of the Gibbs sampler to the analysis of floods. Water Resour Res 35:2825–2832

Alfonso L, Lobbrecht A, Price R (2010) Optimization of water level monitoring network in polder systems using information theory. Water Resour Res 46:W12553

Ben-Zvi A (2009) Rainfall intensity–duration–frequency relationships derived from large partial duration series. J Hydrol 367:104–114

Calsaverini RS, Vicente R (2009) An information-theoretic approach to statistical dependence: copula information. Europ Phys Lett 88(6):3–12

Cannon AJ (2010) A flexible nonlinear modelling framework for nonstationary generalized extreme value analysis in hydroclimatology. Hydrol Process 24:673–685. https://doi.org/10.1002/hyp.7506

Chebana F, Ouarda TBMJ (2011) Multivariate quantiles in hydrological frequency analysis. Environmetrics 22:63–78. https://doi.org/10.1002/env.1027

Chen L, Guo S, Yan B, Liu P, Fang B (2010) A new seasonal design flood method based on bivariate joint distribution of flood magnitude and date of occurrence. Hydrol Sci J 55(8):1264–1280. https://doi.org/10.1080/02626667.2010.520564

Chen L, Singh VP, Guo S, Zhou J (2015) Copula-based method for multisite monthly and daily streamflow simulation. J Hydrol 528:369–384

De Michele C, Salvadori G (2003) A generalized Pareto intensity duration model of storm rainfall exploiting 2-copulas. J Geophys Res 108(D2):1–11

De Michele C, Salvadori G, Canossi M, Petaccia A, Rosso R (2005) Bivariate statistical approach to check adequacy of dam spillway. J Hydrol Eng 10(1):50–55

De Michele C, Salvadori G, Passni G, Vezzoli R (2007) A multivariate model of sea storms using copulas. Coastal Eng 54(10):734–751

Du T, Xiong L, Xu CY, Gippel CJ, Guo S, Liu P (2015) Return period and risk analysis of nonstationary low-flow series under climate change. J Hydrol 527:234–250

El Adlouni S, Ouarda TBMJ, Zhang X, Roy R, Bobée B (2007) Generalized maximum likelihood estimators for the nonstationary generalized extreme value model. Water Resour Res 43: W03410. https://doi.org/10.1029/2005WR004545

Favre AC, Adlouni S, Perreault L, Thiémonge N, Bobée B (2004) Multivariate hydrological frequency analysis using copulas. Water Resour Res 40(1):W01101

Favre AC, Musy A, Morgenthaler S (2002) Two-site modeling of rainfall based on the Neyman-Scott process. Water Resour Res 38(12):1307. https://doi.org/10.1029/2002WR001343

Gauthier TD (2001) Detecting trends using Spearman's rank correlation coefficient. Environmental Forensics 2(4):359–362

Giraldo Osorio JD, García Galiano SG (2012) Non-stationary analysis of dry spells in monsoon season of Senegal River Basin using data from regional climate models (RCMs). J Hydrol 450–451:82–92. https://doi.org/10.1016/j.jhydrol.2012.05.029

Grimaldi S, Serinaldi F (2006a) Asymmetric copula in multivariate flood frequency analysis. Adv Water Resour 29(8):1155–1167

Grimaldi S, Serinaldi F (2006b) Design hyetographs analysis with 3-copula function. Hydrol Sci J 51(2):223–238

Haddad K, Rahman A (2011) Selection of the best fit flood frequency distribution and parameter estimation procedure: a case study for Tasmania in Australia. Stoch Environ Res Risk Assess 25:415

Harmancioglu NB, Yevjevich V (1987) Transfer of hydrologic information among river points. J Hydrol 91:103–111

Hawkes PJ, Gouldby BP, Tawn JA, Owen MW (2002) The joint probability of waves and water levels in coastal engineering design. J Hydraul Res 241–251

Hejazi MI, Cai X, Ruddel Benjamin (2008) The role of hydrologic information to reservoir operations—learning from past releases. Adv Water Resour 31(12):1636–1650

Hosking JRM (1990) L-Moments: analysis and estimation of distributions using linear combinations of order statistics. J R Stat Soc Series B Stat Methodol. 52(1):105–124

Jiang C, Xiong L, Xu CY, Guo S (2015) Bivariate frequency analysis of nonstationary low-flow series based on the time-varying copula. Hydrol Process 29(6):1521–1534. https://doi.org/10.1002/hyp.10288

Joe H (1997) Multivariate models and dependence concepts. Chapman and Hall, London JSTOR, www.jstor.org/stable/2345653

Kao S, Govindaraju RS (2010) A copula-based joint deficit index for droughts. J Hydrol (Amsterdam) 380(1–2):121–134

Kao SC, Govindaraju RS (2007) A bivariate frequency analysis of extreme rainfall with implications for design. J Geophys Res 112(D1):13119

Kao SC, Govindaraju RS (2008) Trivariate statistical analysis of extreme rainfall events via the Plackett family of copulas. Water Resour Res 44:W02415. https://doi.org/10.1029/2007WR006261

Keef C, Svensson C, Tawn JA (2009) Spatial dependence in extreme river flows and precipitation for Great Britain. J Hydrol (Amsterdam) 378(3–4):240–252

Khan S, Ganguly AR, Bandyopadhyay S, Saigal S, Erickson IIIDJ, Protopopescu V, Ostrouchov G (2006) Nonlinear statistics reveals stronger ties between ENSO and the tropical hydrological cycle. Geophys Res Lett 33:L24402. https://doi.org/10.1029/2006-GL027941

Kuhn G, Khan S, Ganguly AR, Branstetter ML (2007) Geospatial temporal dependence among weekly precipitation extremes with applications to observations and climate model simulations in South America. Adv Water Resour 30(12):2401–2423
Li F, Zheng Q (2016) Probabilistic modelling of flood events using the entropy copula. Adv Wat Res 97:233–240, ISSN 0309-1708 https://doi.org/10.1016/j.advwatres.2016.09.016
Li W (1990) Mutual information functions versus correlation functions. J Stat Phys 60:823–837
Molini A, La Barbera P, Lanza LG (2006) Correlation patterns and information flows in rainfall fields. J Hydrol 322(1–4):89–104
Nelsen RB (2006) An introduction to copulas, 2nd edn. Springer-Verlag, New York
Ng WW, Panu US, Lennox WC (2007) Chaos based analytical techniques for daily extreme hydrological observations. J Hydrol 342(1–2):17–41
Renard B, Lang M (2007) Use of a Gaussian copula for multivariate extreme value analysis: some case studies in hydrology. Adv Water Resour 30(4):897–912
Salvadori G, De Michele C (2007) On the use of copulas in hydrology: theory and practice. J Hydrol Eng 12(4):369–380
Salvadori G, De Michele C (2010) Multivariate multiparameter extreme value models and return periods: a copula approach. Water Resour Res 46(10):W10501
Salvadori G, De Michele C, Kottegoda NT, Rosso R (2007) Extremes in nature: an approach using copulas. Springer, New York
Serinaldi F, Bonaccorso B, Cancelliere A, Grimaldi S (2009) Probabilistic characterization of drought properties through copulas. Phys Chem Earth 34(10–12):596–605
Shiau JT (2006) Fitting drought duration and severity with two-dimensional copulas. Water Resour Manage 20(5):795–815
Shiau JT, Wang HY, Tsai CT (2006) Bivariate frequency analysis of floods using copulas. J Am Water Resour Assoc 42(6):1549–1564
Singh VP (1997) The use of entropy in hydrology and water resources. Hydrol Process 11:587–626
Singh VP (2000) The entropy theory as a tool for modeling and decision making in environmental and water resources. J Water Society Am 1:1–11
Singh VP (2011) Hydrologic synthesis using entropy theory: review. J Hydrol Eng 16(5):421–433
Song S, Singh VP (2010) Meta-elliptical copulas for drought frequency analysis of periodic hydrologic data. Stoch Environ Res Risk Assess 24(3):425–444
Steuer R (2006) On the analysis and interpretation of correlations in metabolomic data. Brief Bioinform 7(2):151–158
Steuer R, Kurths J, Daub CO, Weise J, Selbig J (2002) The mutual information: detecting and evaluating dependencies between variables. Bioinformatics 18:231–240
Villarini G, Serinaldi F, Smith JA, Krajewski WF (2009) On the stationarity of annual flood peaks in the continental United States during the 20th century. Water Resour Res 45(8W08417): https://doi.org/10.1029/2008wr007645
Wang QJ (1997) LH moments for statistical analysis of extreme events. Water Resour Res 33(12):2841–2848
Xiao Y, Guo SL, Liu P, Fang B (2008) A new design flood hydrograph method based on bivariate joint distribution. IAHS-AISH Publications 319:75–82
Xiao Y, Guo SL, Liu P, Yan BW, Chen L (2009) Design flood hydrograph based on multi-characteristic synthesis index method. J Hydrol Eng 14(12):1359–1364
Xu Y (2005) Applications of copula-based models in portfolio optimization. Ph.D. Dissertation, University of Miami
Yue S, Ouarda TBMJ, Bobee B (2001) A review of bivariate gamma distribution for hydrological application. J Hydrol 246:1–18
Yue S, Rasmussen P (2002) Bivariate frequency analysis: discussion of some useful concepts in hydrological application. Hydrol Process 16:2881–2898
Zhang L, Singh VP (2006) Bivariate flood frequency analysis using the copula method. J Hydrol Eng 11(2):150–164

Zhang L, Singh VP (2007a) Gumbel Hougaard copula for trivariate rainfall frequency analysis. J Hydrol Eng 12(4):409–419

Zhang L, Singh VP (2007b) Trivariate flood frequency analysis using the Gumbel-Hougaard copula. J Hydrol Eng 12(4):431–439

Zhao N, Lin WT (2011) A copula entropy approach to correlation measurement at the country level. Appl Math Comput 218(2):628–642

Chapter 2
Copula Theory

2.1 Copula Function

The copulas were first introduced by Sklar (1959). Copulas are functions that join or "couple" multivariate distribution functions to their one-dimensional marginal distribution functions (Nelsen 2006). The copula function is capable of exhibiting the structure of dependence between two or more random variables and has recently emerged as a practical and efficient method for modeling the general dependence in multivariate data (e.g., Joe 1997; Nelsen 2006). The advantages of using copulas to model joint distributions are manifold: (1) flexibility in choosing arbitrary marginal and structure of dependence, (2) extension to more than two variables, and (3) separated analysis of marginal distributions and dependence structure (Salvadori et al. 2007; Serinaldi et al. 2009). Hydrological applications of copulas have surged in recent years (e.g. Wang et al. 2010).

2.1.1 Definition

To give a precise definition of copulas, here we restate the Sklar's theorem. If random variables x_1, …, x_n follow an arbitrary marginal distribution function $F_1(x_1)$, …, $F_n(x_n)$, respectively, there then exists a copula, C, that combines these marginal distribution functions to give the joint distribution function, $F(x_1, \ldots, x_n)$ as follows

$$F(x_1, \ldots, x_n) = C\{F_1(x_1), \ldots, F_n(x_n)\} = C(u_1, \ldots, u_n), \quad x_1, \ldots, x_n \in R \tag{2.1}$$

If the marginal distributions $F_i(x_i)$ are continuous, the copula function C is unique. On the contrary, if C is a k-dimensional copula function, F is an

L. Chen and S. Guo, *Copulas and Its Application in Hydrology and Water Resources*, Springer Water, https://doi.org/10.1007/978-981-13-0574-0_2

n-dimensional distribution function and $F_1(x_1)$, …, $F_n(x_n)$ are the respective marginal distributions.

2.1.2 Properties of Copulas

Let $C(u, v)$ be an arbitrary two-dimensional copula function. Then the function C has the following elementary properties (Nelsen 2006).

For every u and v,

$$C(u,0) = C(0,v) = 0 \tag{2.2}$$

and

$$C(u,1) = u \tag{2.3}$$

and

$$C(1,v) = v \tag{2.4}$$

For each u_1 and u_2, v_1 and v_2, if $u_1 \le u_2$ and $v_1 \le v_2$

$$C(u_2,v_2) - C(u_2,v_1) - C(u_1,v_2) + C(u_1,v_1) \ge 0 \tag{2.5}$$

Many families of copulas exist and mainly include the following: (1) meta-elliptical copulas (normal and t), (2) Archimedean copulas (Clayton, Gumbel, Frank, and Ali-Mikhail-Haq), (3) Extreme Value copulas (Gumbel, Husler-Reiss, Galambos, Tawn, and t-EV), and (4) other families (Plackett and Farlie-Gumbel-Morgenstern). Among the various families of copulas, the Archimedean and the meta-elliptical copulas are more popular for hydrologic applications.

2.1.3 Conditional Copulas

Let X and Y be random variables with $U_1 = F_X(x)$ and $U_2 = F_Y(y)$. u_1 and u_2 are specific values. As an example, the conditional distribution function of X given $Y = y$ can be expressed by

$$\begin{aligned} H(X \le x | Y = y) &= C_\theta(u_1 | U_2 = u_2) \\ &= \lim_{\Delta u_2 \to 0} \frac{C_\theta(u_1, u_2 + \Delta u_2) - C_\theta(u_1, u_2)}{\Delta u_2} \\ &= \frac{\partial}{\partial u_2} C_\theta(u_1, u_2) | U_2 = u_2 \end{aligned} \tag{2.6}$$

Similarly, an equivalent formula for the conditional distribution function for Y given $X = x$ can be obtained.

Furthermore, the conditional distribution function of X given $Y \leq y$ can be expressed by

$$H'(X \leq x | Y \leq y) = C_\theta(u_1 | U_2 \leq u_2) = \frac{C_\theta(u_1, u_2)}{u_2} \tag{2.7}$$

Likewise, an equivalent formula for the conditional distribution function for Y given $X \leq x$ can be obtained.

2.2 Archimedean Copulas

Different families of copulas have been proposed and described by Nelsen (2006) and Salvadori et al. (2007). Of all the copula families, the Archimedean family is more desirable for hydrological analyses, because it can be more easily constructed and can be applied whether the correlation among the hydrological variables is positive or negative (Zhang and Singh 2006).

2.2.1 Bivariate Archimedean Copulas

Archimedean copulas are widely used in hydrology, especially the bivariate Archimedean copulas. Previous studies have indicated that copulas perform well for bivariate problems, and in particular, several families of Archimedean copulas, including Gumbel, Frank, and Clayton, have been popular choices for dependence models because of their simplicity and generation properties (Nelson 2006).

2.2.1.1 Gumbel Copula

This family of copulas was defined by Gumbel (1960). It is also a member of the important class of bivariate extreme value copulas (Nelson 1999) and has been widely used in the hydrological analysis of bivariate extreme value. The Gumbel copula C can be written as

$$C(u_1, u_2) = \exp\left\{-[(-\ln u_1)^\theta + (-\ln u_2)^\theta]^{\frac{1}{\theta}}\right\}, \quad \theta \in [1, +\infty) \tag{2.8a}$$

The generator of Gumbel copula is

$$\varphi(t) = (-\ln t)^\theta, \quad \theta \in [1, +\infty) \tag{2.8b}$$

2.2.1.2 Frank Copula

This family was defined by Frank in 1979. A Frank copula is given by

$$C(u_1, u_2) = \frac{1}{\theta}\log(1 + \frac{(e^{\theta u_1} - 1)(e^{\theta u_2} - 1)}{e^{\theta} - 1}), \quad \theta \in (-\infty, +\infty) \tag{2.9a}$$

The generator of Frank copula is

$$\varphi(t) = \ln[\frac{\exp(\theta t) - 1}{\exp(\theta) - 1}], \quad \theta \in (-\infty, +\infty) \tag{2.9b}$$

2.2.1.3 Clayton Copula

A Clayton copula is defined by

$$C(u_1, u_2) = (u_1^{-\theta} + u_2^{-\theta} - 1)^{-\frac{1}{\theta}}, \quad \theta \in (0, +\infty) \tag{2.10a}$$

The generator of Clayton copula is

$$\varphi(t) = t^{-\theta} - 1, \quad \theta \in (0, +\infty) \tag{2.10b}$$

2.2.1.4 Ali-Milhail-Haq Copula

Ali et al. (1978) defined Ali-Milhail-Haq copula as

$$C(u_1, u_2) = \frac{u_1 u_2}{1 - \theta(1 - u_1)(1 - u_2)}, \quad \theta \in [-1, 1] \tag{2.11a}$$

The generator of Ali-Milhail-Haq copula is

$$\varphi(t) = \ln\frac{1 - \theta(1 - t)}{t}, \quad \theta \in [-1, 1] \tag{2.11b}$$

2.2.1.5 Joe Copula

This family was first discussed by Joe (1993). The Joe Copula can be written as

$$C(u, v) = 1 - [(1 - u)^{\theta} + (1 - v)^{\theta} - (1 - u)^{\theta}(1 - v)^{\theta}]^{\frac{1}{\theta}}, \quad \theta \in [1, +\infty) \tag{2.12a}$$

The generator of Joe copula is

$$\varphi(t) = -\ln[1 - (1 - t)^{\theta}], \quad \theta \in [1, +\infty) \tag{2.12b}$$

2.2.2 *Multivariate Archimedean Copulas*

For multi-variables (greater than two), the Archimedean copulas can be divided into symmetric and asymmetric copulas (Chen et al. 2013a, b). Compared with asymmetric ones, the symmetric Archimedean copula in higher dimensions is easily built. However, this copula has only one parameter, which forces that all pairs of variables share the same dependence structure (Chen et al. 2013a, b). To model different dependence structures, Grimaldi and Serinaldi (2006) applied nested classes of the Archimedean copulas. However, the nested method for building multivariate copulas is complicated, and their parameters are estimated by the maximum likelihood method instead of the Kendall tau method.

2.2.2.1 Symmetric Archimedean

The symmetric Archimedean copulas allow modeling dependence in arbitrarily high dimensions with only one parameter, governing the strength of dependence. Joe (1997) defined symmetric Archimedean copulas as

$$C(u) = \varphi^{-1}(\sum_{k=1}^{n} \varphi(u_k)) \tag{2.13}$$

where function $\varphi(u)$, called the generator of the copula, is continuous strictly decreasing function from [0, 1] to [0, ∞) such that $\varphi(0) = \infty$ and $\varphi(1) = 0$ (Serinaldi and Grimaldi 2007), and its inverse φ^{-1} is "completely monotone" on [0, ∞), that is φ^{-1} has derivatives of all orders which alternate in sign (Nelsen 1999; Embrechts et al. 2003).

$$(-1)^k \frac{d^k \varphi^{-1}(t)}{dt^k} \geq 0 \qquad t \in [0, \infty), \tag{2.14}$$

The commonly used trivariate, four-dimensional and multi-dimensional symmetric Archimedean copulas are given in Table 2.1.

Table 2.1 Commonly used trivariate, four-dimensional and multi-dimensional symmetric Archimedean copulas

Copulas	Dimension	Function
Gumbel	3	$C(u_1, u_2, u_3) = \exp\left\{-\left[(-\ln u_1)^\theta + (-\ln u_2)^\theta + (-\ln u_3)^\theta\right]\right\}$
	4	$C(u_1, u_2, u_3, u_4) = \exp\left\{-\left[(-\ln u_1)^\theta + (-\ln u_2)^\theta + (-\ln u_3)^\theta + (-\ln u_4)^\theta\right]\right\}$
	n	$C(u_1, \ldots, u_n) = \exp\left\{-\left[\sum_{i=1}^{n}(-\ln u_i)^\theta\right]^{\frac{1}{\theta}}\right\}$
Frank	3	$C(u_1, u_2, u_3) = \frac{-1}{\theta}\ln\left(1 + \frac{\prod_{i=}^{3}(e^{-\theta u_i}-1)}{(e^{-\theta}-1)^2}\right)$
	4	$C(u_1, u_2, u_3, u_4) = \frac{-1}{\theta}\ln\left(1 + \frac{\prod_{i=1}^{4}(e^{-\theta u_i}-1)}{(e^{-\theta}-1)^3}\right)$
	n	$C(u_1, \ldots, u_n) = -\frac{1}{\theta}\log\left[1 - \frac{\prod_{i=1}^{n}(1-e^{-\theta u_i})}{1-e^{-\theta}}\right]$
Clayton	3	$C(u_1, u_2, u_3) = \left(u_1^{-\theta} + u_2^{-\theta} + u_3^{-\theta} - 2\right)^{-1/\theta}$
	4	$C(u_1, u_2, u_3, u_4) = \left(u_1^{-\theta} + u_2^{-\theta} + u_3^{-\theta} + u_4^{-\theta} - 2\right)^{-1/\theta}$
	n	$C(u_1, \ldots, u_n) = \left(\sum_{i=1}^{n} u_i^{-\theta} - 1\right)^{-\frac{1}{\theta}}$

2.2.2.2 Asymmetric Archimedean Copula

However, correlated hydrological variables can have different bivariate dependence structures. For multi-variables (greater than two), the symmetric Archimedean copula has only one parameter, which forces that all pairs of variables share the same dependence structure. To model the different dependence structures, Grimaldi and Serinaldi (2006) applied nested classes of the Archimedean copulas.

Following Joe (1997), Nelsen (1999), Embrechts et al. (2003) and Whelan (2004), a generalization of Archimedean 2-copulas can be written in the form called "fully nested":

$$\begin{aligned} C(u_1,\ldots,u_n) &= C_1(u_n, C_2(u_{n-1},\ldots,C_{n-1}(u_2,u_1)\cdots)) \\ &= \varphi_1^{-1}(\varphi_1(u_n)+\varphi_1(\varphi_2^{-1}(\varphi_2(u_{n-1})+\cdots \\ &\quad + \varphi_{n-1}^{-1}(\varphi_{n-1}(u_2)+\varphi_{n-1}(u_1))\cdots))) \end{aligned} \tag{2.15}$$

In general, first two variables are coupled by a 2-copula, then this copula is coupled with another variable by a second copula, and so on.

The commonly used trivariate and four-dimensional asymmetric Archimedean copulas are given in Table 2.2.

2.2.2.3 X-Gumbel Copula

Salvadori and Michele (2010) outlined a procedure for introducing a suitable number of extra parameters in a given copula model. In this method, a family of copulas can be defined as

$$C_a(u) = A(u^a)\cdot B(u^{1-a}) = A(u_1^{a_1},\ldots,u_n^{a_n})\cdot B(u_1^{a_1},\ldots,u_n^{a_n}) \tag{2.16}$$

where A and B represents d-copulas; $a = (a_1, \ldots, a_n)$, and $a \in \mathbf{I}$; n represents set of n parameters.

The X-Gumbel model, which is a bivariate version of the symmetric Gumbel model given in a previous section, can be extra parameterized by using Eq. 2.16 as

$$C(u_1,u_2,u_3,u_4) = C_\xi(u_1^{a_1},u_2^{a_2},u_3^{a_3},u_4^{a_4})\cdot C_\chi(u_1^{1-a_1},u_2^{1-a_2},u_3^{1-a_3},u_4^{1-a_4}) \tag{2.17}$$

2.2.3 Application of Archimedean Copulas in Hydrology

Complex hydrological events such as storms, floods, and droughts are often characterized by several correlated random variables. Copulas can model the dependence structure independently of the marginal distributions and allow for

Table 2.2 Commonly used trivariate and four-dimensional asymmetric Archimedean copulas

Copulas	Dimension	Function
Gumbel	3	$C(u_1,u_2,u_3)=\exp\left\{-\left[\left((-\log u_1)^{\theta_2}+(-\log u_2)^{\theta_2}\right)^{\theta_1/\theta_2}+(-\log u_3)^{\theta_1}\right]^{1/\theta_1}\right\}$
	4	$C(u_1,u_2,u_3,u_4)=\exp\left\{-\left[(-\ln u_4)^{\theta_1}+\left(\left((-\log u_1)^{\theta_3}+(-\log u_2)^{\theta_3}\right)^{\theta_2/\theta_3}+(-\log u_3)^{\theta_2}\right)^{\theta_1/\theta_2}\right]^{1/\theta_1}\right\}$
Frank	3	$C(u_1,u_2,u_3)=-\theta_1^{-1}\log\{1-(1-e^{-\theta_1})\theta^{-1}[1-(1-(1-e^{-\theta_2})^{-1}$ $(1-e^{-\theta_2 u_1})(1-e^{-\theta_2 u_2}))^{\theta_1/\theta_2})(1-e^{-\theta_1 u_3})]\}$
	4	$C(u_1,u_2,u_3,u_4)=\frac{-1}{\theta_1}\ln\{1+\frac{1}{e^{-\theta_1}-1}(e^{-\theta_1 u_4}-1)[(1-\frac{1}{1-e^{-\theta_2}}\cdot$ $(1-(1-\frac{1}{1-e^{-\theta_3}}(1-e^{-\theta_3 u_1})(1-e^{-\theta_3 u_2}))^{\theta_2/\theta_3})(1-e^{-\theta_2 u_3}))^{\theta_1/\theta_2}-1]\}$
Clayton	3	$C(u_1,u_2,u_3)=[(u_1^{-\theta_2}+u_2^{-\theta_2}-1)^{\theta_1/\theta_2}+u_3^{-\theta_1}-1]^{-1/\theta_1}$
	4	$C(u_1,u_2,u_3,u_4)=(u_4^{-\theta_1}+((u_1^{-\theta_3}+u_2^{-\theta_3}-1)^{\theta_2/\theta_3}+u_3^{-\theta_2}-1)^{\theta_1/\theta_2}-1]^{-1/\theta_1}$

multivariate distributions with different margins and dependence structures to be built (Dupuis 2007).

Favre (2004) tested four Archimedean copulas on peak flows from the watershed of Peribonka in Quebec, Canada. Zhang (2007) computed the distributions of bivariate rainfall frequency in Amite River, USA by using Archimedean copulas. Grimaldi (2006) analyzed the relationships between peak, volume and duration of a flood frequency by Archimedean copulas. Kao (2008) extended Frank copula to a trivariate one and applied it to the study of the temporal distribution of extreme rainfall events for several stations in Indiana where the estimated parameters lay in the feasible region. Zhang (2012) studied joint probabilities, changing characteristics of extremes and the implications of these changes in Xinjiang by using Archimedean copulas. Gräler et al. (2013) proposed an approach to estimate the expected value of the conditional distribution when the joint density along the level curve is derived, which is developed as conditional expectation combination (CEC) method. Li et al. (2016) used the CEC method to derive the quantiles of flood peak and 7-day volume under different joint return periods, and they found that the bivariate CEC design values have smaller flood volume and larger flood peak than bivariate equivalent frequency combination results in Geheyan reservoir. Xu et al. (2016) derived the general formulae of conditional most likely combination (CMLC) method to describe the dependence between flood peak and volumes using the conditional density function to measure the occurrence likelihood of flood events.

2.3 Meta-Elliptical Copulas

The meta-elliptical copulas were first introduced by Fang et al. (2002), which were extended from the so-called meta-Gaussian distributions constructed by Krzysztofowicz and Kelly (1996), and its properties were examined by Frahm et al. (2003) and Abdous et al. (2005). The meta-elliptical copulas can provide a wide range of positive and negative degrees of joint behavior and model high-dimensional dependence structure with a very simple structure (Kao and Govindaraju 2008). Therefore, they are often employed to solve some high-dimensional problems (Chen et al. 2012).

2.3.1 Meta-Elliptical Copulas

Following Fang (2002), the elliptically contoured distributions are the basic framework of meta-elliptical distributions. A d-dimensional vector Z is said to have an elliptically contoured distribution (or called ECD) with parameters $\mu(d \times 1)$ and $\Sigma(d \times d)$ if it can be expressed in the form

$$Z = \mu + rAu \tag{2.18}$$

where r is a nonnegative random variable; $AA^{\mathrm{T}} = \Sigma$ and A is a $d \times d$ constant matrix; u is a d-dimensional vector which is uniformly distributed and independent of r. If r has a density function, then the density of the vector Z can be written as

$$h(z) = |\Sigma|^{(-1/2)} g((z-\mu)' \Sigma^{(-1)} (z-\mu)) \tag{2.19}$$

where $g(\bullet)$ is a scale function which is controlled by the distribution of r. When a particular $g(\bullet)$ is given, for example, $g(t) \propto e^{-at/2}$ with a is a non-negative real number, then the vector Z can be called multivariate Gaussian. Other common examples of g can be found in Genest (2007).

Suppose X is a d-dimensional random vector which can be expressed in the form

$$Z_i = Q_g^{-1}(F_i(x_i)) \tag{2.20}$$

where Q_g is the CDF of Z and Q_g^{-1} is the inverse of Q_g; and $F_i(x_i)$ is the distribution function of x_i. Then X has a meta-elliptical distribution when its density function is given by

$$h(x_1, x_2, \ldots, x_d) = \Phi(Q_g^{(-1)}(F_1(x_1)), \ldots, Q_g^{(-1)}(F_d(x_d))) \prod_{i=1}^{d} f_i(x_i) \tag{2.21}$$

where $f_i(x_i)$ is the density function of x_i; Φ is the d-variate density weighting function

$$\Phi(x_1, \ldots, x_d) = \frac{|R|^{(-1/2)} g(x^T \Sigma^{(-1)} x)}{\prod_{i=1}^{d} q_g(z_i)} \tag{2.22}$$

where $R = \{\rho_{ij}: \rho_{ii} = 1, -1 < \rho_{ij} < 1;$ for $i \neq j$, $\rho_{ij} = \rho_{ji}; i, j = 1, \ldots, d\}$; and q_g is the PDF of Z.

2.3.2 Structure of Copulas

Two main copulas of meta-elliptical distribution, Meta-Gaussian Copula, and Student t Copula, are discussed in this section. Without loss of generality, we discuss the d-dimensional case.

2.3.2.1 Meta-Gaussian Copula

According to Kelly and Krzyszofowicz (1997), Meta-Gaussian copula's distribution function is given by

$$\begin{aligned} C(u_1, u_2, \ldots, u_d; \Sigma) &= \Phi_\Sigma(\ ^{-1}(u_1), \Phi^{-1}(u_2), \ldots, \Phi^{-1}(u_d)) \\ &= \int_{-\infty}^{\Phi^{-1}(u_1)} \cdots \int_{-\infty}^{\Phi^{-1}(u_d)} \frac{1}{(2\pi)^{\frac{d}{2}}|\Sigma|^{\frac{1}{2}}} \exp\left(-\frac{1}{2}w^T\Sigma^{-1}w\right) dw \end{aligned} \tag{2.23}$$

where Φ^{-1} is the inverse of the standard normal distribution; $\Phi_\Sigma(\Phi^{-1}(u_1), \Phi^{-1}(u_2), \ldots, \Phi^{-1}(u_d))$ multivariate normal distribution function; Σ is a symmetric covariance matrix, and

$$\sum = \begin{pmatrix} 1 & \cdots & \rho_{1d} \\ \vdots & \ddots & \vdots \\ \rho_{d1} & \cdots & 1 \end{pmatrix} \tag{2.24}$$

where

$$\rho_{ij} = \begin{cases} 1, i = j \\ \rho_{ji}, i \neq j \end{cases} (-1 \leq \rho_{ij} \leq 1) \tag{2.25}$$

and w is a d-dimensional integral variable, with the density function

$$\begin{aligned} c(u_1, u_2, \ldots, u_d; \Sigma) &= \frac{\partial^d C(u_1, u_2, \ldots, u_d; \Sigma)}{\partial u_1 \partial u_2 \ldots \partial u_d} \\ &= |\Sigma|^{-1/2} \exp\left[-\frac{1}{2}(\zeta^T\Sigma^{-1}\zeta - \zeta^T\zeta\right] \end{aligned} \tag{2.26}$$

where $\zeta = [\Phi^{-1}(u_1), \Phi^{-1}(u_2), \ldots, \Phi^{-1}(u_d)]^T$.

2.3.2.2 Student T Copula

According to Demarta and Mcneil (2005), the distribution function of Student t copula is given by

$$\begin{aligned} C(u_1, u_2, \ldots, u_d; \Sigma; \nu) &= T_{\Sigma,\nu}(T_\nu^{-1}(u_1), T_\nu^{-1}(u_2), \ldots, T_\nu^{-1}(u_d)) \\ &= \int_{-\infty}^{T_\nu^{-1}(u_1)} \cdots \int_{-\infty}^{T_\nu^{-1}(u_2)} \frac{\Gamma\left(-\frac{\nu+d}{2}\right)}{\Gamma\left(\frac{\nu}{2}\right)} \frac{1}{(\pi\nu)^{\frac{d}{2}}|\Sigma|^{\frac{1}{2}}} \left(1 + \frac{w^T\Sigma^{-1}w}{\nu}\right)^{-\frac{\nu+d}{2}} dw \end{aligned} \tag{2.27}$$

where $T_v^{-1}(\cdot)$ is the inverse of Student t distribution; v is the degree of freedom; $T_{\Sigma,v}\left(T_v^{-1}(u_1), T_v^{-1}(u_2), \ldots, T_v^{-1}(u_d)\right)$ is distribution function of multivariate Student t; and the other symbols are as the same as previously mentioned.

With the density function

$$
\begin{aligned}
c(u_1, u_2, \ldots, u_d; \Sigma; v) &= \frac{\partial^d C(u_1, u_2, \ldots, u_d; \Sigma; v)}{\partial u_1 \partial u_2 \ldots \partial u_d} \\
&= |\Sigma|^{-\frac{1}{2}} \frac{\Gamma\left(\frac{v+d}{2}\right)}{\Gamma\left(\frac{v}{2}\right)} \left[\frac{\Gamma\left(\frac{v}{2}\right)}{\Gamma\left(\frac{v+1}{2}\right)}\right]^d \frac{\left(1 + \frac{\zeta^T \Sigma^{-1} \zeta}{v}\right)^{-\frac{v+d}{2}}}{\prod_{i=1}^d \left(1 + \frac{b_i^2}{v}\right)^{-\frac{v+1}{2}}}
\end{aligned}
\tag{2.28}
$$

where $b_i = T_v^{-1}(u_i)$ and $\zeta = \left[\left(T_v^{-1}(u_1), T_v^{-1}(u_2), \ldots, T_v^{-1}(u_d)\right)\right]^T$.

2.3.3 Applications of Meta-Elliptical Copulas in Hydrology

Genest and Favre (2007) implemented the goodness-of-fit tests to meta-elliptical copulas and analyzed flood peak, volume and duration data of the Romaine River, Canada. Wong et al. (2008) established a joint distribution function of drought intensity, duration, and severity by using Gaussian and Gumbel copulas. Ghosh (2010) showed that meta-elliptical copula was an efficient method to modeling multivariate distribution. Wang et al. (2010) built a trivariate model among volume, duration and peak intensity of extreme rainfall events at 12 stations in Connecticut by using meta-elliptical copula method. Song and Singh (2010) used several meta-elliptical copulas in drought analysis and found that meta-Gaussian and t copula had a better fit. Ma et al. (2013) investigated the drought events in the Weihe river basin and selected the Gaussian and Student t copulas to model the joint distribution among drought duration, severity, and peaks. Chen et al. (2013a, b) measured the correlation between river flows in China and built joint distributions of rivers among several stations by using meta-elliptical copulas. Chen et al. (2013a, b) investigated the dependence structure in different drought states and calculated drought probabilities and return periods based on a four-dimensional meta-elliptical copula for the upper Han River basin in China. Xu et al. (2015) developed a regional drought frequency analysis model based on trivariate copulas by considering the spatio-temporal variations of drought events. Cui et al. (2017) utilized k-means classification and t-copula to demonstrate the regional drought occurrence probability and return period based on trivariate drought properties, i.e., drought duration, severity, and peak.

2.4 Parameter Estimation Method for Copulas

2.4.1 Parameter Estimation Method of Archimedean Copulas

The parameters of symmetric Archimedean copulas can be estimated by the method of moments with the use of the Kendall's correlation coefficient. For asymmetric Archimedean copulas, the inference functions for margins method (IFM) or maximum pseudo-likelihood (MPL) method can be selected.

2.4.1.1 Kendall's Correlation Coefficient

Kendall correlation coefficient is widely used for measures of association for non-normal multivariate distributions. If X_1 and X_2 are two random vectors, Kendall's correlation coefficient τ of them can be written as (Kruskal 1958)

$$\begin{aligned}\tau = {} & P\left[\left(X_{1i} < X_{1j}, X_{2i} < X_{2j}\right) or \left(X_{1i} > X_{1j}, X_{2i} > X_{2j}\right)\right] \\ & - P\left[\left(X_{1i} < X_{1j}, X_{2i} > X_{2j}\right) or \left(X_{1i} > X_{1j}, X_{2i} < X_{2j}\right)\right]\end{aligned} \tag{2.29}$$

Parameters θ and τ have connections with the Archimedean copulas. The relationships between them for four widely used copulas are summarized in Table 2.3.

2.4.1.2 The Maximum Pseudo-likelihood Method

The maximum pseudo-likelihood method is a semiparametric approach, where pseudo-observation values always lie between 0 and 1 ($[0, 1]^n$) (Bezak 2014). The pseudo log-likelihood has the form:

$$L(\theta) = \sum_{k=1}^{n} \log\left\{c_\theta\left(F_{1\mathrm{n}}(X_{1k}), \ldots, F_{pn}(X_{pk})\right)\right\} \tag{2.30}$$

where F_{in} stands for the empirical distribution function of the ith variable; c_θ is copula density which can be calculated as partial derivative of copula functions:

$$c_\theta(u_1, \ldots, u_\mathrm{n}) = \frac{\partial^n C_\theta(u_1, \ldots, u_n)}{\partial u_1 \ldots \partial u_n} \tag{2.31}$$

Table 2.3 Four common families of Archimedean copulas and their generator

Copula	θ
GH Copula	$\tau = 1 - \frac{1}{\theta}$
Frank Copula	$\tau = 1 + \frac{4}{\theta}\left(\frac{1}{\theta}\int_0^\theta \frac{t}{e^t - 1}dt - 1\right)$
Clayton Copula	$\tau = \theta/(\theta + 2)$
Ali-Mikhail-Haq Copula	$\tau = 1 - \frac{2(\theta + (1-\theta)^2 \ln(1-\theta))}{3\theta^2}$

Take the derivative of Eq. 2.30 concerning θ and equating it to zero yields:

$$\frac{1}{n}\frac{\partial}{\partial\theta}L(\theta) = \frac{1}{n}\sum_{k=1}^{n} l_\theta\{\theta, F_{1n}(X_{1k}), \ldots, F_{pn}(X_{pk})\} = 0 \tag{2.32}$$

where L denotes the log-likelihood function; and l_θ denotes the derivative of L concerning parameter θ (Zhang and Singh 2007). Then, we can get the estimator of parameter θ by solving the Eq. 2.32.

2.4.1.3 The Inference Functions for Margins Method

Joe (1997) recommended a parametric two-step procedure known as the inference function for margins (IFM). The IFM method includes two procedures: (1) marginal distributions are computed from the observed values; and (2) the copula dependence parameter, θ, is estimated through the maximization of the log-likelihood function of the copula (Mirabbasi 2012). The log-likelihood has the form:

$$l(\theta) = \sum_{i=1}^{n} \ln c\left(F_1\left(x_1^i; \theta_1\right), \ldots, F_p\left(x_p^i; \theta_p\right); \alpha\right) + \sum_{i=1}^{n}\sum_{j=1}^{n} \ln f_j\left(x_j^i; \theta_j\right) \tag{2.33}$$

First, we can perform n separate estimations, one for each univariate marginal distribution, i.e., obtain

$$\bar{\theta}_j = \arg\max \sum_{i=1}^{n} f_j\left(x_j^i; \theta_j\right) \tag{2.34}$$

for $i = 1, \ldots, n$ and then estimate α given the previous marginal estimates

$$\bar{\alpha} = \arg\max \sum_{i=1}^{n} \ln c\left(F_1\left(x_1^i; \theta_1\right), \ldots, F_p\left(x_p^i; \theta_p\right); \alpha\right) \tag{2.35}$$

2.4.2 Parameter Estimation Method of Meta-Elliptical Copulas

The parameters of meta-elliptical copulas are determined by the linear correlation coefficient ρ, and Σ is a symmetrical correlation matrix, we need only estimate its $d(d-1)/2$ supradiagonal elements (Genest and Favre 2007). According to Hult and Lindskog (2002), the linear correlation coefficient ρ can be computed by

$$\rho_{kl} = \frac{\sigma_{kl}}{\sqrt{\sigma_{kk}\sigma_{ll}}} \tag{2.36}$$

where σ_{kl} is the covariance of samples k and l; and σ_{kk} and σ_{ll} are the variance of sample k and l, respectively.

2.4.2.1 Parameter Estimation Based on Kendall Correlation Coefficient

We can directly find the ρ by Eq. 2.36. However, the correlation coefficient ρ is somewhat complicated for elliptically contoured distributions (Fang 2002), it is more convenient to compute ρ by considering Kendall's correlation coefficient [see, e.g., Nelson (1992, 1998)]. According to Hult and Lindskog (2002), the population value of Kendall's tau is linked to ρ_{kl} through the relation

$$\tau_{kl} = \frac{2}{\pi}\arcsin(\rho_{kl}) \tag{2.37}$$

Thus, the parameters of meta-elliptical copulas can be computed based on the inversion of Kendall's correlation coefficient.

2.4.2.2 The Maximum Pseudo-likelihood Method

The maximum pseudo-likelihood estimation method (MPL) mentioned above can also be used to compute the parameters of Gaussian and Student copulas (Genest 1995; Nadarajah 2005). In the MPL method, the parametric marginal distribution is substituted by empirical or rank-based marginal distribution (Reddy and Ganguli 2012). If X is a vector of observations, the empirical distribution function of it can be calculated by the following expression:

$$\hat{u}_{kt}(X_{kt}) = \frac{1}{N+1}\sum_{i=1}^{N} 1(X_{ki} \le X_{kt}) \tag{2.38}$$

where N is the sample size.

Considering the Gaussian copula, the maximum pseudo-likelihood function is given by

$$L(x_1, x_2, \ldots, x_d; \Sigma) = \prod_{i=1}^{d} c(\hat{u}_{1i}, \hat{u}_{2i}, \ldots, \hat{u}_{di}; \Sigma_i) \tag{2.39}$$

where $c(\cdot)$ is the density of Gaussian copula; $\hat{u}_{di}$ is the empirical marginal distribution value of current variables; Σ_i is the symmetrical covariance matrix of X. Then, build an equation based on Eq. 2.39

$$\sum_{i=1}^{n} \frac{\partial}{\partial \Sigma} \{\ln[c(\hat{u}_{1i}, \hat{u}_{2i}, \ldots, \hat{u}_{di}; \Sigma_i)]\} = 0 \tag{2.40}$$

We can get the estimator of parameter Σ by solving Eq. 2.40. This method can also estimate the parameters of Student copula. The maximum pseudo-likelihood function is given by

$$L(x_1, x_2, \ldots, x_d; \Sigma; v) = \prod_{i=1}^{d} c(\hat{u}_{1i}, \hat{u}_{2i}, \ldots, \hat{u}_{di}; \Sigma_i; v) \tag{2.41}$$

Then let $\partial \ln L / \partial \Sigma = 0$ and $\partial \ln L / \partial v = 0$ as following:

$$\begin{cases} \sum_{i=1}^{n} \frac{\partial}{\partial \Sigma} \{\ln[c(\hat{u}_{1i}, \hat{u}_{2i}, \ldots, \hat{u}_{di}; \Sigma_i, v)]\} = 0 \\ \sum_{i=1}^{n} \frac{\partial}{\partial v} \{\ln[c(\hat{u}_{1i}, \hat{u}_{2i}, \ldots, \hat{u}_{di}; \Sigma_i, v)]\} = 0 \end{cases} \tag{2.42}$$

where $c(\cdot)$ is the density of Student copula. By solving Eq. 2.42, Σ and v can be determined.

2.5 Goodness-of-Fit for Copulas

2.5.1 Fitting Evaluation of Copulas

When evaluating the goodness-of-fit of a model, maybe the most natural idea is plotting a scatter plot of the pairs $(\widehat{F}_i, C_i)$, where $\widehat{F}_i$ and C_i are respectively the empirical and theoretical values (Genest and Favre 2007). Empirical copulas are rank-based, empirically joint cumulative probability measures (Nelsen 2006). For the bivariate case, the empirical copula of the observed data (u_i, v_i) is as follows:

$$F(u_i, v_i) = \frac{1}{N}\sum_{i=1}^{N} I(\frac{D_i}{n+1} \leq u_i, \frac{S_i}{n+1} \leq v_i) \tag{2.43}$$

where N is the sample size; $I(A)$ denotes the indicator variable of the logical expression A and assumes a value of 0 if A is false and 1 if A is true; and the ranks of the ith observed duration and the severity data are represented as D_i and S_i, respectively (Mirabbasi et al. 2012).

By the graphical model selection method proposed by Genest and Rivest (1993), the best-fitting model is the one whose scatter plot is the closest to the 45° diagonal.

The graphical diagnostic is intuitive but not accurate, so the root mean square error (RMSE), Bayesian information criteria (BIC), and Akaike's information criterion (AIC) are three more common methods to measure the fitting biases of various copulas (Ma et al. 2011; Zhang 2005). The AIC can be obtained either by calculating the maximum likelihood or by calculating the mean square error of the model (Zhang and Singh 2006).

RMSE and BIC can be calculated by

$$MSE = \frac{1}{N}\sum_{i=1}^{N} (P_{e_i} - P_i)^2 \tag{2.44}$$

$$RMSE = \sqrt{MSE} \tag{2.45}$$

$$BIC = N\ \ln(MSE) + m\ \ln(N) \tag{2.46}$$

The AIC values related to maximum likelihood values can be calculated by

$$AIC = 2m - 2\ \ln(L) \tag{2.47a}$$

The AIC values related to mean square error can be calculated by

$$AIC = 2m + N\ \ln(MSE) \tag{2.47b}$$

where MSE is the mean square error of the chosen copula model concerning empirical copula; P_{e_i} and P_i is respectively the empirical probability and theoretical probability; m is the number of parameters; and L is the maximized value of the likelihood function for the estimated model. The copula function with the smaller *AIC*, *BIC* and *RMSE* values is the better one. Thus, these three values can play a guiding role in choosing copulas.

2.5.2 Goodness-of-Fit Test for Copulas

To test whether a copula function can adequately describe the dependence between given series, several goodness-of-fit procedures have been proposed. The Kolmogorov-Smirnov test (Kolmogorov 1933; Smirnov 1948) and the Anderson–Darling test (Anderson and Darling 1952) are two common ways to test our hypothesis (Malevergne and Sornette 2003). Computing statistics D_N and A_N^2 by K-S and A-D methods, and comparing them with corresponding critical values at a certain significance level, if the statistics are less than the corresponding critical values, our hypothesis can be accepted. The statistics D_N and A_N^2 can be computed by

$$D_N = \max_{1 \le i \le N} \left[\frac{i}{N} - C(x_i), C(x_i) - \frac{i-1}{N} \right] \tag{2.48}$$

$$A_N^2 = -N - \frac{1}{N} \sum_{i=1}^{N} (2i-1)[\ln C(x_i) + \ln(1 - C(x_{N+1-i}))] \tag{2.49}$$

where N is the number of observations; $C(\cdot)$ is the hypothesized model that needs to be tested; and x_i is observation in increasing order.

2.6 Copula Entropy Theory

2.6.1 Entropy Theory

The Shannon entropy (Shannon 1948) quantitatively measures the mean uncertainty associated with a probability distribution of a random variable and in turn with the random variable itself in concert with several consistency requirements (Kapur and Kesavan 1992). The entropy of a random variable (r.v.) X can be expressed as:

$$H(X) = -\int_0^{\infty} f(x) \log f(x) \mathrm{d}x \tag{2.50}$$

where $f(x)$ is the probability density function of variable X. In this book, we focus on flood flow, so the range of the variable is from 0 to infinite. In fact, the domain can be extended to any real number.

Equation 2.50 defines the univariate continuous entropy or marginal entropy of X. The units of entropy are given by the base of the logarithm, being “nats” for base e and “bits” for base 2. The natural logarithm will be used in this book.

For two random variables X_1 and X_2, the joint entropy can be expressed as

$$H(X_1, X_2) = -\int_0^{\infty}\int_0^{\infty} f(x_1, x_2)\log f(x_1, x_2)\mathrm{d}x_1\mathrm{d}x_2 \tag{2.51}$$

Let X_1, X_2, …, X_d denote the random variable series, the multidimensional joint entropy can be expressed as:

$$H(X_1, X_2, \ldots, X_d) = -\int_0^{\infty}\cdots\int_0^{\infty} f(x_1, x_2, \ldots, x_d)\log[f(x_1, x_2, \ldots, x_d)]\mathrm{d}x_1\,\mathrm{d}x_2\ldots\mathrm{d}x_d \tag{2.52}$$

The mutual information can be expressed as:

$$T(X_1, X_2) = H(X_1) + H(X_2) - H(X_1, X_2) \tag{2.53}$$

Using Eq. 2.50 and 2.51 into Eq. 2.53, the formulation can be written as:

$$\begin{aligned} T(X_1, X_2) &= -\int_0^{\infty} f(x_1)\log f(x_1)\mathrm{d}x - \int_0^{\infty} f(x_2)\log f(x_2)\mathrm{d}x \\ &\quad + \int_0^{\infty}\int_0^{\infty} f(x_1, x_2)\log f(x_1, x_2)\mathrm{d}x_1\,\mathrm{d}x_2 \\ &= -\int_0^{\infty}\int_0^{\infty} f(x_1, x_2)\log f(x_1)\mathrm{d}x_1\,\mathrm{d}x_2 - \int_0^{\infty}\int_0^{\infty} f(x_1, x_2)\log f(x_2)\mathrm{d}x_1\,\mathrm{d}x_2 \\ &\quad + \int_0^{\infty}\int_0^{\infty} f(x_1, x_2)\log f(x_1, x_2)\mathrm{d}x_1\,\mathrm{d}x_2 \\ &= \int_0^{\infty}\int_0^{\infty} f(x_1, x_2)[-\log f(x_1) - \log f(x_2) + \log f(x_1, x_2)]\mathrm{d}x_1\,\mathrm{d}x_2 \\ &= \int_0^{\infty}\int_0^{\infty} f(x_1, x_2)\log\frac{f(x_1, x_2)}{f(x_2)f(x_1)}\mathrm{d}x_1\,\mathrm{d}x_2 \end{aligned} \tag{2.54}$$

2.6.2 Definition of Copula Entropy

Ma and Sun (2011) took into accounts the copula function and entropy theory together and introduced the concept of copula entropy. Calsaverini and Vicente (2009) discuss a couple of consequences yield by connections between the copula and entropy methods, which involve copula entropy.

Copula entropy is defined as the entropy of copula function, which is related to the joint entropy, marginal entropy and mutual information. Let $x \in R_d$ be random variables with marginal functions $F_i(x)$, $U_i = F_i(x)$, $i = 1, 2, \ldots, d$. Then, U_i are uniformly distributed random variables; and u_i will denote a specific value of U_i. The entropy of the copula function is defined as variable CE, which can be expressed as

$$H_C(U_1, U_2, \ldots, U_d) = -\int_0^1 \cdots \int_0^1 c(u_1, u_2, \ldots, u_d)\log(c(u_1, u_2, \ldots, u_d))\mathrm{d}u_1 \ldots \mathrm{d}u_d \tag{2.55}$$

where $c(u_1, u_2, \ldots, u_d)$ is the probability density function of copulas and expressed as $\frac{\partial C(u_1, u_2, \ldots, u_d)}{\partial u_1 \partial u_2 \ldots \partial u_d}$.

For a bivariate case, the CE can be expressed as

$$H_C(U_1, U_2) = -\int_0^1 c(u_1, u_2)\log(c(u_1, u_2))\mathrm{d}u_1\mathrm{d}u_2 \tag{2.56}$$

2.6.3 Relationship Between CE and MI

The purpose of this section is to find a relationship between CE and MI. The joint probability density function of vector random variable X can be defined as Grimaldi and Serinaldi (2006):

$$f(x_1, x_2, \ldots, x_d) = c(u_1, \ldots, u_d)\prod_{i=1}^{d} f(x_i) \tag{2.57}$$

Based on Eq. 2.50, the joint entropy can be expressed as:

$$
\begin{aligned}
H(X_1, X_2, \ldots, X_d) &= -\int_0^\infty \ldots \int_0^\infty f(x_1, x_2, \ldots, x_d) \log[f(x_1, x_2, \ldots, x_d)] \mathrm{d}x_1\, \mathrm{d}x_2 \ldots \mathrm{d}x_d \\
&= -\int_0^\infty \ldots \int_0^\infty c(u_1, \ldots, u_n) \prod_{i=1}^{d} f(x_i) \log[c(u_1, \ldots, u_n) \prod_{i=1}^{d} f(x_i)] \mathrm{d}x_1\, \mathrm{d}x_2 \ldots \mathrm{d}x_d \\
&= -\int_0^\infty \ldots \int_0^\infty c(u_1, \ldots, u_d) \prod_{i=1}^{d} f(x_i) \{\log[c(u_1, \ldots, u_d)] + \sum_{i=1}^{d} \log[f(x_i)]\} \mathrm{d}x_1\, \mathrm{d}x_2 \ldots \mathrm{d}x_d \\
&= -\int_0^\infty \ldots \int_0^\infty c(u_1, \ldots, u_d) \prod_{i=1}^{d} f(x_i) \cdot \log[c(u_1, \ldots, u_d)] \mathrm{d}x_1\, \mathrm{d}x_2 \ldots \mathrm{d}x_d \\
&\quad - \int_0^\infty \ldots \int_0^\infty c(u_1, \ldots, u_d) \prod_{i=1}^{d} f(x_i) \cdot \sum_{i=1}^{d} \log[f(x_i)] \mathrm{d}x_1\, \mathrm{d}x_2 \ldots \mathrm{d}x_d \\
&= A + B
\end{aligned}
\tag{2.58}
$$

where

$$
\begin{aligned}
A = &-\int_0^\infty \ldots \int_0^\infty c(u_1, \ldots, u_d) \prod_{i=1}^{d} f(x_i) \cdot \sum_{i=1}^{d} \log[f(x_i)] \mathrm{d}x_1\, \mathrm{d}x_2 \ldots \mathrm{d}x_d \\
&= -\int_0^\infty \ldots \int_0^\infty f(x_1, x_2, \ldots, x_d) \cdot \sum_{i=1}^{n} \log[f(x_i)] \mathrm{d}x_1\, \mathrm{d}x_2 \ldots \mathrm{d}x_d \\
&= -\int_0^\infty \ldots \int_0^\infty f(x_1, x_2, \ldots, x_d) \cdot \{\log[f(x_1)] + \cdots + \log[f(x_d)]\} \mathrm{d}x_1\, \mathrm{d}x_2 \ldots \mathrm{d}x_d \\
&= -\int_0^\infty \ldots \int_0^\infty f(x_1, x_2, \ldots, x_d) \cdot \log[f(x_1)] \mathrm{d}x_1\, \mathrm{d}x_2 \ldots \mathrm{d}x_d \\
&\ldots - \int_0^\infty \ldots \int_0^\infty f(x_1, x_2, \ldots, x_d) \cdot \log[f(x_d)] \mathrm{d}x_1\, \mathrm{d}x_2 \ldots \mathrm{d}x_d \\
&= -\int_0^\infty \log[f(x_1)] [\int_0^\infty \ldots \int_0^\infty f(x_1, x_2, \ldots, x_d) \cdot \mathrm{d}x_2 \ldots \mathrm{d}x_d] \mathrm{d}x_1
\end{aligned}
$$

$$= \ldots - \int_0^\infty \log[f(x_d)][\int_0^\infty \ldots \int_0^\infty f(x_1, x_2, \ldots, x_d) \cdot \mathrm{d}x_1 \ldots \mathrm{d}x_{d-1}]\mathrm{d}x_d$$
$$= -\sum_{i=1}^{d} \int_0^\infty f(x_i) \log[f(x_i)]\mathrm{d}x_i = \sum_{i=1}^{d} H(X_i) \tag{2.59}$$

Noting that $\mathrm{d}u = \mathrm{d}x \cdot f(x_i)$,

$$B = -\int_0^\infty \ldots \int_0^\infty c(u_1, \ldots, u_d) \prod_{i=1}^{n} f(x_i) \cdot \log[c(u_1, \ldots, u_d)]\mathrm{d}x_1\, \mathrm{d}x_2 \ldots \mathrm{d}x_d$$
$$= -\int_0^\infty \ldots \int_0^\infty c(u_1, \ldots, u_d) \cdot \log[c(u_1, \ldots, u_d)]\mathrm{d}u_1\, \mathrm{d}u_2 \ldots \mathrm{d}u_d = H_C(\mathbf{u}) \tag{2.60}$$

Therefore, the joint entropy can be expressed as the sum of the d univariate marginal entropies and the copula entropy.

$$H(X_1, X_2, \ldots, X_d) = \sum_{i=1}^{d} H(X_i) + H_C(u_1, u_2, \ldots, u_d) \tag{2.61}$$

Equation 2.61 indicates that the joint entropy $H(X_1, X_2, \ldots, X_d)$ is divided into two parts: the sum of the d marginal entropies $H(X_i)$ and the copula entropy $H_C(U_1, U_2, \ldots, U_d)$.

For $d = 2$,

$$H(X_1, X_2) = H(X_1) + H(X_2) + H_C(U_1, U_2) \tag{2.62}$$

From Eq. 2.53,

$$T(X_1, X_2) = H(X_1) + H(X_2) - H(X_1, X_2) = -H_C(X_1, X_2) \tag{2.63}$$

From Eq. 2.63, it can be seen that the mutual information is the negative copula entropy. It is well known that the mutual information can measure the linear and non-linear dependencies. Therefore, the copula entropy can be used to estimate the linear and non-linear dependencies.

Research on copula entropy has received significant attention recently. Calsaverini and Vicente (2009) discussed a couple of consequences yielded by connections between the copula and entropy methods, which involve copula entropy. Zhao and Lin (2011) applied copula entropy models with two and three variables to measure the dependence in stock markets. The advantage of the copula entropy is summarized as follows: (1) it makes no assumptions about the marginal distributions and can be used for higher dimensions, (2) the mutual information can be obtained from the calculation of the copula entropy instead of the marginal or joint entropy, which estimates the MI more directly and avoids the accumulation of systematic bias. Until now the copula entropy method has not been widely used in the hydrological field.

2.6.4 *Calculation of Copula Entropy*

Two methods are proposed to calculate the copula entropy. One is multiple integration method, and the other is Monte Carlo method.

2.6.4.1 Multiple Integration Method

From Eq. 2.55, the copula entropy can be derived using the multiple integration method. First, parameters of the copula function need to be estimated, and then the copula probability density function can be determined. The multiple integration method, proposed by Berntsen et al. (1991), is applied to calculate the multiple integrations. In order to test this multiple integration method, we use the copula probability density function as an integrand. The result of integration should be 1.

2.6.4.2 Monte Carlo Method

For more variables, it may be difficult to calculate multiple integrations. The Monte Carlo method can be used to calculate the copula entropy. For a multivariate vector with support in [0, 1], the copula entropy can be obtained by

$$H_C(u_1, u_2, \ldots, u_d) = -\int\limits_{[0,1]^d} c(U)\ln c(U)\mathrm{d}U = -E[\ln c(U)] \tag{2.64}$$

The copula entropy equals the expected value of $-\ln[c(\mathbf{U})]$, which can be derived by the Monte Carlo method. Similar to the multiple integration method, first the dependence structure and parameters of the copula function need to be determined. M pairs of $\mathbf{u}$ are generated from the determined copula function, and then average values of the $-\ln[c(\mathbf{U})]$ are calculated.

An example for calculating the copula entropy is given as follows. The FORTRAN code is used to do the calculation. For example, we calculate the copula entropy of two variables X and Y. The Gumbel Copula is used to establish the joint distribution of variables X and Y. The parameter of Gumbel copula is 1.5. The Gumbel copula can be described as:

$$C(u_1, u_2) = \exp\{-[(-\ln u_1)^{1.5} + (-\ln u_2)^{1.5}]^{1/1.5}\} \tag{2.65}$$

First, the multiple integration method is used. The integrand function is Eq. 2.65. The multiple integration method, proposed by Berntsen et al. (1991), is applied to calculate the multiple integrations. The value of copula entropy of variables X and Y is—0.166.

Second, the Monte Carlo method is used. According to Eq. 2.64, The copula entropy equals the expected value of $-\ln[c(\mathbf{U})]$. First, the copula function is established as shown in Eq. 2.65. 10,000 pairs of $\mathbf{u}$ are generated from the determined copula function, and then the average values of $-\ln[c(\mathbf{U})]$ are calculated. The value of copula entropy is -0.153.

References

Abdous B, Genest C, Remillard B (2005) Dependence properties of meta-elliptical distributions. In: Statistical modeling and analysis for complex data problems. In: GERAD 25th Anniversary Series. Springer, New York, vol 1, pp 1–15

Ali MM, Mikhail NN, Haq MS (1978) A class of bivariate distributions including the bivariate logistic. J Multivariate Anal 8(3):405–412

Anderson TW, Darling DA (1952) Asymptotic theory of certain "goodness of fit" criteria based on stochastic processes. Ann Math Stat 23:193–212

Berntsen J, Espelid TO, Genz A. (1991) An adaptive algorithm for the approximate calculation of multiple integrals. ACM

Bezak N, Mikoš M, Šraj M (2014) Trivariate frequency analyses of peak discharge, hydrograph volume and suspended sediment concentration data using copulas. Water Resour Manag 28(8):2195–2212

Calsaverini RS, Vicente R (2009) An information-theoretic approach to statistical dependence: copula information. Eur Phys Lett 88(6):3–12

Chen L, Singh VP, Guo S et al (2013a) Drought analysis using copulas. J Hydrol Eng 18(7): 797–808

Chen L, Singh VP, Guo S (2013b) Measure of correlation between river flows using the copula-entropy method. J Hydrol Eng 18(12):1591–1606

Chen L, Singh VP, Guo S, Hao Z, Li T (2012) Flood coincidence risk analysis using multivariate copula functions. J Hydrol Eng 17(6):742–755

Cui H, Zhang J, Yao F (2017) Multivariate drought frequency estimation using copula method in southwest china. Theoretical & Applied Climatology 127(3-4): 1-15

Demarta S, Mcneil AJ (2005) The t copula and related copulas. Int Stat Rev 73(1):111–129

Dupuis DJ (2007) Using Copulas in Hydrology: Benefits, Cautions, and Issues. J Hydrol Eng 12(4):381–393

Embrechts P, Lindskog F, Mcneil A (2003) Modelling dependence with copulas and applications to risk management. Elsevier

Fang HB, Fang KK, Kotz Samuel (2002) The Meta-elliptical distributions with given marginal. J Multivar Anal 82(1):1–16

Favre AC, Adlouni SE, Perreault L et al (2004) Multivariate hydrological frequency analysis using copulas. Water Resour Res 40(1):290–294

Frahm G, Junker M, Szimayer A (2003) Elliptical copulas: applicability and limitations. Stat Probab Lett 63:275–286

Genest C, Favre AC, Béliveau J et al (2007) Meta-elliptical copulas and their use in frequency analysis of multivariate hydrological data. Water Resour Res 43(9):223–236
Genest C, Ghoudi K, Rivest LP (1995) A semiparametric estimation procedure of dependence parameters in multivariate families of distributions. Biometrika 82(3):543–552
Genest C, Rivest LP (1993) Statistical inference procedures for bivariate Archimedean copulas. J Am Stat Assoc 88(423):1034–1043
Ghosh S (2010) Modelling bivariate rainfall distribution and generating bivariate correlated rainfall data in neighboring meteorological subdivisions using copula. Hydrol Process 24(24):3558–3567
Gräler B, van den Berg Vandenberghe S, Petroselli A, Grimaldi S, De Baets B, Verhoest N (2013) Multivariate return periods in hydrology: a critical and practical review focusing on synthetic design hydrograph estimation. Hydrol Earth Syst Sci 17(4):1281–1296
Grimaldi S, Serinaldi F (2006) Asymmetric copula in multivariate flood frequency analysis. Adv Water Resour 29(8):1155–1167
Gumbel EJ (1960) Multivariate extreme distribution. Bull Int Stat Inst 39(2):471–475
Hult H, Lindskog F (2002) Multivariate extremes, aggregation and dependence in elliptical distributions. Adv Appl Probab 34(3):587–608
Joe H (1997) Multivariate models and multivariate dependence concepts. Chapman & Hall
Joe H (1993) Parametric families of multivariate distributions with given margins. J Multivar Anal 46(2):262–282
Kao SC, Govindaraju RS (2008) Trivariate statistical analysis of extreme rainfall events via the Plackett family of copulas. Water Resour Res 44(2):W02415
Kapur JN, Kesavan HK (1992) Entropy optimization principles and their applications. Academic Press
Kelly KS, Krzysztofowicz R (1997) A bivariate meta-Gaussian density for use in hydrology. Stohastic Hydrology and Hydraulics 11:17-31
Kolmogorov A (1933) Sulla determinazione empirica di una legge distribuzione(distribution? G Ist Ital Attuari 4:83–91
Kruskal WH (1958) Ordinal measures of association. J Am Stat Assoc 53(284):814–861
Krzysztofowicz R, Kelly KS (1996) A meta-Gaussian distribution with specified marginals. Technical report, University of Virginia
Li T, Guo S, Liu Z, Xiong L, Yin J (2016) Bivariate design flood quantile selection using copulas. Hydrol Res. https://doi.org/10.2166/nh.2016.049
Ma J, Sun Z (2011) Mutual information is copula entropy. Tsinghua Sci Technol 16(1):51–54
Ma M, Song S, Ren L et al (2013) Multivariate drought characteristics using trivariate Gaussian and student t copulas. Hydrol Process 27(8):1175–1190
Malevergne Y, Sornette D (2003) Testing the Gaussian copula hypothesis for financial assets dependences. Quant Financ 3(4):231–250
Mirabbasi R, Fakheri-Fard A, Dinpashoh Y (2012) Bivariate drought frequency analysis using the copula method. Theor Appl Climatol 108(1–2):191–206
Nelsen RB. (1998) Concordance and Gini's measure of association. Journal of Nonparametric Statistics 9(3):227-238
Nadarajah S, Kotz S (2005) Information matrices for some elliptically symmetric distributions. SORT 29(1)
Nelsen BR (2006) An introduction to copulas. Springer, New York, USA
Nelsen RB (1992) On measures of association as measures of positive dependence. Stat Probab Lett 14:269–274
Nelson BR (1999) An introduction to copulas. Springer, New York, USA
Whelan Niall (2004) Sampling from Archimedean copulas. Quant Financ 4(3):339–352
Reddy MJ, Ganguli P (2012) Application of copulas for derivation of drought severity–duration–frequency curves. Hydrol Process 26(11):1672–1685
Salvadori G, Michele CD (2010) Multivariate multiparameter extreme value models and return periods: a copula approach. Water Resour Res 46(10):219–233
Salvadori G, Michele CD, Kottegoda NT et al (2007) Extremes in nature. Springer, Netherlands

Serinaldi F, Grimaldi S (2007) Fully nested 3-copula: procedure and application on hydrological Data. J Hydrol Eng 12(4):420–430

Serinaldi F, Bonaccorso B, Cancelliere A, et al. (2009) Probabilistic characterization of drought properties through copulas. Physics & Chemistry of the Earth Parts A/b/c 34(10-12):596-605

Shannon Claude E (1948) A mathematical theory of communication. Bell Syst Tech J 27(3): 379–423. https://doi.org/10.1002/j.1538-7305.1948.tb01338.x

Sklar A (1959) Fonctions de répartition à n dimensions et leurs marges. Publ Inst Stat Univ Paris 8:229–231

Smirnov N (1948) Table for estimating the goodness of fit of empirical distributions. Ann Math Stat 19:279–281

Song S, Singh VP (2010) Meta-elliptical copulas for drought frequency analysis of periodic hydrologic data. Stochast Environ Res Risk A 24(3):425–444

Wang XJ, Gebremichael M, Yan J (2010) Weighted likelihood copula modeling of extreme rainfall events in Connecticut. J Hydrol 390:1–2

Wong G, Lambert MF, Metcalfe AV (2008) Trivariate copulas for characterisation of droughts. J ANZIAM 49

Xu C, Yin J, Guo S, Liu Z, Hong X (2016) Deriving design flood hydrograph based on conditional distribution: a case study of Danjiangkou reservoir in Hanjiang basin. Math Probl Eng. https://doi.org/10.1155/2016/4319646

Xu K, Yang D, Xu X, Lei H (2015) Copula based drought frequency analysis considering the spatio-temporal variability in Southwest China. J Hydrol 527:630–640

Zhang L (2005) Multivariate hydrological frequency analysis and risk mapping. LSU Doctoral Dissertations. 1351.https://digitalcommons.lsu.edu/gradschool_dissertations/1351

Zhang L, Singh VP (2006) Bivariate rainfall frequency distributions using Archimedean copulas. J Hydrol 332(1):93–109

Zhang L, Singh VP (2007) Gumbel-Hougaard copula for trivariate rainfall frequency analysis. J Hydrol Eng 12(4):409–419

Zhang Q, Singh VP, Li J, Jiang F, Bai Y (2012) Spatio-temporal variations of precipitation extremes in Xinjiang, China. J Hydrol 434–435(2):7–18

Zhao N, Linb WT (2011) A copula entropy approach to correlation measurement at the country level. Appl Math Comput 218(2):628–642

Chapter 3
Copula-Based Flood Frequency Analysis

3.1 Introduction

Flood frequency analysis is a constant concern in the hydrological practice. The sizing of bridges, culverts and other facilities, the design capacities of levees, spillways and other control structures, and reservoir operation or management depend upon the estimated magnitude of various design flood values (ASCE 1996). Nowadays, the general methodology based on the univariate distribution is to derive the fitted distribution representing the probability of an annual maximum flood being exceeded (USWRC 1981; MWR 2006).

As the duration of gauged record rarely exceeds 50 years, estimates corresponding to high return period obtained from the systematic data alone are subject to large sampling errors. Furthermore, the existence of a cyclic variation over periods longer than the duration of the records might well introduce further bias (Leese 1973; Stedinger and Cohn 1986; Guo and Cunnane 1991). Therefore, to overcome the problem of relatively short data series for frequency analysis, the need to augment the flow record with historical is widely acknowledged in the hydrological community. Several methods for incorporating historical information into flood frequency studies have been suggested, including historically weighted moments, maximum likelihood, probability weighted moments and L-moments (USWRC 1982; Guo and Cunnane 1991; Hosking 1995).

The hydrologic extreme values and critical thresholds derived from complex hydrological events for engineering design are usually obtained from single site characteristics (e.g., annual maximum peak discharge). Therefore, conventional hydrological frequency analysis has also mainly focused on one characteristic value and univariate distributions that cannot provide a complete description of hydrologic events with multi-characteristics. Many hydrological frequency problems, such as design flood hydrograph that includes flood peak and flood volumes, should be solved by the multivariate distributions (Dupuis 2007; Xiao et al. 2008, 2009).

L. Chen and S. Guo, *Copulas and Its Application in Hydrology and Water Resources*, Springer Water, https://doi.org/10.1007/978-981-13-0574-0_3

In this chapter, the multivariate frequency analysis has been carried out. One of the main difficulties in the multivariate quantile estimation is how to choose the proper combinations of design values of the concerned random variables for a given multivariate return period of hydrologic structure design. Take the bivariate case (peak discharge Q and flood volume W) as an example. The combinations can differ greatly regarding their values: moving along the multivariate quantile curve to an asymptote, one of the two variables will approach its marginal value, while the other tends to increase indefinitely (for unbounded random variables). Chebana and Ouarda (2011) proposed the decomposition of the level curve into a naive part (tail) and the proper part (central); they assumed that the naive part was composed of two segments starting at the end of each extremity of the proper part. Salvadori et al. (2011) introduced two basic design realizations, i.e., component-wise excess design realization and most-likely design realization. Li et al. (2016) used the conditional expectation combination method to derive the quantiles of flood peak and 7-day volume under different JRPs, and they found that the bivariate design values have smaller flood volume and larger flood peak than bivariate equivalent frequency combination results.

3.2 Annual Maximum Flood Frequency Analysis Based on Copula

Annual maximum (AM) flood series can be characterized by flood occurrence dates and flood magnitudes. The marginal distribution of flood occurrence dates, peak discharges, and flood volumes are established.

3.2.1 Margin Distribution of AM Flood Occurrence Dates

The AM flood occurrence dates can be described by the directional statistics (DS) method. The date firstly should be converted to the angle of a circle by

$$\alpha_i = D_i \frac{2\pi}{L} \quad 0 \le \alpha_i \le 2\pi \tag{3.1}$$

where L is the length of flood season; D_i is the flood occurrence date.

The x and y coordinates of the flood dates described by the angles is determined by

$$(a_i, b_i) = (\cos\, \alpha_i, \sin\, \alpha_i) \tag{3.2}$$

$$\bar{a} = \sum_{i=1}^{n} \cos x_i / n \tag{3.3}$$

$$\bar{a} = \sum_{i=1}^{n} \sin x_i / n \tag{3.4}$$

where n is the sample size.

The mean direction of the circular data (denoted by $\bar{\theta}$) is estimated by

$$\bar{\theta} = \begin{cases} \arctan \bar{b}/\bar{a} & \bar{a} > 0, \quad \bar{b} > 0 \\ 2\pi + \arctan \bar{b}/\bar{a} & \bar{a} > 0, \quad \bar{b} < 0 \\ \pi + \arctan \bar{b}/\bar{a} & \bar{a} < 0 \\ \pi/2 & \bar{a} = 0, \quad \bar{b} > 0 \\ 3\pi/2 & \bar{a} = 0, \quad \bar{b} < 0 \\ \mathit{unkown} & \bar{a} = 0, \quad \bar{b} = 0 \end{cases} \tag{3.5}$$

A measure of the variability of the flood occurrences about the mean date is determined by defining the mean resultant vector as:

$$\bar{r} = \sqrt{\bar{a}^2 + \bar{b}^2} \quad 0 \le r \le 1 \tag{3.6}$$

where $\bar{r}$ describes the dispersion measure (Black and Werritty 1997).

Since the distribution of dates is on a circle, rather than along a line, the use of the normal distribution is no longer appropriate. Therefore, the von Mises distribution is introduced and used to describe seasonal data with a single peak.

Fisher (1993) termed the von Mises distribution as the "natural" analog of the normal distribution for seasonal data with a single peak. It is the most commonly used and has some similar characteristics to the normal distribution (Mardia 1972). The probability density function of von Mises distribution is given by:

$$f(x) = \frac{1}{2\pi I_0(\kappa)} \exp[\kappa \cos(x - \mu)] \; 0 \le x \le 2\pi, \; 0 \le \mu \le 2\pi, \; \kappa \ge 0 \tag{3.7}$$

It is symmetric and unimodal, with a mean direction at μ and the dispersion given by a concentration parameter $\kappa = A^{-1}(r)$• $A^{-1}(r)$ is the inverse function of A•$I_0(\kappa)$ is the modified Bessel function of order zero. For large values of κ, the distribution is concentrated around the mean. When $\kappa = 0$, the density gives the uniform distribution on [0, 2].

3.2.2 *Margin Distribution of AM Flood Peaks and Volumes*

For the AM flood series, the Pearson type III (P-III) has been recommended by MWR (2006) as a uniform procedure for flood frequency analysis in China. The PDF of the P-III distribution is given in Table 1.1 of Chap. 1.

3.2.3 Bivariate Distribution of AM Flood Occurrence Dates and Magnitudes

For estimating the design flood, the bivariate joint distributions of AM flood occurrence dates and magnitudes (or flood peaks and volumes) need to be built. Every joint distribution can be written regarding a copula and its univariate marginal distributions. The copula is a function that links univariate marginal distribution functions to construct a multivariate distribution function. The definition and establishment of copulas can be seen in Chap. 2. The Gumbel copula is used to establish the joint distribution in this section.

3.2.4 Case Study

As an illustrative example, the Geheyan reservoir is selected as a case study. The Geheyan reservoir is a key control and multi-purpose water resources engineering project in the Qingjiang Basin, which is one of the main tributaries of the Yangtze River in China. The basin encompasses an area of 17,000 km^2 with the annual average rainfall 1500 mm. The annual average discharge and runoff at dam site are 393 m^3/s and 124 × 108 m^3 (from 1951 to 2005), respectively. The flood season lasts for five months from 1 May to 30 September (153 days).

3.2.4.1 Computation of Empirical Probability

The empirical probabilities can be computed by the Gringorten plotting–position formula

$$P(j) = \frac{j - 0.44}{n + 0.12} \tag{3.8}$$

where $P(j)$ is the cumulative frequency, indicating the probability that a given value is less than the jth smallest observation in the data set of n observations.

Observed joint probabilities are computed based on the same principle as in the case of a single variable. A two-dimensional table is constructed first in which the variables X and Y are arranged in ascending order. The joint cumulative frequency (non-exceedance joint probability) is then given by (Yue et al. 1999):

$$F(t_k, q_j) = P(X \leq t_k, Y \leq q_j) = \frac{\sum_{m=1}^{k} \sum_{l=1}^{j} n_{m,l} - 0.44}{n + 0.12} \tag{3.9}$$

3.2.4.2 Evaluation Criteria

A Chi-Square Goodness-of-fit test (χ^2), mean *Rbias* and *RRMSE* are selected to test the fitting descriptive ability of flood frequency curve, which can be calculated by

$$\chi^2 = \sum_{i=1}^{n} \left(P_{the}(i) - P_{emp}(i)\right)^2 / P_{emp}(i) \tag{3.10}$$

$$Rbias = \frac{1}{n}\sum_{i=1}^{n} \left(\hat{Q}(i) - Q(i)\right)/Q(i) \tag{3.11}$$

$$RRMSE = \sqrt{\frac{1}{n}\sum_{i=1}^{n}\left(\frac{\hat{Q}(i) - Q(i)}{Q(i)}\right)^2} \tag{3.12}$$

where P_{the} and P_{emp} are the theoretical and empirical frequencies; and $\hat{Q}(i)$ and $Q(i)$ are the estimated and observed values, respectively.

3.2.4.3 Conditional Probability

The parameters of Von Mises and P-III distribution are estimated by L-moments method for given AM flood series of occurrence dates, peak discharges or volumes, respectively. A Chi-Square Goodness-of-fit test is performed to test the assumption, H_0, that the flood occurrence dates and magnitudes follow the Von Mises and P-III distributions. Table 3.1 shows that the assumption cannot be rejected at the 0.5% significance level. It is shown that the values of *Rbias* and *RRMSE* are very small, which mean that the marginal distribution can fit data set very well.

Table 3.2 lists the conditional probability of $P(X > x_p|Y > y_{1\%})$ given x_p. Under the condition of annual maximum flood magnitude $Y > y_{1\%}$, the probability corresponding occurrence date after May 27 is 98.45%, the probability of annual maximum flood occurred during May 27 to 29 is (98.45−29.86%) = 68.59%, and during July 18 to 29 is (81.16−75.29%) = 5.87%.

Table 3.1 The goodness of fit and χ^2 test statistics

Index	Rbias	RRMSE	χ^2	c	$\chi^2_{0.995}(N-c-1)$
Von Mises	−4.378	0.982	0.253	2	82.001
P-III	0.254	0.327	0.903	3	80.747
Bivariate			4.400	6	76.969

Table 3.2 Conditional probability of X given $Y > y_{1\%}$

P (%)	0.01	0.1	1	10	20	30	40	50	70	90	99
x_p (Arc)	6.28	6.27	6.09	4.48	3.71	3.27	2.92	2.62	2.02	1.11	0.17
Dates	9/29	9/28	9/25	8/17	7/29	7/18	7/10	7/2	6/18	5/27	5/4
CP (%)	0.84	6.17	29.86	65.42	75.29	81.16	85.47	88.93	94.36	98.45	99.87

Note CP means the conditional probabilities $P(X > x_p | Y > y_{1\%})$

3.2.4.4 Fitting Marginal Distributions

The marginal distribution frequency curves of flood peaks and 7-day flood volumes are shown in Fig. 3.1, in which the line represents the theoretical distribution, and the crossing represents the empirical probabilities. Figure 3.1 indicates that these theoretical distributions can fit the observed data reasonably well.

The Gumbel copula is used to model the dependence between the extreme maximum annual flood peaks and 7-day flood volume in this study. The probability plot of joint distribution is shown in Fig. 3.2, in which the Gumbel copula can fit the empirical bivariate distribution very well.

3.3 Copula-Based Flood Frequency Considering Historical Information

Flood events consist of flood peaks and flood volumes that are mutually correlated and need to be described by multivariate analysis methods, of which the copula functions are most desirable ones. Until now, the multivariate flood frequency

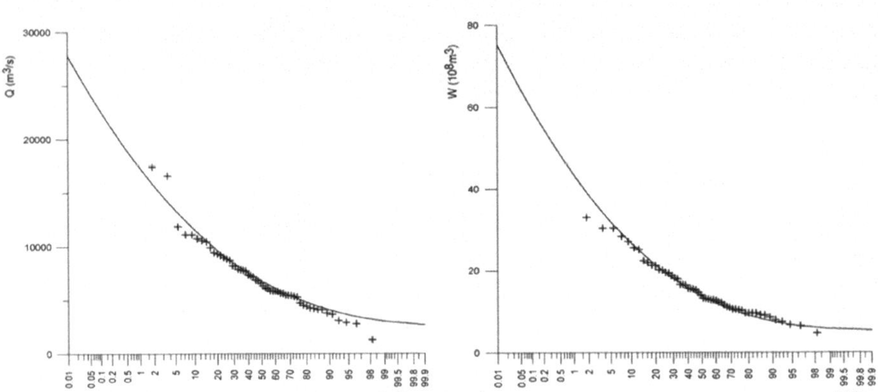

(a) Flood peak (b) 7-day flood volume

Fig. 3.1 Probability curves of flood peak and 7-day flood volume

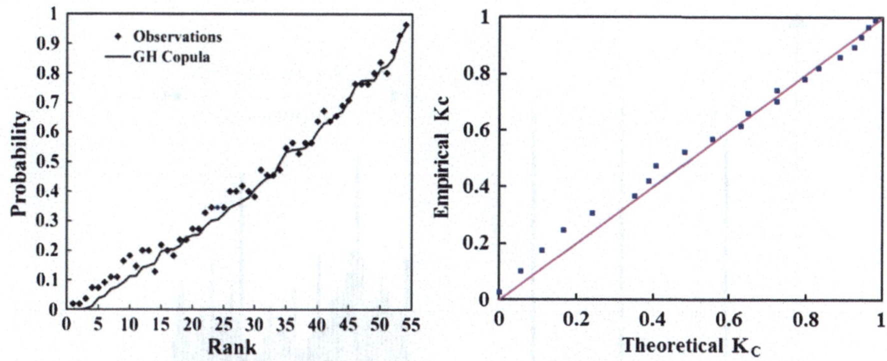

(a) Bivariate probability distribution

(b) Kendall's function results

Fig. 3.2 Comparison of observed and theoretical bivariate probability distribution

analysis methods based on copulas doesn't consider the historical flood information. This may underestimate or overestimate the flood quantiles or conditional probabilities corresponding to high return periods, especially when the length of gauged record data series is relatively short.

3.3.1 Maximum Likelihood Estimation for Censored Samples

In certain sampling situations, the exact values of a proportion of the sample are unknown, although their range may be specified. Usually, the range consists of all points above or below a threshold level. Under these circumstances, the sample is said to be censored. Censored samples occur, for example, when instruments are not calibrated for measurements above or below a certain level. Both historical data and recent flood data (i.e., systematic record) may give rise to censored samples, but because the censoring is generally above a threshold in the former and below in the latter, they must be treated separately (Leese 1973).

Censored-sample maximum likelihood estimators were initially developed by Hald (1949) and Cohen (1976) for the normal and lognormal distributions. They were subsequently adapted by Leese (1973), Condie and Lee (1982), and Stedinger and Cohn (1986) for common case in hydrology where one have both a censored-sample historical flood record and also a systematic gaged record. The maximum likelihood estimation method for type-I censoring is described as follows.

In the annual maximum flood series of Fig. 3.3, there is a total of g known floods. Of these, k is known to be the k largest in the period of n years. The n year period contains within it a systematic record (recently gauged data) of s years ($s \leq n$) length. Of the k largest floods, c occurred during the systematic record

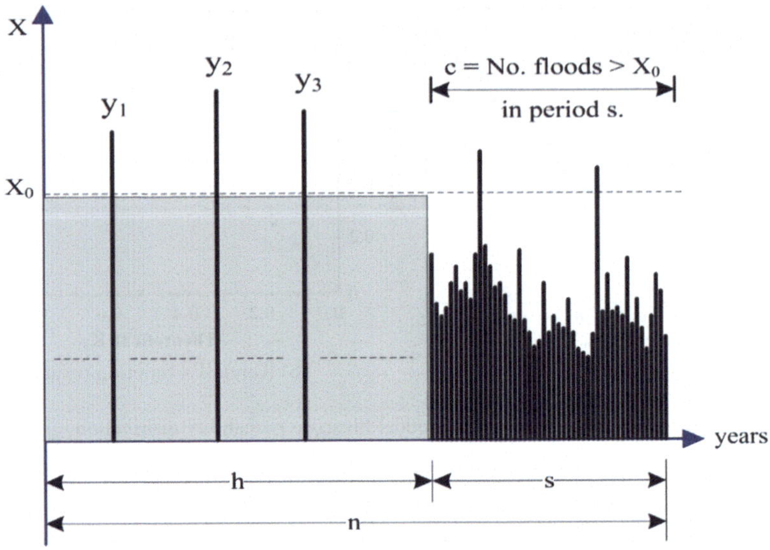

Fig. 3.3 Sketch of the annual maximum flood series when historical floods are available. Notations: s—the length of the systematic record; h—the length of the pre-gauging period; y_1, y_2, y_3—historical flood events; X_0—perception threshold

($c \leq k$ and $c < s$, and also $g = s+k-c$). Assume a fixed threshold X_0 exceeded by the k largest floods and not exceeded by any of the remaining n–k floods, recorded or not (i.e., the k values which exceed X_0 form a type I censored sample). It is also noted that the m ($m = k-c$) floods in the pre-gauging period h ($h = n-s$) are known as they are included in the k values which exceed X_0, and it is assumed that no other floods exceeded the threshold during that period.

Let f_X and F_X denote the probability density function (PDF) and the cumulative distribution function (CDF) of variable X, respectively. The resulting likelihood function for the whole sample of $s+m$ known and $h-m$ unknown values is given by (Leese 1973; Condie 1986; Stedinger and Cohn 1986; Guo and Cunnane 1991)

$$l(\boldsymbol{\alpha}) = \prod_{i=1}^{s+m} f_X(x_i)[\int_{-\infty}^{X_0} f_X(x)dx]^{h-m} \tag{3.13}$$

where α is the parameter vector of f_X and F_X.

Since c flood events exceeding the perception threshold X_0 occur among the systematic data (analogously to the sketch in Fig. 3.3), the c events are virtually removed from the period s and are treated as historical data (Bayliss and Reed 2001). Then, Eq. 3.13 can be expressed as

$$l(\boldsymbol{\alpha}) = \prod_{i=1}^{s-c} f_X(x_i) \prod_{j=1}^{k} f_X(y_j) [\int_{-\infty}^{X_0} f_X(x)dx]^{h-m} \tag{3.14}$$

where $x_i(i = 1, 2 \ldots s - c)$ denotes the systematic data less than the threshold X_0 and $y_i(j = 1, 2 \ldots k)$ denotes the k ($k = m+c$) largest floods exceeding the threshold X_0; $\prod_{i=1}^{s-c} f_X(x_i)$ and $\prod_{j=1}^{k} f_X(y_j)$ are the likelihood functions of $s-c$ systematic records and the k largest floods, respectively; and $[\int_{-\infty}^{X_0} f_X(x)dx]^{h-m}$ represents the likelihood function for the $h-m$ unknown values, which has been defined and applied by Leese (1973), Condie (1986), Stedinger and Cohn (1986), and Guo and Cunnane (1991).

The log-likelihood function for the univariate distribution can be expressed as

$$L(\boldsymbol{\alpha}) = \sum_{i=1}^{s-c} \log f_X(x_i) + \sum_{j=1}^{k} \log f_X(y_j) + (h - m) \log F_X(X_0) \tag{3.15}$$

The maximum likelihood estimates are those values of α that maximize Eq. 3.15.

3.3.2 *Bivariate Flood Frequency Analysis with Historical Information*

The conventional flood frequency analysis incorporation with historical information is based on univariate distribution. To overcome the shortcomings of univariate frequency analysis, a multivariate copula-based flood frequency analysis model that considers historical information was proposed and discussed by Li et al. (2013). As the historic flood events occurred hundreds of years ago, the durations of them are hard to measure or investigate. There is no publication or any gauged record related to the duration samples of historical floods. Besides, the perception threshold of flood duration is also difficult to fix for maximum likelihood estimation. Thus, only the distribution of flood peak and volume with historical information is studied.

3.3.3 *Inference Function for Margins Method*

In classical statistics, the inference function for margins (IFM) method was first defined as a terminology by McLeish and Small (1988). Compared with other estimation methods, the IFM method is the preferred fully parametric method for multidimensional parameter estimation because it is close to maximum likelihood (ML) in approach and is easier to implement (Joe and Xu 1996; Joe 1997). Comparisons of various types have been made in Xu (1996) for some multivariate models which suggest that the IFM method is highly efficient compared to maximum likelihood. Similar comparisons have also been made by Joe (1997), (2005) and the derived conclusions are: (1) the ML estimation is much more time-consuming than IFM method, (2) the IFM method allows one to do inference and modelling starting with univariate and lower-dimensional margins, (3) there is some robustness against misspecification of the dependence structure and also there should be more robustness against outliers or perturbations of the data, compared with the ML method; and (4) the IFM rather than the ML method avoids the sparseness problem to a certain degree, especially if parameters can all be estimated from univariate and bivariate likelihoods. Therefore, the IFM method is selected and described briefly as follows:

Under the assumption that the marginal distributions are continuous with probability density functions $f_X(x;\boldsymbol{\alpha}_1)$ and $f_Y(y;\boldsymbol{\alpha}_2)$, the joint PDF then becomes

$$f_{X,Y}(x,y;\boldsymbol{\alpha}_1,\boldsymbol{\alpha}_2,\theta) = c_\theta[F_X(x;\boldsymbol{\alpha}_1),F_Y(y;\boldsymbol{\alpha}_2)]f_X(x;\boldsymbol{\alpha}_1)f_Y(y;\boldsymbol{\alpha}_2) \tag{3.16}$$

where F_X and F_Y are univariate CDFs with respective parameter vectors $\boldsymbol{\alpha}_1$, $\boldsymbol{\alpha}_2$, and c_θ is the density of C_θ parametrized by a parameter θ, defined as

$$c_\theta(u,v) = \frac{\partial^2 C_\theta(u,v)}{\partial u \partial v} \tag{3.17}$$

For the observed bivariate series $(x_1, y_1),\ldots, (x_s, y_s)$ with a sample size s, we can consider the two log-likelihood functions for the univariate marginal distribution, i.e.

$$L_1(\boldsymbol{\alpha}_1) = \sum_{i=1}^{s} \log f_X(x_i;\boldsymbol{\alpha}_1) \tag{3.18a}$$

$$L_2(\boldsymbol{\alpha}_2) = \sum_{i=1}^{s} \log f_Y(y_i;\boldsymbol{\alpha}_2) \tag{318b}$$

and the log-likelihood function for the joint distribution,

$$L(\theta,\boldsymbol{\alpha}_1,\boldsymbol{\alpha}_2) = \sum_{i=1}^{s} \log f_{X,Y}(x_i,y_i;\boldsymbol{\alpha}_1,\boldsymbol{\alpha}_2,\theta) \tag{3.19}$$

The IFM method consists of two separate optimizations of univariate likelihoods, followed by an optimization of multivariate likelihood as a function of the dependence parameter vector. More specifically,

(a) The log-likelihoods $L_1(\boldsymbol{\alpha}_1)$ and $L_2(\boldsymbol{\alpha}_2)$ of the two univariate marginal distributions are separately maximized by Eq. 3.18a, 318b to get estimates $\hat{\boldsymbol{\alpha}}_1$ and $\hat{\boldsymbol{\alpha}}_2$;
(b) The function $L(\theta, \hat{\boldsymbol{\alpha}}_1, \hat{\boldsymbol{\alpha}}_2)$ is maximized over θ to get $\hat{\theta}$ in Eq. 3.19.

That is, under regularity conditions, $(\hat{\boldsymbol{\alpha}}_1, \hat{\boldsymbol{\alpha}}_2, \hat{\theta})$ is the solution of

$$(\partial L_1/\partial \boldsymbol{\alpha}_1, \partial L_2/\partial \boldsymbol{\alpha}_2, \partial L/\partial \theta) = 0 \tag{3.20}$$

This procedure is computationally simpler than that of estimating all parameters $\boldsymbol{\alpha}_1, \boldsymbol{\alpha}_2, \theta$ simultaneously in Eq. 3.19.

3.3.4 *Modified IFM Method with Incorporation of Historical Information*

Since the current IFM method can only be used for systematic data series, a modified IFM (MIFM) method with an incorporation of historical and paleological information is proposed and described as follows.

Let x_i and y_i (i = 1,..., s-c) respectively denote the systematic data of marginal distributions (flood peak and volume); g_j and p_j (j = 1,..., k) respectively denote the k largest floods of marginal distributions (flood peak and volume) with the same years of occurrence. Of the k largest floods, c occurred during the systematic record and m occurred during the pre-gauging period h ($k = m+c$ and $h = n - s$); X_0 (or Y_0) is the fixed threshold of margin exceeded by the k largest flood peaks (or volumes) and not exceeded by any of the remaining $n - k$ flood peaks (or volumes). Furthermore, let f_x, and f_y denote the univariate marginal PDFs, and F_x, and F_y denote the univariate marginal CDFs of variables X and Y, respectively. f_{XY} denotes the joint PDF.

Referring to Eq. 3.14, the likelihood function with historical floods for joint distributions can be described as

$$\begin{aligned} l(\theta, \boldsymbol{\alpha}_1, \boldsymbol{\alpha}_2) &= \prod_{i-1}^{s-c} f_{XY}(x_i, y_i) \prod_{j-1}^{k} f_{XY}(g_j, p_j) [\int_{-\infty}^{X_0} \int_{-\infty}^{Y_0} f_{XY}(x, y) dx dy]^{h-m} \\ &= \prod_{i=1}^{s-c} f_{XY}(x_i, y_i) \prod_{j=1}^{k} f_{XY}(g_j, p_j) \{ C_\theta [F_X(X_0), F_Y(Y_0)] \}^{h-m} \end{aligned} \tag{3.21}$$

Then, the log-likelihood function for joint distribution can be expressed as:

$$\begin{aligned}L(\theta, \boldsymbol{\alpha}_1, \boldsymbol{\alpha}_2) = &\sum_{i=1}^{s-c} \log c_\theta[F_X(x_i), F_Y(y_i)] + \sum_{j=1}^{k} \log c_\theta[F_X(g_j), F_Y(p_j)] \\ &+ (h-m)\log C_\theta[F_X(X_0), F_Y(Y_0)] + \sum_{i=1}^{s-c} \log f_X(x_i) + \sum_{j=1}^{k} \log f_X(g_j) \\ &+ \sum_{i=1}^{s-c} \log f_Y(y_i) + \sum_{j=1}^{k} \log f_Y(p_j)\end{aligned} \tag{3.22}$$

In which, the two log-likelihood functions for the univariate marginal distribution are

$$L_1(\boldsymbol{\alpha}_1) = \sum_{i=1}^{s-c} \log f_X(x_i) + \sum_{j=1}^{k} \log f_X(g_j) \tag{3.23}$$

$$L_2(\boldsymbol{\alpha}_2) = \sum_{i=1}^{s-c} \log f_Y(y_i) + \sum_{j=1}^{k} \log f_Y(p_j) \tag{3.24}$$

Similar to the IFM method, the MIFM method also consists of two separate procedures:

(a) The log-likelihoods $L_1(\boldsymbol{\alpha}_1)$ and $L_2(\boldsymbol{\alpha}_2)$ are separately maximized by Eqs. 3.23 and 3.24 to get estimates $\hat{\boldsymbol{\alpha}}_1$ and $\hat{\boldsymbol{\alpha}}_2$;
(b) The function $L(\theta, \hat{\boldsymbol{\alpha}}_1, \hat{\boldsymbol{\alpha}}_2)$ is maximized by Eq. 3.22 over θ to get $\hat{\theta}$.

As a consequence, the precious historical information is used to estimate not only the parameters of marginal distributions but also the dependence parameters of joint distribution that is based on the correlation of the marginal distributions. The more additional information of marginal distribution provides, the more precise dependence structure will be obtained.

3.3.5 *Case Study*

The Three Gorges reservoir (TGR) in China is selected as an illustrative example. The basin area of TGR is one million km^2, and the annual average discharge and runoff volume at the dam site are 14,300 m^3/s and 4510 $\times$ 10^8 m^3, respectively. The TGR located on middle reaches of the Yangtze River is the largest water conservancy project in the world, with a normal pool level at an elevation of 175 m. The total storage capacity of the TGR is 393 $\times$ 10^8 m^3, of which 221.5 $\times$ 10^8 m^3

is flood control storage, and 165 × 10^8 m^3 is the conservation regulating storage volume. With 26 hydro-generators installed, the mean annual electricity output of the TGR reaches up to 847 × 10^8 kW•h. The TGR also plays a key role in the flood prevention of Yangtze River basin which is the richest area in China (Li et al. 2010).

3.3.5.1 Systematic Record and Historical Floods

The annual maximum peak discharge (Q), 3-day flood volume (W_3), and 15-day flood volume (W_{15}) are available with a systematic record of 128 years (1882–2009, i.e., no systematic data are formally gauged before 1882). Besides the systematic observations, a lot of historical flood events had been investigated by CWRC (Changjiang Water Resources Commission) in the last century for the design of the Three Gorges Project. The gathered information from gauging authority records, historical documents, archives, flood marks and stone inscriptions showed the concrete positions of high water stages recorded. As a result, the eight largest historical floods since 1153 were quantificationally evaluated by CWRC and other relevant units (CWRC 1996).

As the same notations defined previously, the length of the systematic observations is unequivocally given: s = 128 years; since no extraordinary flood occurred during the systematic record, c = 0 and k = m; for the joint distribution of flood peak (Q) and 3-day flood volume (W_3), k = m = 8; for the joint distribution of flood peak and 15-day flood volume (W_{15}), k = m = 3; the perception thresholds of peak discharge, 3-day flood volume and 15-day flood volume are X_{0Q} = 80,000 m^3/s, X_{0w3} = 200 × 10^8 m^3 and X_{0w15} = 780 × 10^8 m^3, respectively; and the pre-gauging period, h = 730 (i.e. from 1153 to 1882). These data settings are also listed in Table 3.3.

3.3.5.2 Parameter Estimation for Marginal Distributions

The empirical probabilities of univariate discontinuous series can be computed by Weibull formula recommended by MWR (2006)

$$P_i = P(x \geq x_i) = \begin{cases} P_h(i) = \frac{i}{n+1} & i = 1, \cdots, k \\ P_s(i) = P_h(k) + (1 - P_h(k)) \times \frac{i}{s-c+1} & i = 1, \cdots, s-c \end{cases} \tag{3.25}$$

Table 3.3 Data settings for the modified IFM method

Variables	Threshold X_0/Y_0	h	s	k	m
Q (m^3/s)	80,000	730	128	8	8
W_3 (10^8 m^3)	200				
Q (m^3/s)	80,000	730	128	3	3
W_{15} (10^8 m^3)	780				

where P_i represents the exceedance probability; $P_h(i)$ is the empirical probabilities of historical floods for $i = 1,\ldots, k$; $P_s(i)$ is the empirical probabilities of systematic data for $i = 1,\ldots, s-c$; and the meanings of n, k, s, c are the same as those defined in Fig. 3.3.

The parameters of the P-III marginal distributions estimated by the first stage of the MIFM method in Eqs. 3.23 and 3.24 are listed in Table 3.4. A Chi-Square Goodness-of-fit test is performed to test the assumption, H_0, that the flood magnitudes follow the P-III distribution. Table 3.5 shows that the assumption cannot be rejected at the 5% significance level. The marginal distribution frequency curves of flood peak and flood volumes are drawn in Fig. 3.4, in which the line represents the theoretical distribution, the crossings and circles represent systematic record and historical flood data, respectively. Figure 3.4 indicates that all the theoretical distributions can fit the observed data reasonably well.

3.3.5.3 Empirical Joint Probabilities of Dependence Flood Variables

Empirical (observed) joint probabilities of flood peak (Q) and volume (W) are computed in a manner analogous to that for a univariate variable. A two-dimensional table is constructed in which the variable X and Y are arranged in descending order. The joint probabilities (exceedance) of k historical floods and s-c systematic data are empirically computed separately, which are expressed as

$$F(x_i, y_i) = P(X \geq x_i, Y \geq y_i) = \begin{cases} P_h(i) = \dfrac{\sum_{l=1}^{i}\sum_{p=1}^{i} N_{lp}}{n+1} & i = 1,\ldots,k \\ P_s(i) = P_h(k) + (1 - P_h(k)) \times \dfrac{\sum_{l=1}^{i}\sum_{p=1}^{i} M_{lp}}{s-c+1} & i = 1,\ldots,s-c \end{cases} \tag{3.26}$$

where $F(x_i, y_i)$ is obtained by arranging the number of (x_i, y_i) by either x_i or y_i; $P_h(i)$ is the empirical joint probabilities of historical floods and N_{lp} is the number of (x_i, y_i) counted as $x_j \geq x_i$ and $y_j \geq y_i$, $i = 1,\ldots, k$, $1 \leq j \leq i$; $P_s(i)$ is the empirical

Table 3.4 Estimated parameters of P-III marginal distributions for flood peak and volumes by MIFM

Variables	$\hat{\alpha}$	$\hat{\beta}$	$\hat{\delta}$
Q (m^3/s)	11.11	0.0003	17066.7
W_3 (10^8 m^3)	11.89	0.1348	39.7
W_{15} (10^8 m^3)	18.26	0.0463	118.22

Table 3.5 Hypothesis test results of P-III marginal distributions for flood peak and volumes

Variables	$\chi_{0.05}$	Chi-Square statistics, χ^2
Q (m^3/s)	7.815	4.924
W_3 (10^8 m^3)	9.488	5.048
W_{15} (10^8 m^3)	7.815	4.110

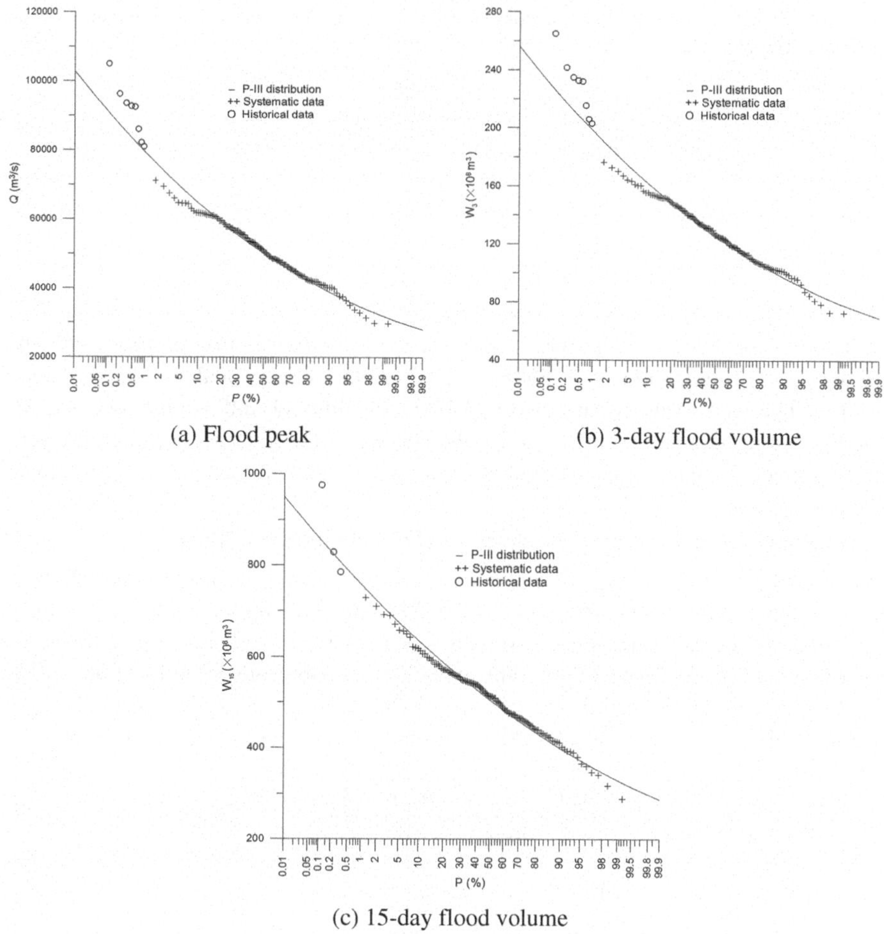

(a) Flood peak

(b) 3-day flood volume

(c) 15-day flood volume

Fig. 3.4 P-III distributions fitted to flood peak and volumes with historical information

joint probabilities of systematic data and M_{lp} is the number of (x_i, y_i) counted as $x_j \geq x_i$ and $y_j \geq y_i$, $i = 1,\ldots, s-c$, $1 \leq j \leq i$; and n is the total length of the analyzed time period ($n = s+h$).

3.3.5.4 Identification of Copula

The parameters of marginal distributions are estimated in the first stage of MIFM method. The dependence parameter θ is obtained by maximizing the log-likelihood function of the joint distribution. For Gumbel copula, the estimation results are $\theta = 16.2524$ for the joint distribution of flood peak and 3-day flood volume, and $\theta = 3.2977$ for that of flood peak and 15-day flood volume. For Student copula, the

Table 3.6 RMSE of Gumbel and student's copulas and upper TDC estimated by parametric and nonparametric methods

Variables	RMSE		λ_U of copula		$\hat{\lambda}_U^{LOG}$	$\hat{\lambda}_U^{SEC}$	$\hat{\lambda}_U^{CFG}$
	Gumbel	Student	Gumbel	Student			
Q (m^3/s) W_3 (10^8 m^3)	0.0262	0.0874	0.9564	0.8954	0.9442	0.9511	0.9482
Q (m^3/s) W_{15} (10^8 m^3)	0.0413	0.2149	0.7661	0.5262	0.7218	0.7618	0.7109

estimation results are ($\theta = 0.9947$, $v = 6$) for the joint distribution of flood peak and 3-day flood volume, and ($\theta = 0.8598$, $v = 5$) for that of flood peak and 15-day flood volume. The root mean square errors (*RMSE*) of Gumbel and Student copulas are listed in Table 3.6. The comparison results show that the Gumbel copula represents the bivariate distribution of correlated flood peak and volumes better than that of Student copula.

The upper tail dependence coefficients (TDC) of Gumbel copula ($\lambda_U = 2 - 2^{1/\theta}$) and student's t copula ($\lambda_U = 2t_{v+1}\left(-\sqrt{(v+1)(1-\theta)/(1+\theta)}\right)$) are computed by the estimated parameters and listed in Table 3.6. The upper TDC can also be estimated by the nonparametric estimation, which is a much more general as no assumption is made about copula and marginal distributions (Poulin et al. 2007). The Log, Sec and CFG estimators of upper TDC (Coles et al. 1999; Joe et al. 1997; Poulin et al. 2007; Frahm et al. 2005) are respectively determined as follows.

$$\hat{\lambda}_U^{LOG} = 2 - \frac{\log C_n((n-k)/n, (n-k)/n)}{\log((n-k)/n)}, \quad 0<k<n \tag{3.27}$$

$$\hat{\lambda}_U^{SEC} = 2 - \frac{1 - C_n((n-k)/n, (n-k)/n)}{1-(n-k)/n}, \quad 0<k<n \tag{3.28}$$

$$\hat{\lambda}_U^{CFG} = 2 - 2\exp\left[\frac{1}{n}\sum_{i=1}^{n}\log\left(\sqrt{\log\frac{1}{U_i}\log\frac{1}{V_i}}\Big/\log\frac{1}{\max(U_i, V_i)^2}\right)\right] \tag{3.29}$$

in which

$$C_n(u, v) = \frac{1}{n}\sum_{i=1}^{n}\mathbf{I}\left(\frac{R_i}{n+1} \le u, \frac{S_i}{n+1} \le v\right) \tag{3.30}$$

where $C_n(u, v)$ is the empirical copula, I denote the indicator function, R_i and S_i are the ranks of block maxima x_i and y_i, respectively. $\{(U_1, V_1), \ldots, (U_n, V_n)\}$ denote random sample obtained from the copula C.

The nonparametric estimation results of upper TDC are calculated and also listed in Table 3.6. The comparison results of Table 3.7 show that the upper TDC of Gumbel copula is much closer to the nonparametric estimation results than that of

Table 3.7 Parameters of marginal distributions and copula estimated by different data and methods

Variables	IFM				MIFM			
	P-III			Copula	P-III			Copula
	$\hat{\alpha}$	$\hat{\beta}$	$\hat{\delta}$	$\hat{\theta}$	$\hat{\alpha}$	$\hat{\beta}$	$\hat{\delta}$	$\hat{\theta}$
Q (m^3/s)	13.72	0.0004	16933.3	15.1545	11.11	0.0003	17066.7	16.2524
W_3 (10^8 m^3)	15.75	0.1736	36.28		11.89	0.1348	39.7	
Q (m^3/s)	13.72	0.0004	16933.3	3.0962	11.11	0.0003	17066.7	3.2977
W_{15} (10^8 m^3)	22.15	0.0541	102.38		18.26	0.0463	118.22	

student copula. This indicates that Gumbel copula reproduces better the observed tail dependence coefficient, and the extreme behavior of Gumbel copula is more similar to that of the sample. Therefore, the Gumbel copula is used to model the dependence between the extreme maximum annual flood peak and volumes in this study.

3.3.5.5 Copula-Based Conditional Distributions

The conditional flood distributions with historical flood data can be easily derived if the copula-based bivariate flood distribution is constructed. For instance, the conditional distributions for flood volume given that the peak discharge exceeding a certain threshold q_{x0} can be expressed as

$$P(W \le w | Q > q_{X0}) = \frac{P(W \le w, Q > q_{X0})}{P(Q > q_{X0})} = \frac{F_Y(w) - C_\theta[F_X(q_{X0}), F_Y(w)]}{1 - F_X(q_{X0})} \quad (3.31a)$$

$$P(W > w | Q > q_{X0}) = \frac{P(W > w, Q > q_{X0})}{P(Q > q_{X0})} = \frac{1 - F_X(q_{X0}) - F_Y(w) + C_\theta[F_X(q_{X0}), F_Y(w)]}{1 - F_X(q_{X0})} \quad (3.31b)$$

where F_x and F_Y represent the marginal distributions, and θ represents the dependence parameter of the bivariate distribution.

Likewise, the conditional distribution functions for peak discharge given that the flood volumes exceeding a certain threshold W_{Y0} can be expressed as

$$P(Q \le q | W > w_{Y0}) = \frac{P(Q \le q, W > w_{Y0})}{P(W > w_{Y0})} = \frac{F_X(q) - C_\theta[F_X(q), F_Y(w_{Y0})]}{1 - F_Y(w_{Y0})} \quad (3.32a)$$

$$P(Q > q | W > w_{Y0}) = \frac{P(Q > q, W > w_{Y0})}{P(W > w_{Y0})} = \frac{1 - F_X(q) - F_Y(w_{Y0}) + C_\theta[F_X(q), F_Y(w_{Y0})]}{1 - F_Y(w_{Y0})} \tag{3.32b}$$

The historical floods, which usually occurred as extraordinary events, may help exposit the correlation of variables with high return period. As a consequence, the incorporation of historical information into bivariate frequency analysis can provide better insight into the dependence structure of variables. The conditional probabilities accounting for historical floods can provide more comprehensive and adequate information, which is useful in evaluating the flood prevention capability.

3.3.5.6 Comparative Study and Discussions

The comparative study and discussions of MIFM and IFM methods are conducted in this section. First, the parameters of marginal distributions (Q, W_3, and W_{15}) and copulas are estimated by IFM and MIFM methods, respectively. Table 3.7 shows that the different data and methods lead to different parameter estimation results of both marginal distributions and copula. Second, the quantiles of flood peak (Q), 3-day flood volume (W_3) and 15-day flood volume (W_{15}) are estimated by univariate distribution (Chinese design flood guidelines), MIFM and IFM methods, respectively.

The Relative Errors (RE) of T-year quantile estimator are calculated by

$$\text{RE} = \frac{\hat{X}_T - X_T}{X_T} \times 100\% \tag{3.33}$$

where X_T is the univariate quantile estimated by univariate distribution (Chinese design flood guidelines) with an incorporation of historical information; $\hat{X}_T$ represents the bivariate quantiles estimated by MIFM method with an incorporation of historical information or by IFM method using systematic records alone.

The relative errors (*RE*) of flood peak, 3-day flood volume, and 15-day flood volume are calculated and listed in Tables 3.8, 3.9, and 3.10, respectively. The results of these tables indicate that the bivariate quantiles estimated by MIFM

Table 3.8 Comparison of quantile Q estimated by univariate and bivariate distributions

T (years)	Univariate quantile Q_T (m^3/s)	MIFM		IFM	
		$\hat{Q}_T$ (m^3/s)	*RE* (%)	$\hat{Q}_T$ (m^3/s)	*RE* (%)
10,000	102,900	103,100	0.19	95,900	−6.80
1000	91,700	91,900	0.22	86,400	−5.78
100	79,400	79,700	0.38	75,800	−4.53
Mean relative error			0.26		−5.70

Table 3.9 Comparison of quantile W_3 estimated by univariate and bivariate distributions

T (years)	Univariate quantile W_{3T} (10^8 m^3)	MIFM		IFM	
		$\hat{W}_{3T}$ (10^8m^3)	RE (%)	$\hat{W}_{3T}$ (10^8m^3)	RE (%)
10,000	255.9	256.3	0.16	246.0	−3.87
1000	228.4	228.9	0.22	220.8	−0.33
100	198.0	198.6	0.30	193.0	−2.53
Mean relative error			0.23		−3.24

Table 3.10 Comparison of quantile W_{15} estimated by univariate and bivariate distributions

T (years)	Univariate quantile W_{15T} (10^8 m^3)	MIFM		IFM	
		$\hat{W}_{15T}$ (10^8 m^3)	RE (%)	$\hat{W}_{15T}$ (10^8 m^3)	RE (%)
10,000	950.3	958.2	0.83	924.4	−2.73
1000	859.5	868.1	1.00	842.5	−1.97
100	757.9	767.8	1.31	750.7	−0.95
Mean relative error			1.05		−1.88

approach is much closer to the univariate quantiles than that estimated by IFM method. The quantiles estimated by IFM method are much smaller than that of Chinese design flood guidelines. The mean relative errors are equal to −5.70, −3.24, and −1.88% for flood peak, 3-day flood volume, and 15-day flood volume, respectively.

3.4 Bivariate Design Flood Quantile Selection Using Copulas

To derive the feasible range, a boundary identification method is suggested, which is inspired by the ideas of Chebana and Ouarda (2011) and Volpi and Fiori (2012). Li et al. (2016) estimated the bivariate feasible ranges of flood peak and flood volume suitable for combination in the critical level curve. Two combination methods for estimating unique bivariate flood quantiles, i.e., the EFC method and the CEC method, are proposed based on the assumption of the relationship between u and v (or q and w).

3.4.1 Bivariate Return Period

In the conventional univariate analysis, flood events of interest are often defined by return periods. In the bivariate domain, however, it is still discussed by the community as to which method is most suitable to transform the joint exceedance probability to a bivariate joint return period (JRP). Different JPRs estimated by copula function have been developed for the case of a bivariate flood frequency analysis. Eight types of possible joint events were presented by Salvadori and De Michele (2004) using "OR" and "AND" operators, of which, two cases are of the greatest interest in hydrological applications (Shiau et al. 2006; Salvadori and De Michele 2004):

(1) (OR case) either $Q > q$ or $W > w$, i.e.,

$$E_{or} = \{Q > q \, \text{or} \, W > w\} \tag{3.34}$$

(2) (AND case) both $Q > q$ and $W > w$, i.e.,

$$E_{and} = \{Q > q \, \text{and} \, W > w\} \tag{3.35}$$

In simple words: for E_{or} to happen it is sufficient that either peak discharge Q or flood volume W (or both) exceed given thresholds; instead; for E_{and} to happen it is necessary that both Q and W are larger than prescribed values. Thus, two different JRPs can be defined accordingly (De Michele et al. 2005):

$$T_{or} = \frac{\mu}{P[Q > q \, or \, W > w]} = \frac{\mu}{1 - F(q, w)} \tag{3.36}$$

$$T_{and} = \frac{\mu}{P[Q > q \, and \, W > w]} = \frac{\mu}{1 - F_Q(q) - F_W(w) + F(q, w)} \tag{3.37}$$

where μ is the mean inter-arrival time between two consecutive events (in the case of annual maxima μ = 1 year), and $F(q, w) = P(Q \leq q, W \leq w)$.

The Kendall JRP was introduced by Salvadori and De Michele (2004) to identify the univariate critical threshold in a multivariate context, which is given by:

$$\theta_t = \frac{\mu_T}{1 - K_C(t)} \tag{3.38}$$

where K_C is the Kendall's distribution function associated with the joint cumulative distribution function of the copula's level curves: $K_C(\text{t}) = P[C(\text{u}, \text{v}) \leq t]$. It allows for the calculation of the probability that a random point (u, v) in the unit square has a smaller (or larger) copula value than a given critical probability level t. In other words, it is related to the probability of occurrence of an event in the area over the copula level curve of value t.

Different definitions of the multivariate return period are available in the literature, based on regression analysis, bivariate conditional distributions, survival Kendall distribution function, and structure performance function. For instance, some studies have focused on a structure-based return period for the design and or risk assessment of hydrological structures in a bivariate environment (Volpi and Fiori 2014).

A comprehensive review of the JRP estimation methods was given by Volpi and Fiori (2014).

The OR return period given in Eq. 3.36 has been extensively applied in multivariate hydrological frequency analysis (e.g., Shiau et al. 2006; Salvadori and De Michele 2004; Chebana and Ouarda 2011; Volpi and Fiori 2012; Li et al. 2013). In this study, we focus on the OR case for quantile estimation in a bivariate context.

3.4.2 *Feasible Range Identification for Bivariate Quantile Curve*

The critical level curve, as shown in Fig. 3.5, was defined as a bivariate quantile curve by Chebana and Ouarda (2011). As previously stated, for the case of OR return period, the function that describes the level curve for any given return period T or critical probability level p has two asymptotes, $q = q_p$ and $w = w_p$, where $q_p = F_Q^{-1}(p)$ and $w_p = F_W^{-1}(p)$ are the quantiles of the marginal distribution for the given probability level p. According to Eq. 3.36 in the bivariate case, the choice of an appropriate return period T or a critical probability level p for hydraulic structure design will lead to the infinite combinations of flood peak and volume. However, all the bivariate flood events with the same value of T or p along the level curve differ greatly not only in terms of their quantile values, but also in terms of their probability of occurrence, which is measured by the joint probability density function (PDF), i.e., $f(q, w)$, evaluated along the critical level curve (Volpi and Fiori 2012). Meanwhile, different combinations of Q and W are generally not equivalent from a practical point of view, although they all satisfy the flood prevention standards. The boundaries (see points B and C in Fig. 3.5) for selection of design flood peak and volume are necessary in the case that the flood combinations are outside the boundaries with unrealistically low occurrence probabilities.

Chebana and Ouarda (2011) proposed a method to decompose the quantile curve in Fig. 3.5 into a naive part (i.e., the subset BC) and a proper part (outside subset

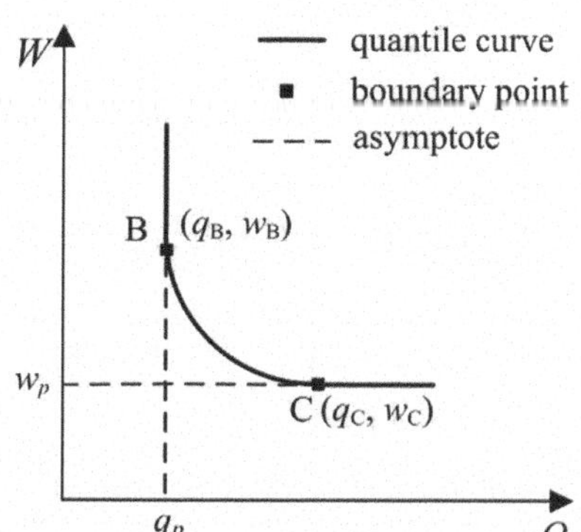

Fig. 3.5 Bivariate quantile curve with a critical probability level p

BC). They assumed that the naive part is composed of two segments starting at the end of each extremity of the proper part. They also suggested selecting these boundary points according to the empirical version or as close as to the asymptotes (the naive part). Volpi and Fiori (2012) defined the distance of each point along the quantile curve in Fig. 3.5 from its vertex as a random variable (s) and derived its PDF. The boundary points of the quantile curve are identified with a chosen percentage in the probability of the events. They also proposed a way of decomposition of the quantile curve into the naive part and proper part. However, the procedure presented by Volpi and Fiori (2012) is difficult to apply in the curvilinear coordinate system [$s(x, y)$, $n(x, y)$] or to derive the expression of a random variable (s). To overcome these limitations, an approach to identify the boundary points (i.e., B and C) of the quantile curve is developed. A new density function $\varphi(q)$ is used to measure the relative likelihood of flood events, which is a non-curvilinear variable in the procedure.

To derive the new density function with a chosen probability level to decompose the quantile curve, a joint distribution of annual maximum flood peak (Q) and flood volume (W) should be built by copula functions. The joint distribution function $F(q, w)$ can be expressed in terms of its marginal functions and $F_W(w)$ by using an associated dependence function C, $F(q, w) = C[F_Q(q), F_W(w)]$.

It is found that flood peak and volumes are usually upper-tailed dependent variables and the Gumbel copula can reproduce best the observed tail dependence coefficient (e.g., Poulin et al. 2007) Therefore, the Gumbel copula is taken as an example to illustrate the developed boundary identification method because of its easy expression and wide applications (Li et al. 2013).

For the Gumbel copula function, the relationship of joint distribution $C_\theta(u, v)$ and bivariate return period T can be expressed as (μ = 1 for annual maxima flood series):

$$C_\theta(u, v) = \exp\{-[(-\ln u)^\theta + (-\ln v)^\theta]^{1/\theta}\} = 1 - \frac{1}{T} \quad (3.39)$$

where θ is the dependence parameter of the Gumbel copula, $u = F_Q(q)$, $v = F_W(w)$.

Thus, the relationship between u and v with the given bivariate return period T can be derived as:

$$v = \exp\left\{-[(-\ln u)^\theta - (-\ln(1 - \frac{1}{T}))^\theta]^{1/\theta}\right\} \quad (3.40)$$

Replacing $u = F_Q(q)$, and $v = F_W(w)$into the above equation yields:

$$F_W(w) = \exp\left\{-[(-\ln F_Q(q))^\theta - (-\ln(1 - \frac{1}{T}))^\theta]^{1/\theta}\right\} = \eta(F_Q(q)) \quad (3.41)$$

in which, $\eta(x) = \exp\left\{-[(-\ln x)^\theta - (-\ln(1 - \frac{1}{T}))^\theta]^{1/\theta}\right\}$

Thus, the relationship between Q and W with the fixed bivariate return period T can be derived as:

$$w = F_W^{-1}(v) = F_W^{-1}(\eta(F_Q(q)) = \varsigma(q) \tag{3.42}$$

where $F_W^{-1}(v)$ is the inverse CDF of flood volume W. The above equation reveals that W can be derived by Q if the bivariate return period T is fixed.

It should be noted that other copulas with more complicated formulas sometimes may be needed. For the Frank copula, Clayton copula and several two-parameter copulas, the implicit expression for describing the relationship between Q and W in Eqs. 3.39 to 3.42 can be derived. For copulas with more complicated expressions, the numerical method should be applied. For example, the unique value of w could be obtained with given q by a trial and error method.

After obtaining the corresponding relationship of the values of w and q for the flood events along the critical level curve, the bivariate joint PDF of w and q can be expressed according to Sklar's theory as (Nelsen 2006):

$$f(q,w) = c_\theta(F_Q(q), F_W(w)) \cdot f_Q(q) \cdot f_W(w) \tag{3.43}$$

where $f_Q(q)$ and $f_W(w)$ are univariate PDFs of flood peak and volume, respectively, and $c_\theta(u,v)$ is the density of $C_\theta(u,v)$ and defined as:

$$c_\theta = \frac{\partial^2 C_\theta(u,v)}{\partial u \partial v} \tag{3.44}$$

Referring to Eqs. 3.41 and 3.42, the bivariate joint PDF of flood peak and volume can be finally described as the function of the single random variable of flood peak Q for the fixed bivariate return period T, i.e.,

$$f(q,w) = c_\theta(F_Q(q), \eta(F_Q(q))) \cdot f_Q(q) \cdot f_W(\varsigma(q)) \tag{3.45}$$

According to Eq. 3.45, there is a curve that can describe the relationship between joint PDF $f(q, w)$and flood peak Q for a given bivariate return period T or a critical probability level p. Assume that the area between the curve of $f(q, w)$ and the horizontal axis of flood peak Q is A, i.e.,

$$A = \int_{q_p}^{+\infty} f(q,w)dq = \int_{q_p}^{+\infty} c(F_Q(q), \eta(F_Q(q))) \cdot f_Q(q) \cdot f_W(\varsigma(q))dp \tag{3.46}$$

where q_p represents univariate design value of flood peak, i.e., $q_p = F_Q^{-1}(p)$, which is chosen as the lower bound of flood peak in the estimation of the bivariate design flood values.

As $f(q, w)$ is a joint density function of q and w, area A does not equal to 1 if only q is taken as an integral variable (i.e., $A \neq 1$). A new density function $\varphi(q)$ over the area A which has proper density characters is constructed and expressed as follows:

$$\varphi(q) = \frac{f(q, w)}{A} = \frac{f(q, w)}{\int_{q_p}^{+\infty} f(q, w) dq} \tag{3.47}$$

Obviously, there is a one-to-one correspondence between the density function $\varphi(q)$ and bivariate PDF $f(q, w)$. The density function $\varphi(q)$ varies with the horizontal axis and $\int_{q_T}^{+\infty} \varphi(q) dq = 1$.

As previously stated, the bivariate design flood combinations near the upper and lower bounds of the quantile curve have lower occurrence probability than that near the middle of the quantile curve. As a consequence, the bivariate PDF $f(q, w)$ of bivariate design flood combination near the upper and lower bounds of quantile curve is smaller than that near the middle of the quantile curve. The density function $\varphi(q)$ has the same property as the bivariate PDF $f(q, w)$. As the design flood peak (or flood volume) varies from the lower bound, i.e., (q_p) to infinitely great, the density function $\varphi(q)$ increases to the maximum value and then decreases gradually, as shown in Fig. 3.6. The vertex of the density function $\varphi(q)$ describing the full dependence (Chebana and Ouarda 2011; Volpi and Fiori 2012) between peak and volume has the highest density. In other words, this is the most likely bivariate design flood event.

Once the density function $\varphi(q)$ along Q is defined by Eq. 3.43, we can evaluate the lower and upper bounds that contain $\varphi(q)$ with probability of $1-\varepsilon$, for a given probability level ε. The quantiles of lower and upper bounds (q_{B} and q_{C}) are specified respectively by (Volpi and Fiori 2012):

$$\int_{q_p}^{q_B} \varphi(q) dq = \alpha_1 \tag{3.48}$$

$$\int_{q_p}^{q_C} \varphi(q) dq = 1 - \alpha_2 \tag{3.49}$$

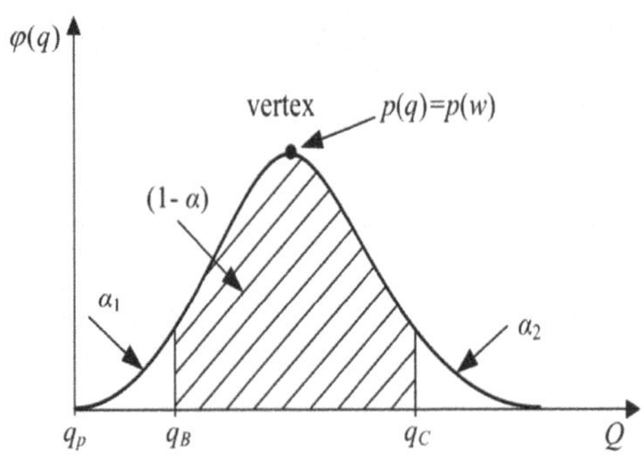

Fig. 3.6 Relationship between density function $\varphi(q)$ and flood peak Q

where $\alpha_1 + \alpha_2 = \varepsilon$. The lower and upper bounds q_B and q_C identify a feasible range on the quantile curve, bounded by the points of coordinates $(q_B, \zeta(q_B))$ and $(q_C, \zeta(q_C))$, that excludes the ε percentage in the probability of the critical events. The probability levels α_1 and α_2 can be arbitrarily chosen, taking account of the specific problem under investigation (Volpi and Fiori 2012).

3.4.3 *Bivariate Flood Quantile Selection*

For a given bivariate return period T, there are countless combinations of u and v that satisfy Eq. 3.39. To derive the design values of flood peak q and flood volume w, the unique combination of u and v (or q and w) should be determined. Hence besides Eq. 3.39, one more equation that can establish the relationship between u and v (or q and w) is necessary. Two combination methods were proposed to derive the quantiles of flood peak and flood volume for given multivariate return periods, and they are now outlined.

3.4.3.1 Equivalent Frequency Combination Method

With a given bivariate return period T, we assume that the flood peak and flood volume have the same probability of occurrence, i.e., $u = v$ (or $F_Q(q) = F_W(w)$). This assumption is usually taken as a uniform procedure for the derivations of design flood values and design flood hydrograph in China (MWR 2006; Xiao et al. 2008, 2009; Chen et al. 2010). Then, the design frequency of bivariate equivalent frequency combination can be obtained by jointly solve the equation $u = v$ and Eq. (3.39).

Taking the Gumbel copula for example, the relationship between u and v with the given bivariate return period T is described in Eq. 3.39. Based on the assumption that $u = v$, the probabilities of occurrence of flood peak and volume (i.e., u and v) can be estimated by the solution of the following equation.

$$u = v = (1 - \frac{1}{T})^{\varsigma} \tag{3.50}$$

where $\varsigma = 2^{-\frac{1}{\theta}}$, and θ is the dependence parameter of the Gumbel copula.

Consequently, the design value of bivariate equivalent frequency combination can be derived by the inverse function of marginal distributions:

$$q = F_Q^{(-1)}(u) \tag{3.51a}$$

$$w = F_W^{(-1)}(v) \tag{3.51b}$$

3.4.3.2 Conditional Expectation Combination Method

Since the flood peak Q and flood volume W are dependent variables, one may wish to predict the value of W based on an observed value of Q. Let $g(Q)$ be a predictor, i.e., $g \in N = \{\text{all Borel functions } g \text{ with } E[g(Q)]^2 < \infty$ Each predictor is assessed by the "mean squared prediction error" $E[W - g(Q)]^2$. The conditional expectation $E(W|Q)$ is the best predictor of W in the sense that

$$E[W - E(W|Q)]^2 = \min_{g \in N} E[W - g(Q)]^2 \tag{3.52}$$

Herein, during a flood event, when the flood peak $Q = q$ takes place; the conditional expectation $E(w|q)$ is used to estimate the value of flood volume, which can be derived by

$$E(w|q) = \int_{-\infty}^{+\infty} w f_{W|Q}(w) dw \tag{3.53}$$

where $f_{W|Q}(w)$ is the density function of the conditional CDF $F_{W|Q}(w)$ and defined as (Zhang and Singh 2006).

$$f_{W|Q}(w) = \frac{f(q, w)}{f_Q(q)} = \frac{c_\theta(u, v) f_Q(q) f_W(w)}{f_Q(q)} = c_\theta(u, v) f_W(w) \tag{3.54}$$

Hence, Eq. 3.53 can be expressed by

$$E(w|q) = \int_{-\infty}^{+\infty} w f_{W|Q}(w) dw = \int_{-\infty}^{+\infty} w c_\theta(u, v) f_W(w) dw = \int_0^1 F_W^{-1}(v) c_\theta(u, v) dv \tag{3.55}$$

where $F_W^{-1}(\cdot)$ is the inverse CDF of W.

Then, the flood peak q and $E(w|q)$ will be the conditional expectation combination if the following equations are satisfied

$$\begin{cases} u = F_Q(q) \\ v = F_W[E(w|q)] \\ \frac{1}{1 - C_\theta(u,v)} = T \end{cases} \tag{3.56}$$

The above equation can be solved by trial and error method with different values of q.

3.4.4 *Case Study*

3.4.4.1 Bivariate Quantile Curve and Feasible Range Identification

The return period of design flood of Geheyan reservoir, i.e., $T = 1000$-year, is selected as the bivariate return period and $T = 200$-year is also chosen for comparison. The bivariate quantile curves of the two return periods are shown in Fig. 3.7. Even if the Gumbel copula model is symmetric, the probability density function $\varphi(q)$ is not symmetrical due to the difference in the marginal distributions.

The upper and lower bounds on the level curve are estimated numerically by solving Eqs. 3.48 and 3.49, and assuming for simplicity (although other assumptions are possible) $\alpha_1 = \alpha_2 = \varepsilon/2$, with $\varepsilon = 0.05$. The upper and lower bounds are denoted as B_1 and C_1, respectively, in Fig. 3.7. It is found that the bounds are close to the horizontal asymptote (i.e., $w_7 = 61.49 \times 10^8$ m^3 for $T = 1000$ and $w_7 = 50.23 \times 10^8$ m^3 for $T = 200$) and vertical asymptote (i.e., $q_p = 22{,}800$ m^3/*s* for $T = 1000$ and $q_p = 19{,}300$ m^3/*s* for $T = 200$) due to the small value assumed for the probability level ε. The upper and lower bounds are also calculated by the boundary identification method proposed by Volpi and Fiori (2012). The results are also presented in Table 3.11, and the derived bounds are denoted as B_2 and C_2, as shown in Fig. 3.7. It is shown that the bounds estimated by the proposed method and that proposed by Volpi and Fiori (2012) are very similar.

3.4.4.2 Estimation of Bivariate Flood Quantiles

The bivariate EFC and CEC methods are used to estimate flood peak and 7-day flood volume quantiles with return periods of $T = 1000$ and $T = 200$ years,

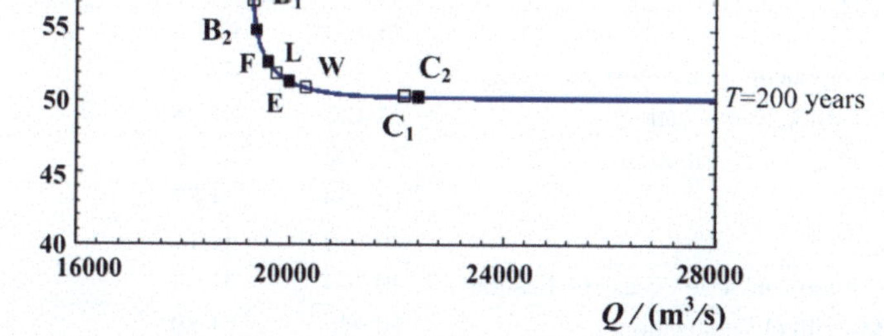

Fig. 3.7 Bivariate quantile curve of joint distribution of flood peak and 7-day flood volume

Table 3.11 Comparison of the lower and upper bounds of the quantile curve

Boundary identification method	Return period	Lower bound		Upper bound	
		Q_p(m^3/s)	W_7 (10^8 m^3)	Q_p(m^3/s)	W_7 (10^8 m^3)
Volpi and Fiori (2012)	1000	22,930	65.84	26,080	61.54
	200	19,350	50.27	22,460	55.86
Li et al. (2016)	1000	23,000	65.76	26,100	61.52
	200	19,400	54.49	22,500	50.26

respectively. For comparison, the univariate flood quantiles (called marginal quantiles by Chebana and Ouarda 2011) are estimated by marginal distributions, assuming that the univariate return periods (T_Q and T_W) are equal to the bivariate return period (i.e., $T_Q = T_W = T$). The univariate flood quantiles can be obtained from the equations $q = F_Q^{-1}(p) = F_Q^{-1}(1 - \frac{1}{T})$ and $w = F_W^{-1}(p) = F_W^{-1}(1 - \frac{1}{T})$. The results of the component-wise excess realization and the most likely realization proposed by Salvadori et al. (2011) are also estimated. The estimation results of bivariate and univariate quantiles are listed in Table 3.12. It is shown that the design values of bivariate quantiles are larger than those of univariate quantiles. The quantiles estimated by the four bivariate event selection methods are also shown in Fig. 3.7, and the estimation points of the EFC method are denoted as point E, while the quantiles estimated by the CEC method are denoted as point F. For the results of selection approaches proposed by Salvadori et al. (2011), the events of component-wise excess realization are denoted as point *W*, and the events of most likely realization are denoted as point *L*. From Fig. 3.7, we find that the joint design values estimated by the four event-selection methods are within the feasible regions. Consequently, the two proposed methods and selection approaches proposed by Salvadori et al. (2011) can be selected as an option of deriving unique flood quantiles, and they can satisfy the inherent law of hydrologic events and have a statistical basis to some degree. It can be seen from Table 3.12 and Fig. 3.7 that

Table 3.12 Design flood values and corresponding highest water levels estimated by bivariate quantile combinations and univariate distribution

T	Method	Q_p(m^3/s)	W_7 ($\times 10^8$ m^3)	Z_{max} (m)
1000	EFC	23,390	63.09	202.97
	CEC	23,420	62.98	202.92
	Component-wise excess realization	23,510	62.78	202.90
	Most-likely realization	23,400	63.05	202.95
	Univariate distribution	22,800	61.49	202.58
200	EFC	19,800	51.87	198.10
	CEC	20,130	51.11	197.79
	Component-wise excess realization	20,200	51.03	197.59
	Most-likely realization	19,940	51.50	197.82
	Univariate distribution	19,300	50.23	197.30

the estimated events of the EFC method and that of the most likely realization are similar. The bivariate EFC results have larger flood volume and smaller flood peak than bivariate CEC results. As well, the results estimated by the component-wise excess realization have larger flood peak and smaller flood volume than the other three methods.

3.4.4.3 Design Flood Hydrograph Based on Joint Distribution

The two combination methods are applied to derive the design flood hydrograph (DFH), and the resulting highest reservoir water level is selected as an index to evaluate the effects of different hydrological loads on the structure. The DFH for a dam is the flood of suitable probability and magnitudes adopted to ensure safety of the dam in accordance with appropriate design standards. The annual maximum flood hydrograph of 1997, which has a high peak and large volume with a posterior-peak shape, is selected as a typical flood hydrograph (TFH). The DFH with bivariate combinations is amplified from a TFH by the following method (Xiao et al. 2008):

$$DFH(t) = (TFH(t) - Q_{TFH}) \times (w/DT - q)/(W_{TFH}/DT - Q_{TFH}) + q \quad (3.57)$$

where $DFH(t)$ and $TFH(t)$ are the flood discharges of the DFH and TFH for time t respectively; Q_{TFH} is flood peak discharge of TFH; W_{TFH} is 7-day flood volume of TFH for flood duration DT; q and w are flood peaks and 7-day flood volumes of bivariate design flood combination, respectively. Nevertheless, other DFH generation methods based on flood peak and volume are also available and can be applied with the bivariate design value combinations.

The DFHs of 1000-year and 200-year return periods are constructed, respectively, with the bivariate EFC method and bivariate CEC method as shown in Fig. 3.8. It is found in Fig. 3.8 that only a few differences exist between the DFHs estimated by the EFC and CEC methods. This is because that the differences between the bivariate design values vary within a small range. Volpi and Fiori (2012) found that the feasible range on a p-level curve strongly depends on the correlation coefficient of Q and W. In the limiting case of full dependence, the level curve reduces to its vertex and the width of the feasible range tends to 0 (Volpi and Fiori 2012). Since the Kendall correlation coefficient between flood peak and 7-day volume in Geheyan reservoir equals to 0.66, the differences of quantiles estimated by EFC and CEC methods are relatively small in this case study.

The DFH rescaled by univariate distribution design values and two realizations proposed by Salvadori et al. (2011) is also derived from TFH by Eq. 3.59. These DFHs are routed through the Geheyan reservoir with initial water level (flood control limiting water level, 192.2 m). The corresponding highest reservoir water levels (Z_{max}) are calculated and are listed in Table 3.12.

It is shown in Table 3.12 that the design values of flood peak and 7-day flood volume obtained by univariate distribution method are both smaller than those

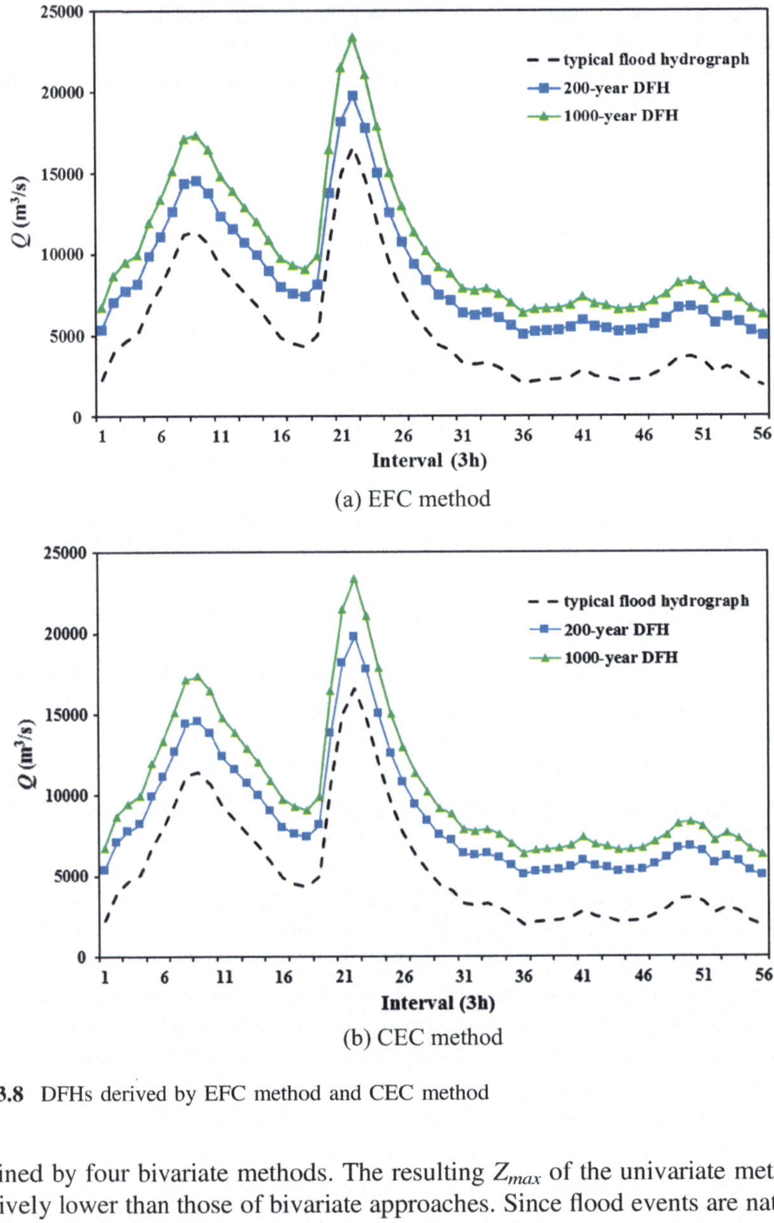

Fig. 3.8 DFHs derived by EFC method and CEC method

obtained by four bivariate methods. The resulting Z_{max} of the univariate method is relatively lower than those of bivariate approaches. Since flood events are naturally multivariate phenomena and flood peak and flood volume are mutually correlated, the quantiles estimated by bivariate distribution are more rational than these by univariate distribution (Chebana and Ouarda 2011).

The comparison results listed in Table 3.12 also show that Z_{max} obtained by bivariate EFC method is larger than that obtained by the other three bivariate methods, while the component-wise excess method reaches the lowest Z_{max}. The results of Z_{max} calculated by most-likely realization are a little lower than those of the EFC method, and the CEC method obtains a slightly higher Z_{max} than the component-wise excess method. Comparing the results of 200-year and 1000-year return period, it is found that the differences among the four bivariate methods decrease as the return period increases. The water level reaches 202.97 m by the EFC method and is slightly higher than other methods for the 1000-year return period. Since the Geheyan reservoir has a large amount of flood control storage with annual regulation ability, the design flood volume is relatively more important than peak discharge for flood prevention safety. As a consequence, the bivariate EFC method with slightly larger 7-day flood volume is safer for reservoir design than other methods.

3.5 Conclusion

According to the bivariate joint distribution of annual maximum flood occurrence dates and magnitudes, flood peaks and volumes, a flood frequency analysis model with an incorporation of historical floods are established based on GH copula. Modified inference functions for the margins (MIFM) method and the quantile curve boundary identification method are developed. The following conclusions are drawn from this Chapter:

(1) The Von Mises and Pearson Type III distributions can fit observed data series very well. The goodness-of-fit tests indicate a good agreement between observed and theoretical probabilities for both marginal and joint distributions.
(2) The proposed MIFM method may reduce the uncertainties of parameter estimation in flood frequency analysis, since the historical floods have been taken into account.
(3) The quantile combination methods provide a simple but effective way for bivariate quantile estimation with given bivariate return period. The results illustrate that the joint design values estimated by the two proposed combination methods are within the feasible regions, and the equivalent frequency combination method perform satisfactorily.

References

ASCE (American Society of Civil Engineers) (1996) Hydrology handbook. In: ASCE manuals and reports on engineering practices no. 28. American Society of Civil Engineers, New York, USA
Bayliss AC, Reed DW (2001) The use of historical data in flood frequency estimation. Report to MAFF.CEH Walingford
Black AR, Werritty A (1997) Seasonality of flooding: a case study of North Britain. J Hydrol 195:1–25

Chebana F, Quarda TBMJ (2011) Multivariate quantiles in hydrological frequency analysis. Environmetrics 22:441–455
Chen L, Guo SL, Yan BW, Liu P, Fang B (2010) A new seasonal design flood method based on bivariate joint distribution of flood magnitude and date of occurrence. Hydrol Sci J 55(8):1264–1280
Cohen AC (1976) Progressively censored sampling in the three parameters log-normal distribution. Technometrics 18(1):99–103
Coles S, Heffernan J, Tawn J (1999) Dependence measures for extreme value analysis. Extremes 2(4):339–365
Condie R (1986) Flood samples from a three-parameter lognormal population with historical information: the asymptotic standard error of estimate of the T-year flood. J Hydrol 85: 139–150
Condie R, Lee K (1982) Flood frequency analysis with historical information. J Hydrol 58:47–62
CWRC (Changjiang Water Resources Commission) (1996) Hydrologic inscription cultural relics in Three Gorges Reservoir. Science Press, Beijing (in Chinese)
De Michele C, Salvadori G, Canossi M, Petaccia A, Rosso R (2005) Bivariate statistical approach to check adequacy of dam spillway. J Hydrol Eng 1:50–57
Dupuis DJ (2007) Using copulas in hydrology: benefits, cautions, and issues. J Hydrol Eng 12 (4):381–393
Fisher NI (1993) Statistical analysis of circular data. Cambridge University Press, Cambridge
Frahm G, Junker M, Schmidt R (2005) Estimating the tail dependence coefficient: properties and pitfalls. Insur Math Econ 37(1):80–100
Guo SL, Cunnane C (1991) Evaluation of the usefulness of historical and paleological floods in quantile estimation. J Hydrol 129:245–262
Hald A (1949) Maximum likelihood estimation of the parameters of a normal distribution which is truncated at a known point. Skand Aktuarietidskrift 32(1/2):119–134
Hosking JRM (1995) The use of L-moments in the analysis of censored data". In: Balakrishnan N (ed) Recent advances in life-testing and reliability. CRC Press, Boca Raton, Fla, pp 545–564
Joe H (1997) Multivariate models and dependence concepts. Chapman & Hall, London
Joe H (2005) Asymptotic efficiency of the two-stage estimation method for copula-based models. J Multivariate Anal 94:401–419
Joe H, Xu JJ (1996) The estimation method of inference functions for margins for multivariate models. Technical Report no. 166, Department of Statistics, University of British Columbia
Leese MN (1973) Use of censored data in the estimation of Gumbel distribution parameters for annual maximum flood series. Water Resour Res 9(6):1534–1542
Li T, Guo S, Chen L, Guo J (2013) Bivariate flood frequency analysis with historical information based on Copula. J Hydrol Eng 18(8):1018–1030
Li T, Guo S, Liu Z, Xiong L, Yin J (2016) Bivariate design flood quantile selection using copulas. Hydrol Res. https://doi.org/10.2166/nh.2016.049
Li X, Guo SL, Liu P, Chen G (2010) Dynamic control of flood limited water level for reservoir operation by considering inflow uncertainty. J Hydrol 391:124–132
Mardia KV (1972) Statistics of directional data. Academic Press, London
McLeish DL, Small CG (1988) The theory and applications of statistical inference functions. Lecture Notes in Statistics, 44. Springer-verlag, New York
MWR (Ministry of Water Resources) (2006) Regulation for calculating design flood of water resources and hydropower projects. Chinese Water Resources And Hydropower Press, Beijing (in Chinese)
Nelsen RB (2006) An introduction to copulas, 2nd edn. Springer, New York
Poulin A, Huard D, Favre AC, Pugin S (2007) Importance of tail dependence in bivariate frequency analysis. J Hydrol Eng 12(4):L394–L403
Salvadori G, De Michele C, Durante F (2011) Multivariate design via Copulas. Hydrol Earth Syst Sci Discuss. 8:5523–5558

Salvadori G, De Michele C (2004) Frequency analysis via copulas: theoretical aspects and applications to hydrological events. Water Resour Res 40:W12511. https://doi.org/10.1029/2004WR003133

Shiau JT, Wang HY, Tsai CT (2006) Bivariate frequency analysis of floods using copulas. J Am Water Resour Assoc 42(6):1549–1564

Stedinger JR, Cohn TA (1986) The value of historical and paleoflood information in flood frequency analysis. Water Resour Res 22(5):785–793

USWRC (US Water Resources Council) (1981) Guidelines for determining flow frequency, Bulletin 17B. D. C, Washington

USWRC (US Water Resources Council) (1982) Guidelines for determining flood flow frequency, Bull. 17B (revised), U.S. Gov. Print. Off., Washington, D. C

Volpi E, Fiori A (2012) Design event selection in bivariate hydrological frequency analysis. Int Assoc Sci Hydrol 57(8):1506–1515

Volpi E, Fiori A (2014) Hydraulic structures subject to bivariate hydrological loads: return period, design, and risk assessment. Water Resour Res 50(2):885–897

Xiao Y, Guo SL, Liu P, Yan B, Chen L (2009) Design flood hydrograph based on multi characteristic synthesis index method. J Hydrol Eng 14(12):1359–1364

Xiao Y, Guo SL, Liu P, Fang B (2008) A new design flood hydrograph method based on bivariate joint distribution. In: Chen XH, Chen YD, Xia J, Zhang H (eds) Hydrological sciences for managing water resources in the asian developing world, IAHS Press, IAHS Publications 319, Wallingford, pp 75–82

Xu JJ (1996) Statistical Modelling and inference for multivariate and longitudinal discrete response data. Ph.D. thesis, Department of Statistics, University of British Columbia

Yue S, Quarda TBMJ, Bobée B, Legendre P, Bruneau P (1999) The Gumbel mixed model for flood frequency analysis. J Hydrol 226:88–100

Zhang L, Singh VP (2006) Bivariate flood frequency analysis using the copula method. J Hydrol Eng 11(2):150–164

Chapter 4
Copula-Based Seasonal Design Flood Estimation

4.1 Introduction

Since the rain-producing systems vary with season, the river flood is usually characterized as seasonality. Seasonal fluctuations are a significant source of variability in runoff records. However, seasonality is often overlooked when evaluating flood risk due to the use of annual value for defining extreme values. The phrase "1 in 100 years" flood does not inform whether a given extreme value is more likely to come from one season over another. The oversight of seasonality is also common to the peak-over-threshold method, even though this method is capable of obtaining more than one extreme value per year (Michael et al. 2007).

In the reservoir operation, the water level of the reservoir should be limited below the flood control water level (FCWL) during flood season to offer adequate storage for flood control. The current FCWL, which plays a key role in the flood prevention and floodwater utilization, is mainly determined according to the design floods estimated from annual maximum flood series while neglecting the seasonal information. This results in over-standard for flood prevention and a waste of floodwater in most of the years. Therefore, the design flood guideline in China stresses the importance of classifying the annual maximum floods caused by different generating mechanisms (MWR 1993). For floodwater utilization, it's very valuable to use the seasonal flood information in flood frequency analysis o operate the reservoir more effectively during flood seasons without enhancing the flood prevention risk. How to reasonably and optimally design seasonal floods that reflects seasonal variations poses a challenge to hydrologists and engineers nowadays. It is a very important and urgent issue in the management of reservoirs in China (Guo et al. 2004; Fang et al. 2007).

The conventional flood frequency analysis methods are based on univariate distributions, mainly concentrated on the analysis of peak discharge or flood volume series. For seasonal design flood, statistical analysis of the flood occurrence dates is also very useful and important. Generally, the annual maximum flood often

L. Chen and S. Guo, *Copulas and Its Application in Hydrology and Water Resources*, Springer Water, https://doi.org/10.1007/978-981-13-0574-0_4

occurs in main flood season, and median or small floods occur in other ones. The flood occurrence date is also a random variable and follows a particular distribution, which is different from that of flood magnitude. Thus, seasonal design flood should consider both the dates and magnitudes of flood events that may be described by a bivariate joint distribution. Chen et al. (2010) proposed a new seasonal design flood method, which considers dates of flood occurrence and magnitudes of the peaks (runoff) based on copula function. Their results show that the proposed method can satisfy the flood prevention standard, and provide more information about the flood occurrence probabilities in each sub-season. Yin et al. (2017) used three bivariate flood quantile selection methods, namely equivalent frequency combination (EFC) method, conditional expectation combination (CEC) method and conditional most likely combination (CMLC) method, to estimate unique seasonal design flood to meet the needs in engineering. Results showed that the CMLC method is more rational in physical realism and recommended for estimating the seasonal design floods, which can provide rich information as the references for flood risk assessment, reservoir scheduling, and management.

4.2 Review of Seasonal Design Flood Methods

The issue of seasonal flood frequency analysis was identified by Creager et al. as early as in 1951. The aim of the seasonal design flood is to determine the relationship between hydrograph and return period in each season. Two current seasonal design flood methods: one was suggested by Chinese design flood guideline (MWR 1993), other was proposed by Singh et al. (2005). These two methods are referred as Chinese method and Singh's method in this study and reviewed as follows.

4.2.1 Chinese Method

The seasonal maximum (SM) flood series extract the maximum peak discharge (or runoff volumes) from each season during each year of record. The seasonal T-year design flood is obtained by fitting a particular distribution, such as P-III distribution used in China (MWR 1993).

In this method, the annual maximum values Y can be described as:

$$Y = \{Y_1, Y_2, \ldots, Y_s\} \tag{4.1}$$

where $Y_1, Y_2, \ldots, Y_s$ are the seasonal maximum flood series; and s is the number of the sub-seasons.

According to the Eq. (4.1), the extreme value distribution of the annual maximum flood series can be defined as:

$$F_T(y) = F_1(y)F_2(y)\cdots F_S(y) \tag{4.2}$$

where $F_T(y)$ is the distribution of the annual maximum flood series; and $F_1(y),\ldots,F_s(y)$ are the distributions of seasonal maximum flood series (Waylen and Woo 1982). For a fixed value *y*, Eq. 4.2 shows that $F_T(y)$ will always be less than or equal to the smallest of the $F_i(y)$, since each of the latter values must always be in the range [0, 1]. In other words, the annual frequency curve must always lie on or above the highest of the seasonal frequency curves on a common probability paper, i.e., *T*-year seasonal design floods are always less than the annual design floods (Durrans et al. 2003).

The Chinese flood prevention standard is defined by annual return period $T{,}T = 1/(1 - F_T(y))$, while the Chinese design flood guideline assumes that the seasonal design frequency is equal to the annual design frequency, namely $F_T(y) = F_1(y_1) = F_2(y_2) = 1 - 1/T$(MWR 1993). Assuming two sub-seasons, take the 100-year design flood for example. If $F_1(y) = F_2(y)$, i.e., in the case of identical distribution, Eq. 4.2 leads to $F_T(y) = (F_1(y))^2 = 0.98$. This means that when the seasonal design method is used, the combined frequency of them cannot reach the annual prevention standard. If the combination frequency must reach to the annual design frequency, at least one of the seasonal frequencies must exceed the annual frequency analysis. Therefore, the current Chinese seasonal design flood method cannot satisfy the flood prevention standard.

4.2.2 Singh's Method

Annual maximum flood series are formed by extracting the annual maximum peak discharge (or runoff volumes) from each year of record. If *n* is the number of recorded years and n_i is the number of annual maxima that occur in the *i*th season, then $n = \sum_{i=1}^{s} n_i$ (Durrans et al. 2003).

This method can be described as follows: considering that the occurrence of a flood event $B = \{Y > y\}$ must be associated with one of the events $\{A_i\}$, *i* = 1, …, s. $\{A_i\}$ means the annual maximum flood that occurs during the *i*th season.

The exceedance frequency $P(y, A_i)$ of seasonal design flood is defined as:

$$P(y, A_i) = P(y|A_i)P(A_i) \tag{4.3}$$

where $P(y|A_i)$ is the exceedance probability that an annual flood maximum occurring in the *i*th season. $P(A_i)$ is the probability of an annual maximum occurring in the *i*th season, *i* = 1, …, *s*.

This method has been described by Thomas et al. (1998), who pointed out that its use is valid for both independent and dependent seasonal flood distributions. Singh et al. (2005) applied this method to estimate design flood from a nonidentically distributed series and provided an estimation procedure for practical use.

The sum of the probabilities of seasonal design flood is given by:

$$P(Y \geq y) = \sum_{i=1}^{s} P(B \cap A_i) = \sum_{i=1}^{s} P(A_i)P(y|A_i) \tag{4.4}$$

Equation 4.4 is the total probability law and expresses the frequency distribution of the annual maximum flood as the sum of the frequency distribution of those annual maximum floods that are conditioned on the maxima occurring in the ith season with the probability weight $P(A_i)$ (Singh et al. 2005).

Assuming the annual maxima occurring in different seasons are identically distributed, the conditional frequency distribution $P(y|A_i)$ is free of A_i, then Eq. 4.4 leads to

$$P(y) = P(y_0)\sum_{i=1}^{s} P(A_i) = P(y_0) \tag{4.5}$$

where $P(y_0)$ is a fixed frequency distribution indicating that the overall annual maxima are identically distributed. Equation 4.5 shows the validity of Eq. 4.4 which can satisfy the flood prevention standard (Singh et al. 2005).

The flood frequency distribution $P(y|A_i)$ should be estimated from those observed values of the Y_i flood series that are picked as the annual maximum floods. For some drier season, there may be few or even no samples to be drawn. It is not accurate and reliable to use these data series for calculation. Equation 4.4 suffers in practice from the fact that n_i for one season will usually be considerably smaller than n_i for another season. Because of this, the reliability with which each conditional distribution in Eq. 4.4 may be estimated will vary from season to season. Furthermore, since n_i for any season will always be less than or equal to n, this approach essentially limits the lengths of the record samples (Durrans et al. 2003).

4.3 A New Seasonal Design Flood Method

The sampling methods, flood seasonality identification methods, and the copula functions are introduced and discussed. The von Mises distribution is used to describe the flood occurrence dates, while the P-III distribution or exponential distribution (Ex) is selected as marginal distribution for annual maximum flood series or peak-over-threshold samples, respectively. A new seasonal design flood method is described as follows.

4.3.1 Sampling Method

Sampling methods play an important role in flood frequency analysis. The annual maximum (AM), seasonal maximum (SM) and peaks-over-threshold (POT) sampling methods are used and compared in this section.

The POT sampling method is also widely used in flood frequency analysis because more information can be obtained compared with that of the AM or SM sampling method. To guarantee the independence of the samples, the flood peaks are selected by two criteria suggested by Institute of Hydrology of UK (IH 1999): (1) two peaks have to be separated by at least three times the average time to rise, in which the average time to rise is determined from the synthetic records as 2 days and is kept constant throughout the study; and (2) The minimum discharge between two peaks has to be less than two-thirds of the discharge of the first of the two peaks.

4.3.2 Identification of Seasonality

The whole flood season is usually divided into three sub-seasons, and these sub-seasons are defined as the pre-flood season, main flood season and post-flood season (MWR 1993; Ngo et al. 2007).

Several types of approaches for detecting flood seasonality have been proposed. One type of approaches is to segment flood season regarding climatological and river basin physiographic characteristics by analyzing the rain-producing system (Black and Werrity 1997; Singh et al. 2005). The other type is to segment flood season by using visual identification based on some measurements of flood seasonality. Ouarda et al. (1993) proposed two variations of a graphical method for identification of river flood season from peaks-over-threshold (POT) data. Cunderlik et al. (2004) used the relative frequency (RF) method and directional statistics (DS) method to identify the seasonality. The RF method is based on counting the number of events in each season, to allow comparisons between records, expressing these counts as a percentage of the total number of events in each record (Black and Werritty 1997; Cunderlik et al. 2004; Ouarda et al. 2006). The DS method describes the seasonality by defining the mean day of the flood (directional mean) and the flood variability measure. The DS, RF, and POT methods are compared in this study.

4.3.3 Seasonal Design Flood Estimation

The season design flood can be characterized by flood occurrence dates and flood magnitudes. In this section, first, the marginal distribution of flood occurrence dates and flood magnitudes are established.

4.3.3.1 Margin Distribution of Flood Occurrence Dates

The von Mises distribution introduced in Chap. 3. Three can only be used for unimodal distribution. Since the annual maximum floods may be generated by different mechanisms, the flood occurrence data series often obey a multimodal distribution. Thus, a mixed von Mises distribution which can describe the multimodal character is comprised of a finite mixture of von Mises distributions. The probability density function for a mixture of N von Mises distributions (vM-pdf) takes the following form:

$$f_X(x) = \sum_{i=1}^{N} \frac{p_i}{2\pi I_0(\kappa_i)} \exp[\kappa_i \cos(x - \mu_i)] \\ 0 \le x \le 2\pi,\ 0 \le \mu_i \le 2\pi,\ \kappa_i \ge 0 \tag{4.6}$$

where p_i is the mixing proportion, μ_i is the mean direction, and κ_i is the concentration parameter.

Various methods can be used to estimate the 3 N parameters on which the mixture of N vM-pdfs depends (Carta and Ramírez 2007). The least squares (LS) method is used in this book, in which the 3 N unknown parameter values can be estimated by minimizing the sum of the squares of the deviations between the experimental data and the calculated value (Carta et al. 2008).

4.3.3.2 Margin Distribution of Flood Magnitudes

For the AM flood series, the P-III distribution has been recommended by MWR (1993) as a uniform procedure for flood frequency analysis in China. The formula of P-III distribution is given in Table 1.1.

The classical use of the POT sampling method comprises the assumptions of a Poisson-distributed number of threshold exceedances and exponentially distributed peak exceedances (Lang et al. 1999). The probability density function of 2-parameter Ex distribution is given in Table 1.1 as well. For POT flood series, the 1-parameter Ex distribution is used by setting the parameter $\gamma = 0$

$$f_{YPOT}(y) = \frac{1}{\alpha} e^{(-y/\alpha_0)} \tag{4.7}$$

where α_0 is a parameter of the Ex distribution.

The parameters of P-III and Ex distributions are estimated by L-moments method (Hosking and Wallis 1997).

4.3.3.3 Bivariate Distribution of Flood Occurrence Dates and Magnitudes

For estimating seasonal design flood, the bivariate joint distributions of flood occurrence dates and magnitudes need to be built. Every joint distribution can be written regarding a copula and its univariate marginal distributions. The copula is a function that links univariate marginal distribution functions to construct a multivariate distribution function. The definition and establishment of copulas can be seen in Chap. 2. The Gumbel–Hougaard, Frank, Clayton, and Ali-Mikhail–Haq copulas are used to establish the joint distribution.

4.3.3.4 Seasonal Design Flood Estimation

Seasonal design flood is related to the flood dates X and magnitudes Y and follows a two-dimensional distribution $F(x, y)$. Assuming all floods occur during whole flood season, the annual exceedance probability can be defined as:

$$P(X \leq t, Y > q) = F_X(t) - F(t, q) \tag{4.8}$$

where t is the last day of the flood season, and q is a specific discharge value. $F_X(t)$ is the marginal distribution function of t.

$F(t, q)$ is the joint distribution of the flood peak which occurs before the date t with the value less than or equal to the discharge q, and can be described by

$$\begin{aligned}
F(t,q) &= \int_{-\infty}^{q} \int_{-\infty}^{t} f(x,y)dxdy \\
&= \int_{-\infty}^{q} \int_{-\infty}^{t} f_X(x) f_Y(y|x)dxdy \\
&= \int_{-\infty}^{t} f_X(x) \int_{-\infty}^{q} f_Y(y|x)dydx \\
&= \int_{-\infty}^{t} f_X(x)F(q|x)dx \\
&= \int_{0}^{t} F(q|x)dF_X(x) \\
&= \sum_{i=1}^{s} P(Y \leq q | x_i < x < x_{i+1}) F_X(x_i < x < x_{i+1})
\end{aligned} \tag{4.9}$$

$$\begin{aligned}
P(X \le t, Y > q) &= F_X(t) - \sum_{i=1}^{s} P(Y \le q|\Delta x_i) F_X(x_i < x < x_{i+1}) \\
&= \sum_{i=1}^{s} F_X(x_i < x < x_{i+1}) \\
&\quad - \sum_{i=1}^{s} P(Y \le q|\Delta x_i) F_X(x_i < x < x_{i+1}) \\
&= \sum_{i=1}^{s} F_X(x_i < x < x_{i+1})(1 - P(Y \le q|x_i < x < x_{i+1})) \\
&= \sum_{i=1}^{s} F_X(x_i < x < x_{i+1}) P(Y \ge q|x_i < x < x_{i+1})
\end{aligned} \tag{4.10}$$

where $f_X(x)$ is the marginal density function of variable x; $f(x, y)$ is two-dimensional density function; $f_Y(y|x)$ and $F_Y(y|x)$ are the conditional probability density and distribution function of y; and x_i represents a segmentation point. If s equals the number of the sub-seasons, then Eq. 4.10 is as the same as that of Eq. 4.4. It is also indicated from Eq. 4.10 that the seasonal design flood frequency curves are located below the annual one.

If $F_X(x_{i-1} < x < x_i)$ is replaced with $P(A_i)$, the exceedance probability for the seasonal design flood frequency $P(q, A_i)$ is defined as:

$$\begin{aligned}
P(q, A_i) &= P_i(Y_i > q|A_i)P(A_i) \\
&= P(x_{i-1} \le X \le x_i, Y_i > q) \\
&= F(x_i) - F(x_{i-1}) - F(x_i, q) + F(x_{i-1}, q)
\end{aligned} \tag{4.11}$$

where x_{i-1} and x_i are the segmentation points. Equation 4.11 indicates that seasonal design flood is related to bivariate joint distributions. The seasonal design flood frequency $P(q, A_i)$ can be described by the probability weight $P(A_i)$ and the conditional frequency distribution $P(q|A_i)$. Since the range of the $P(q|A_i)$ is from 0 to 1, the value of $P(q, A_i)$ is restricted within $P(A_i)$.

4.4 Case Study

4.4.1 Identification of Flood Seasonality

The Geheyan reservoir is selected as a case study. Fifty-four year (1954–2004) discharge records are used to analyze seasonal design flood. For the POT sampling method, the threshold value with 3500 m^3/s is selected, which corresponds to a mean of 2.57 exceedances per year.

In the Qingjiang basin, floods frequently occur in summer from June to early August when the monsoon fronts advance from south to north or in the fall from late August to early October when the fronts withdraw from north to south. Although both summer and fall floods result from frontal rains, their hydrological characteristics are distinctly different because the intensity of the rain-producing system is varied with seasons (Singh et al. 2005). The statistical analysis results of 10-day rainfall data are listed in Table 4.1. It can be seen from Table 4.1 that most of the rainfall occurs from late June to middle July, whereas in other time of the flood season a relatively small amount of rainfall is received. Seasonal variation of trends is that flood events begin to increase from late June and decrease in late July. Therefore, those two periods might be the segmentation points.

The DS, RF and POT methods are used to describe the flood seasonality, and the results of these methods are listed in Table 4.2. It can be seen that the RF method has the shortest main flood season (from June 21 to July 20) because of the clustering of dates of flood occurrences into ten days. Ouarda et al. (2006) pointed out that the seasonality method based on the peaks-over-threshold (POT) approach lead to the best results. However, the result of POT method also has the shortest main flood season (from June 26 to July 26) at the Qingjiang basin as shown in Table 4.2. Compared with POT method, the result of DS approach has a 5-day difference for each sub-season.

In order to ensure the flood control safety, the results of DS method are chosen, since it has the longest main flood season from June 21 to July 31. The mean day of the flood (directional mean) is on July 2. The flood occurrence dates sampled by

Table 4.1 Statistical analysis results of 10-day rainfall data of the Geheyan basin

Date		≥60 mm	50–59 mm	40–49 mm	30–39 mm	Mean values (mm)	Percentage (%)
May	Early	1	2	3	9	58	6.26
	Mid.	1	1	4	11	58	6.19
	Late	0	2	2	13	61	6.51
June	Early	3	3	2	18	61	6.56
	Mid.	4	0	1	9	60	6.40
	Late	4	2	18	13	85	9.14
July	Early	3	6	14	11	86	9.20
	Mid.	7	7	11	9	84	9.03
	Late	1	2	6	7	60	6.42
August	Early	2	2	4	6	60	6.42
	Mid.	1	1	4	9	59	6.27
	Late	3	1	5	8	56	5.96
Sept.	Early	0	2	7	11	49	5.30
	Mid.	4	3	5	9	58	6.25
	Late	0	2	2	3	38	4.08

Table 4.2 Results of three methods for identification of the seasonality

Methods	The pre-flood season	The main flood season	The post-flood season
DS	May 1–June 20	June 21–July 31	Aug. 1–Sept. 30
RF	May 1–June 20	June 21–July 20	July 21–Sept. 30
POT	May 1–June 25	June 26–July 26	July 27–Sept. 30

POT method are translated into a location on the circumference of a circular drawn in Fig. 4.1. It can be seen from Fig. 4.1 that the flood events are mainly centered June 20 to July 31, and the interval time of the adjacent flood events is obviously shorter in this period.

In summary, the flood season of the Qingjiang basin can be divided into three sub-seasons, i.e., the pre-flood season, main flood season and post-flood season. Based on the analysis results above, the pre-flood season is from May 1 to June 20; the main flood season is from June 21 to July 31, and the post-flood season is from August 1 to Sept. 30.

4.4.2 Computation of Empirical Frequency

The empirical probabilities can be computed by Eqs. 3.8 and 3.9.

Fig. 4.1 Application of DS method for flood occurrence dates based on POT samples

4.4.3 Bivariate Distribution

The joint distribution is established for the AM and POT samples respectively, and the estimated parameters of the margin distribution and joint distribution are listed in Table 4.3. Some statistical tests are used for margin and joint distributions. A chi-square goodness-of-fit test is performed to test the assumption H_0 that the flood magnitude follows P-III distribution or Ex distribution. A Kolmogorov–Smirnov (K-S) test is used to test the assumption H_0 that the flood occurrence dates follow mixed von Mises distribution. The results shown in Table 4.4 indicate that these assumptions cannot be rejected at the 5% significant level. The fitted frequency histograms of the flood occurrence date by the mixed von Mises distribution for POT sample series are drawn in Fig. 4.2. The margin distribution frequency curves of flood occurrence dates and magnitudes are shown in Figs. 4.3 and 4.4, respectively, of which the line represents the theoretical distribution, and the crossing means empirical probability of observations. Figures 4.3 and 4.4 indicate that all the theoretical distributions can fit the observed data reasonably well, although there are some uncertainties in the dataset itself.

Four widely used copulas, namely the Gumbel-Hougaard, the Ali-Mikhail-Haq, the Frank and the Clayton are compared and discussed. The root mean square error (RMSE) and Akaike's information criterion (AIC) are used to identify the most appropriate copula distribution (Zhang and Singh 2006). The equation for RMSE can be expressed by

Table 4.3 Estimated parameters of the marginal and joint distributions for peak discharges

Sampling method	P-III or Ex distribution			Mixed von Mises distribution			Joint distribution
				μ_i	κ_i	pi	θ
	α	β	δ	1.03	4.80	0.17	
AM	2.520	0.001	2 467	2.81	3.28	0.70	1.82
				5.66	27.43	0.13	
	λ	α_0		0.62	16.52	0.11	
POT	2.58	2.361		2.72	1.61	0.72	0.83
				5.60	3.76	0.17	

Table 4.4 Hypothesis test for margin distributions

Samples	P-III or Ex distribution		Mixed von Mises distribution	
	$\chi^2_{0.95}$	Chi-squared statistics χ^2	$D_{n,0.95}$	K-S statistics D_n
AM	7.815	1.800	0.180	0.089
POT	12.592	2.758	0.115	0.047

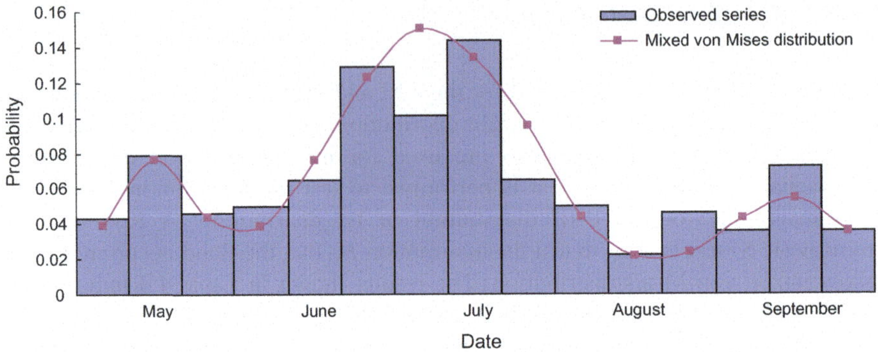

Fig. 4.2 Fitted frequency histograms of flood occurrence dates by the mixed von Mises distribution for POT samples

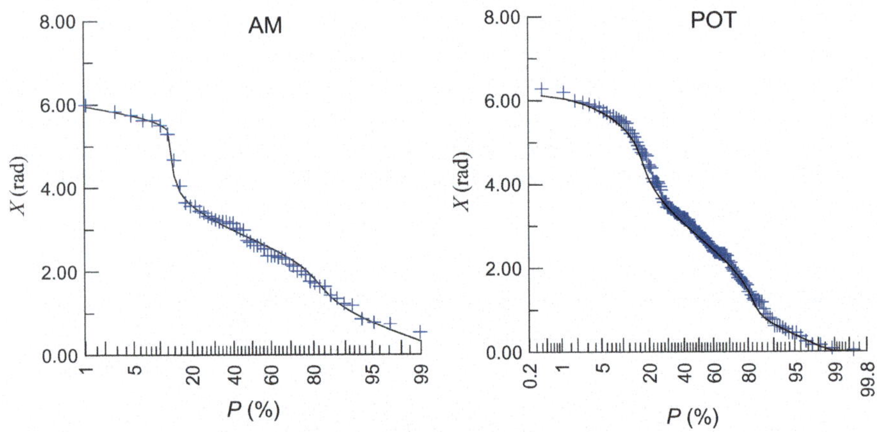

Fig. 4.3 Frequency curves of flood occurrence dates based on AM and POT samples

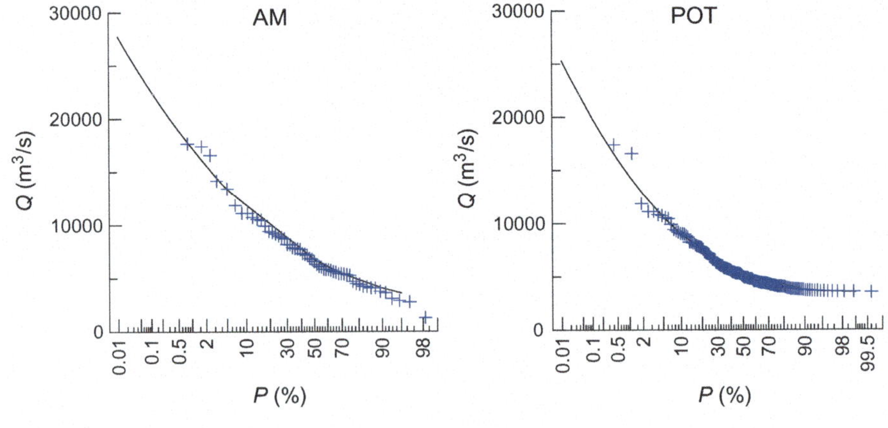

Fig. 4.4 Frequency curves of flood magnitudes based on AM and POT samples

$$RMSE = \sqrt{\frac{1}{N}\sum_{i=1}^{N}\left(x_{the}(i) - x_{emp}(i)\right)^2} \quad (4.12)$$

where N represents the number of observations; and $x_{the}(i)$ and $x_{emp}(i)$ denote the ith calculated and observed values, respectively.

The Akaike information criterion (AIC), developed by Akaike (1974), is used to identify the appropriate probability distribution. The AIC can be obtained either by calculating the maximum likelihood or by calculating the mean square error of the model (Zhang and Singh 2006). The AIC values related to maximum likelihood values can be expressed by

$$AIC = 2k - 2\ \ln(L) \quad (4.13a)$$

The AIC values related to mean square error can be expressed by

$$AIC = 2k + N\ \ln(MSE) \quad (4.13b)$$

where k is the number of parameters in the statistical model; L is the maximized value of the likelihood function for the estimated model; and $MSE = RMSE^2$.

The *RMSE* and *AIC* values related to *MSE* (Eq. 4.13b) for different copulas are listed in Table 4.5. The best family is the one which has the minimum *RMSE* and *AIC* values. It can be seen that Frank and Clayton family fit the empirical joint probabilities better than Gumbel and Ali-Mikhail–Haq. No obvious difference exists between Frank family and Clayton family.

The empirical joint probabilities of the combinations of flood occurrence dates and flood peak magnitudes are plotted versus theoretical probabilities as shown in Fig. 4.5, which shows that no significant difference between empirical and theoretical joint probabilities can be detected.

It may be concluded that the proposed bivariate joint distribution is suitable to represent the flood occurrence dates and magnitudes at the Geheyan reservoir basin. The joint distribution of the AM flood series is shown in Fig. 4.6.

Table 4.5 The RMSE and AIC values for different copula functions

Family		AM			POT	
	θ	*RMSE*	*AIC*	θ	*RMSE*	*AIC*
Gumbel–Hougaard	1.24	0.042	−318	1.10	0.031	−942
Ali-Mikhail–Haq	0.70	0.080	−249	0.37	0.045	−838
Frank	1.82	0.038	−329	0.83	0.028	−970
Clayton	0.49	0.039	−326	0.20	0.028	−970

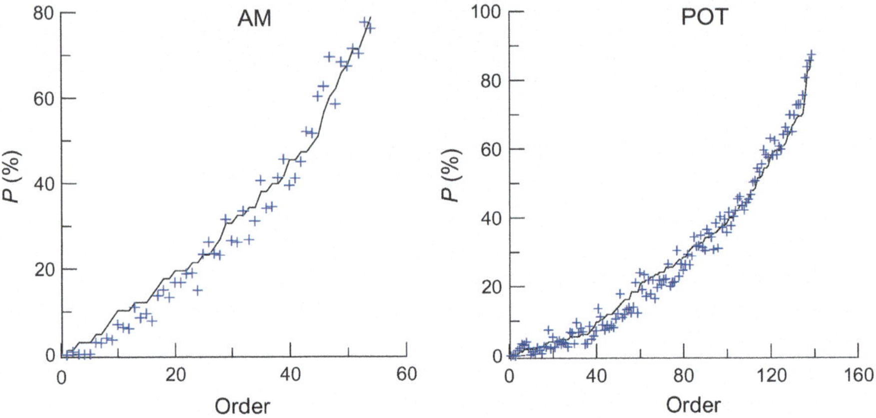

Fig. 4.5 Joint distribution and empirical probabilities of the observed combinations based on AM and POT samples

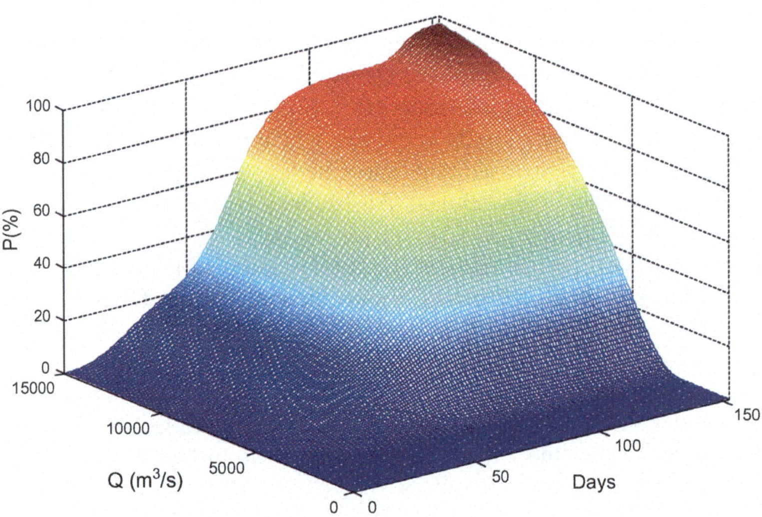

Fig. 4.6 Joint distribution of the flood occurrence dates and magnitudes based on AM samples

4.4.4 Seasonal Design Flood Estimation

The seasonal design floods in the pre-flood, main flood and post-flood sub-seasons are calculated by Eq. 4.11. The curve fitting method that based on minimizing the sum of the squares of the deviations between the observed values obtained from a plotting position formula and theoretical values calculated by Eq. 4.11 for each sub-season is used. An objective function of curve fitting method is given by

$$\text{Min } G(q_j) = \sum_{i=1}^{s} \sum_{j=1}^{N_i} (P(j) - P(q_j, A_i))^2 \tag{4.14}$$

where N_i is the number of the observed data in the ith sub-season; q_j is the observed data in the ith sub-season. $P(j)$ is the cumulative frequency calculated by Eq. 3.9.

The Quasi-Newton method is used to optimize above objective function, and the estimated parameters of von Mises distribution and the seasonal design flood values for AM and POT samples are listed in Tables 4.6 and 4.7 respectively. Figure 4.7 shows that the theoretical curve of seasonal design floods can fit the observational data well.

The relations between the seasonal and annual frequency curves are shown in Fig. 4.8. The seasonal design flood frequency curves are rational, from the point of view that they are lower than the annual design flood frequency curve. Furthermore, the relations between the annual and seasonal design flood frequency curves must be obeyed the Eq. 4.4 or 4.10, which is also taken as a criterion to test the rationality of the seasonal design flood. A goodness-of-fit test for observed and

Table 4.6 Estimated parameters of the Mixed von Mises distribution for AM and POT samples

AM			POT		
μ_i	κ_i	p_i	μ_i	κ_i	p_i
1.03	4.80	0.19	0.61	14.11	0.12
2.81	3.28	0.71	2.77	1.96	0.73
5.66	27.43	0.10	5.61	5.17	0.15

Table 4.7 Comparisons of annual maximum design flood with seasonal design floods estimated by different methods (m^3/s)

Methods	Return period (year)	Annual design	Design values					
			Pre-flood season		Main flood season		Post-flood season	
Chinese method	1,000	22,800	18,700	(−17.98%)	22,200	(−2.63%)	20,500	(−10.09%)
	200	18900	15,090	(−20.31%)	18,383	(−2.92%)	15,890	(−16.08%)
	100	17,400	13,500	(−22.41%)	16,800	(−3.45%)	14,500	(−16.67%)
Singh's method	1,000	22,800	18,384	(−19.37%)	27,298	(19.73%)	15,018	(−34.13%)
	200	18,900	15,206	(−19.69%)	22,883	(20.85%)	13,395	(−29.26%)
	100	17,400	14,282	(−17.92%)	20,626	(18.54%)	12,545	(−27.90%)
Proposed method AM	1,000	22,800	20,300	(−10.96%)	24,000	(5.26%)	22,200	(−0.03%)
	200	18,900	15,400	(−18.52%)	19,200	(1.59%)	17,300	(−0.08%)
	100	17,400	14,200	(−18.39%)	18,100	(4.02%)	16,200	(−0.07%)
POT	1,000	22,044	21,040	(−4.55%)	22,750	(3.20%)	22,042	(−0.01%)
	200	18,270	17,013	(−6.88%)	19,216	(5.18%)	17,986	(−1.55%)
	100	16,606	15,608	(−6.01%)	17,315	(4.27%)	16,604	(−0.01%)

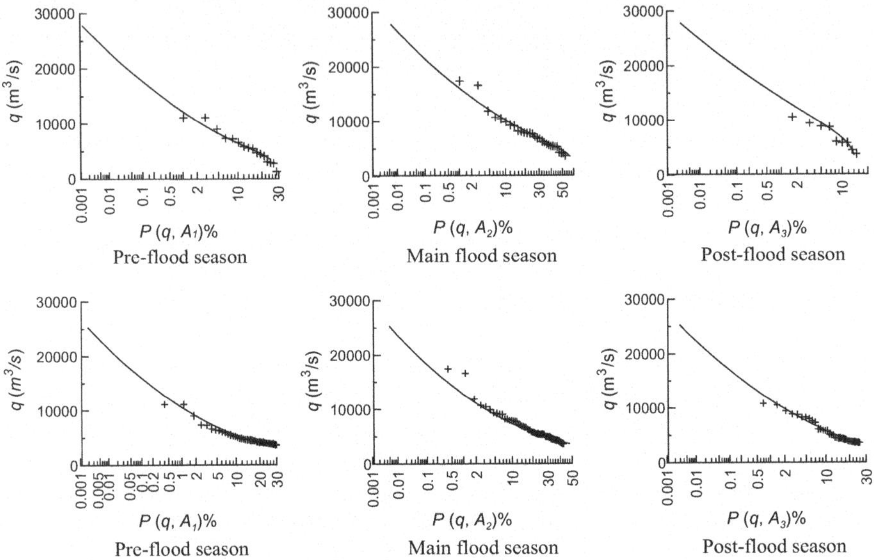

Fig. 4.7 Frequency curves of sub-season design floods based on AM and POT samples

calculated data are shown in Fig. 4.9, in which the line represents the annual theoretical probabilities derived by summing up the seasonal probabilities calculated by Eqs. 4.10 and 4.11, and the crossings represent the empirical probabilities.

1000-year seasonal design floods are estimated by Eq. 4.11, and the results based on AM series are shown in Fig. 4.10. It shows that a surface formed in a three-dimensional Cartesian coordinate system, indicating that various combinations of seasonal design floods can be obtained for a given return period T. As the increase of either or both of two seasonal design flood values, another design flood values will be decreased.

4.4.5 Comparisons of Different Methods

The current seasonal design flood method used in China assumes that the design frequency in each sub-season is identical. In accordance with this hypothesis and the demand of satisfying the flood prevention standards, the seasonal design frequencies must obey the following rules

$$P_X = P_Y = P_Z = P' \tag{4.15}$$

where P_X, P_Y, and P_Z are the design frequencies of the pre-flood season, main flood season and post-flood season, respectively. If the annual maximum flood series is

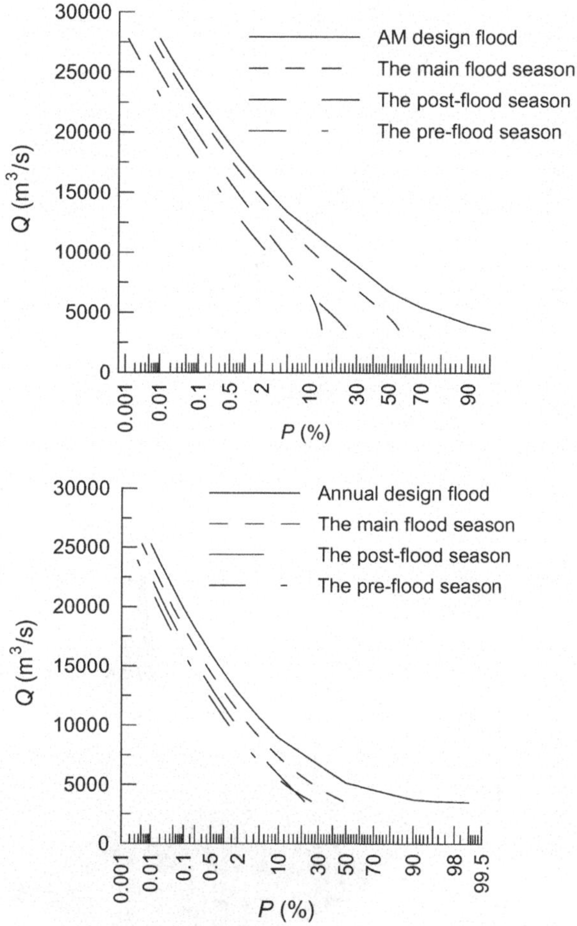

Fig. 4.8 The relations between seasonal and annual design flood frequency curves based on AM and POT samples

used, then P' equals $1/(3T)$. If the POT samples are used, the annual return period needs to be converted to the exceedance probability of the POT method as follows (Rosbjerg 1993):

$$P(Q \geq Q_T) = \frac{1}{\lambda T} \tag{4.16}$$

The comparisons of the annual maximum and seasonal design floods estimated by different methods are given in Table 4.7, where the relative error describes the deviation between the annual design values and seasonal design values. Table 4.7 implies that the seasonal design flood values based on the seasonal maximum series are underestimated in all sub-seasons. Since all of the seasonal design values are less than the annual ones, the current seasonal design flood method used in China is unable to satisfy the flood prevention standards.

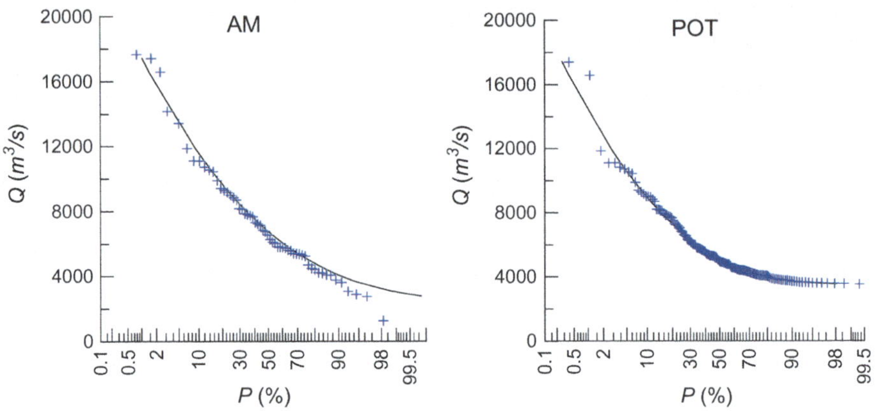

Fig. 4.9 Rational tests of the seasonal design floods based on AM and POT samples

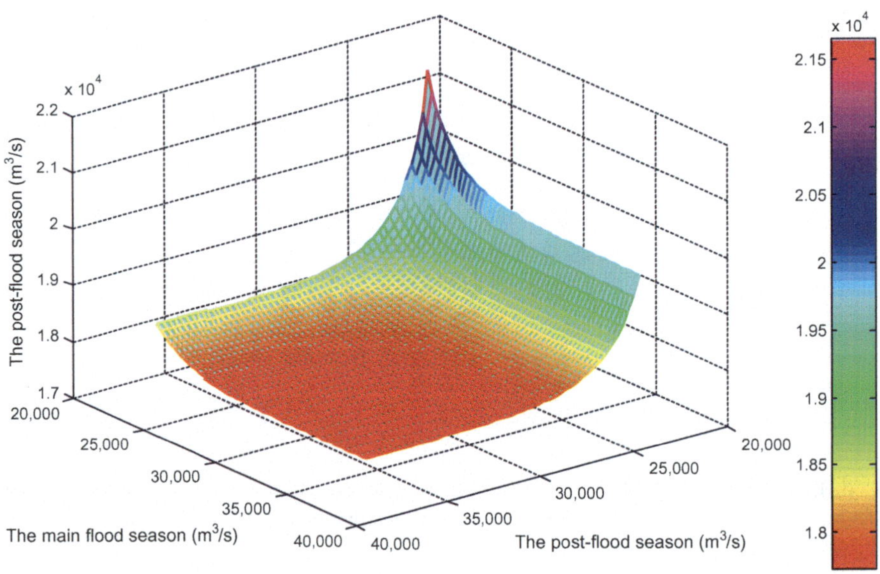

Fig. 4.10 1000-year seasonal design flood peak discharges with different combinations

For the seasonal design flood method suggested by Singh et al. (2005), the sample size of the pre-flood season, main flood season and post-flood season are 16, 20 and 19, respectively. The design values in the pre-flood and post-flood season are lower than those calculated by the Chinese seasonal design flood method. On the other hand, the design values in the main flood season are much higher than that of the annual maximum design floods. The annual and seasonal frequency curves based on Singh's method are drawn in Fig. 4.11. It is shown that the seasonal

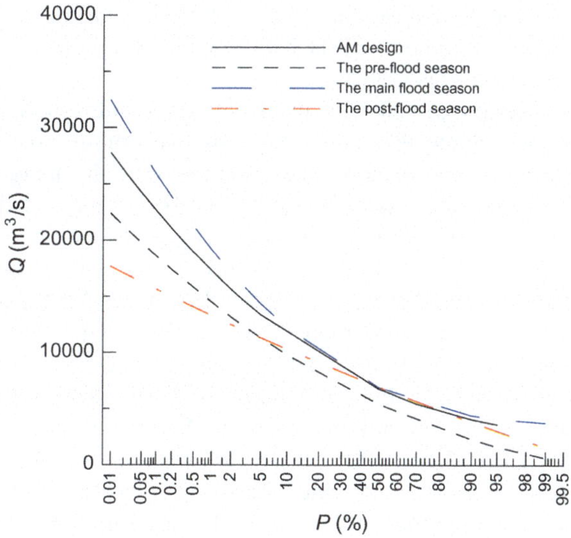

Fig. 4.11 Comparisons of frequency curves calculated by Singh's method with the annual maximum design method

frequency curve of the main flood season is above the annual maximum one. The seasonal frequency curve in the pre-flood season is higher than that in the post-flood season. Actually, the flood in the post-flood season is much larger than that in the pre-flood season. The mean values of the annual maximum flood peak in the pre-flood and post-flood season are 5792 m^3/s and 7060 m^3/s, respectively. The reason for these unreasonable results may be mainly due to that the sample series for some sub-seasons are too short for flood frequency analysis.

The seasonal design floods calculated by the proposed method are also listed in Table 4.7. It indicates that the seasonal design flood values based on AM samples are much higher than the annual maximum design flood values in the main flood season. The design values in the other sub-seasons are less than their corresponding annual maximum ones but greater than those calculated by current seasonal design flood methods. The T-year design flood values calculated by the POT samples are also listed in Table 4.7. It is shown that the seasonal design values of the POT samples in the main flood season exceed the T-year annual design values and less than them in other sub-seasons. The results based on two sampling methods demonstrate that the seasonal design floods estimated by the proposed method are slightly greater than annual design floods in the main flood season and less than them in other flood sub-seasons. Furthermore, the seasonal design floods calculated by Eqs. 4.10 and 4.11 can meet the flood prevention standards.

The design values of the POT samples is less than that of the AM flood series in the main flood season, whereas the values of POT series are larger than AM series in the pre-flood season. The reason for this is mainly due to that the discrepancy exists between the P-III distribution and Ex distribution. For example, 1000-year design values based on the POT and AM samples are equal to 22,044 m^3/s and 22,800 m^3/s, respectively. Compared with the 24% flood occurrence probability for

AM samples, it is about 30% for POT samples in the pre-flood season. No significant difference exists between the results of the POT and AM series in the post-flood season.

The design values of the 1-day maximum runoff volume W_{1d}, 3-day maximum runoff volume W_{3d}, and 7-day maximum runoff volume W_{7d} are also calculated by the proposed method. The parameters of the margin and joint distributions are estimated and listed in Table 4.8. 1000-year and 100-year design flood runoff volumes for each sub-season are calculated by Eq. 4.11 and listed in Table 4.9.

The seasonal FCWL is obtained by the flood hydrograph routing method based on the design flood hydrograph (DFH). One of the methods to derive the DFH is the typical flood hydrograph (TFH) method which has been widely used by practitioners (Nezhikhovsky 1971; Yue et al. 2002). The flood hydrograph with the highest peak or biggest volume is usually selected as a TFH. The DFH is constructed by multiplying each discharge ordinate of the TFH by an amplifier. The TFH of 1979, 1997 and 1998 were selected for the pre-flood season, the main flood season and the post-flood season, respectively. The peak and volume-amplitude (PVA) method is used to derive DFH (MWR 1993; Xiao et al. 2009), and the results are shown in Fig. 4.12. The design flood hydrographs are routed through the reservoir, and the seasonal design FCWL values are determined. They are 201.2, 192.1 and 200.1 m in the pre-flood season, main flood season and post-flood season, respectively.

Table 4.8 Estimated parameters of the marginal and joint distributions for runoff volumes

	P-III distribution			Mixed von Mises distribution			Joint distribution
	α	β	δ	μ_i	κ_i	p_i	θ
W_1				1.03	7.22	0.10	
	2.195	0.646	1.700	2.83	3.55	0.78	2.28
				5.67	35.59	0.11	
W_3				0.93	7.00	0.10	
	1.582	0.231	3.423	2.71	2.95	0.79	1.39
				5.50	8.88	0.11	
W_7				0.60	1.45	0.04	
	1.644	0.156	5.267	2.70	3.22	0.85	2.26
				5.55	4.35	0.12	

Table 4.9 Estimated design runoff volumes by the proposed method (billion m^3)

T	1000			200			100		
Days	Pre-flood	Main flood	Post-flood	Pre-flood	Main flood	Post-flood	Pre-flood	Main flood	Post-flood
W_1	13.55	17.71	16.61	10.64	14.9	13.77	9.38	13.67	12.51
W_3	31.27	37.45	32.93	24.99	31.3	26.68	22.23	28.62	23.93
W_7	44.78	60.15	55.89	34.38	49.97	45.60	29.89	45.53	41.08

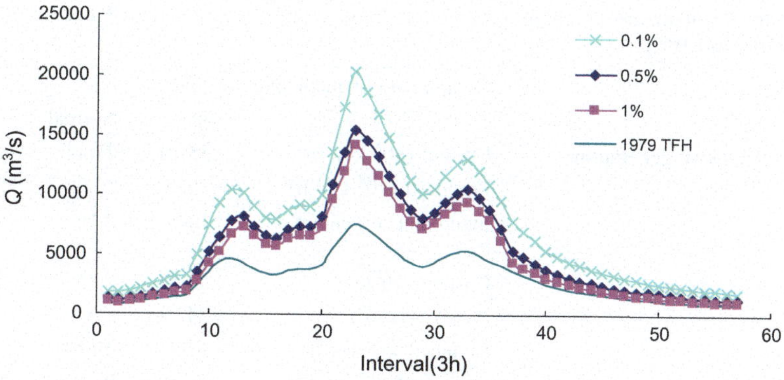

(a) The pre-flood season

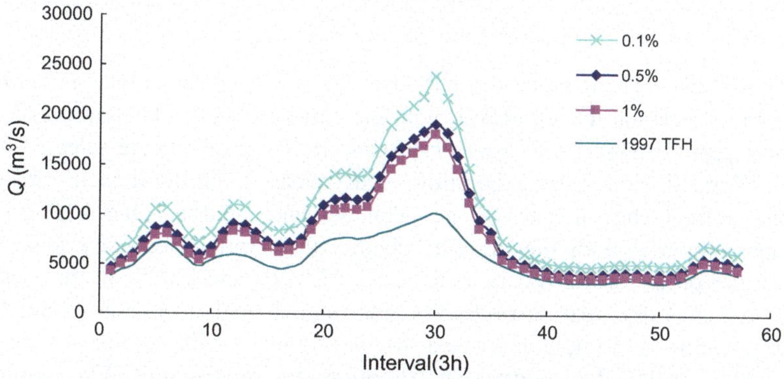

(b) The main flood season

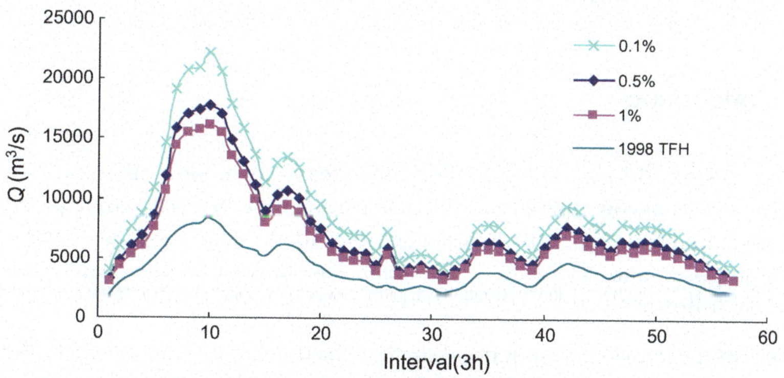

(c) The post-flood season

Fig. 4.12 The derived DFH of the Geheyan reservoir by PVA method

Table 4.10 Comparisons of energy output and flood water utilization rates based on different seasonal design FWCL

Index	Comparison of methods	Year		
		Wet	Normal	Dry
Annual electricity generation	Current (10^8 kW h)	35.23	27.44	23.21
	Proposed (10^8 kW h)	35.89	28.04	23.68
	Increment of generation (%)	1.87	2.19	2.02
Annual spill release	Current ($10^8 m^3$)	25.57	6.85	3.83
	Proposed ($10^8 m^3$)	24.01	5.34	1.75
	Reduced spill release (%)	6.04	10.14	55.24
Flood water resources utilization rate	Current (%)	82.57	93.02	95.70
	Proposed (%)	83.29	94.56	98.05

The daily discharge data set from 1951 to 2004 is used to analyze and compare the benefit of seasonal design FCWL with the current scheme. Three representative years, wet year (1964), normal year (1985) and dry year (2001) are selected for the analysis. Annual electricity generation, spill release and flood water resources utilization at the Geheyan reservoir are calculated and listed in Table 4.10. It can be seen that compared with the current scheme, the annual electricity generations based on the proposed FCWL is increased 1.87, 2.19 and 2.02% in the wet year, normal year and dry year respectively. The annual spill release is reduced. The flood water utilization rate is increased from 82.57 to 83.29% for the wet year, and from 95.70 to 98.05% for the dry year. Therefore, the proposed FCWL can increase energy output and flood water utilization rate.

4.5 Conclusion

Seasonal design floods, which reflect the seasonal flood variation, are very important for reservoir operation and management. A bivariate joint distribution based on copula function, which considers the flood occurrence dates and magnitudes is proposed and established. The main conclusions of this chapter are summarized as follows:

(1) The current seasonal design flood method used in China cannot satisfy the flood prevention standards. Although the Singh's method based on the annual maximum series can meet these standards, the estimated design floods have large errors due to the short length of sample series.
(2) Compared with two current seasonal design flood methods, the proposed method that considers both flood occurrence dates and flood magnitudes is

much more rational in the physical mechanism and can satisfy flood prevention standards in China.

(3) The proposed method can increase energy output and flood water utilization rate and provides a new way for seasonal flood estimation.

References

Akaike H (1974) A new look at the statistical model identification. IEEE Trans Autom Control AC 19(6):716–722

Black AR, Werritty A (1997) Seasonality of flooding: a case study of north Britain. J Hydrol 195:1–25

Carta JA, Ramírez P (2007) Analysis of two-component mixture Weibull statistics for estimation of wind speeds using mixture of von Mises distribution. Renew Energy 32:518–531

Carta JA, Bueno C, Ramírez P (2008) Statistical modeling of directional wind speeds using mixtures of von Mises distributions: case study. Energy Convers Manage 49:897–907

Chen L, Guo S, Yan B, Liu P, Fang B (2010) A new seasonal design flood method based on bivariate joint distribution of flood magnitude and date of occurrence. Hydrol Sci J 55(8):1264–1280

Creager WP, Kinnison HB, Shifrin H, Snyder FF, Williams GR, Gumbel EJ, Matthes GH (1951) Review of flood frequency methods: final report of the subcommittee of the joint division committee on floods. ASCE Transactions 116:1220–1230

Cunderlik JM, Ouarda TBMJ, Bobée B (2004) Determination of flood seasonality from hydrological records. Hydrol Sci J 49(3):511–526

Durrans SR, Eiffe MA, Thomas WO, Goranflo HM (2003) Joint seasonal/annual flood frequency analysis. J Hydrol Eng ASCE 8(4):181–189

Fang B, Guo SL, Wang SX, Liu P, Xiao Y (2007) Non-identical models for seasonal flood. Hydrol Sci J 52(5):974–991

Guo SL, Zhang HG, Chen H, Peng DZ, Liu P, Pang B (2004) Reservoir flood forecasting and control system in China. Hydrol Sci J 49(6):959–972

Hosking JRM, Wallis JR (1997) Regional frequency analysis: an approach based on L-moments. Cambridge University Press, London

Institute of Hydrology (IH) (1999) Flood estimation handbook, vol 3. Inst Hydro. Wallingford, UK

Lang M, Ouarda TBMJ, Bobée B (1999) Towards operational guidelines for over-threshold modeling. J Hydrol 225(3–4):103–117

Michael L, Andrew M, Martin L (2007) Frequency analysis of rainfall and streamflow extremes accounting for seasonal and climatic partitions. J Hydrol 348:135–147

Ministry of Water Resources (MWR) (1993) Regulation for calculating design flood of water resources and hydropower projects. Chinese Water Resour. Hydrop. Press, Beijing, China (in Chinese)

Nezhikhovsky RA (1971) Channel network of the basin information. Hydrometeorological, Leningrad, Russia

Ngo LL, Madsen H, Rosbjerg D (2007) Simulation and optimization modelling approach for operation of the Hoa Binh reservoir. Vietnam J Hydrol 336:269–281

Ouarda TBMJ, Ashkar F, EJabi N (1993) Peaks over threshold model for seasonal flood variations. In: Kuo CY (ed) Engineering hydrology, Proceedings of international symposium, pp 341–346, ASCE, Reston, VA

Ouarda TBMJ, Cunderlik JM, St-hilaire A, Barbet M, Brunear P, Bobée B (2006) Data-based comparison of seasonality-based regional flood frequency methods. J Hydro 330:329–339

Rosbjerg D (1993) Partial duration series in water resources. Technical University of Denmark, Denmark
Singh VP, Wang SX, Zhang L (2005) Frequency analysis of nonidentically distributed hydrologic flood data. J Hydrol 307:175–195
Thomas W, O. Jr, Crockett KL, Johnson A(1998) Discussion of techniques for analysis of ice-jam flooding. In: Proceedings of Association of State Floodplain Managers
Waylen P, Woo MK (1982) Prediction of annual floods generated by mixed processes. Water Resour Res 18(4):1283–1286
Xiao Y, Guo SL, Liu P, Yan BW, Chen L (2009) Design flood hydrograph based on multi-characteristic synthesis index method. J Hydrol Eng 14(12):1359–1364
Yin JB, Guo SL, Liu ZJ, Xiong F (2017) Bivariate seasonal design flood estimation based on copulas. J Hydrol Eng 22(12). https://doi.org/10.1061/(asce)he.19435584.0001594
Yue S, Quarda TBMJ, Bobée B, Legendre P, Bruneau P (2002) Approach for describing statistical properties for flood hydrograph. J Hydrol Eng 7:147–153
Zhang L, Singh VP (2006) Bivariate flood frequency analysis using the copula method. J Hydrol Eng 11(2):150–164

Chapter 5
Drought Analysis Using Copulas

5.1 Introduction

A drought is a natural hazard that results from a deficiency of precipitation as compared with the expected or normal amount, which can translate into the insufficient amount of water to meet the demands of human activities and the environment (Estrela and Vargas 2012). It occurs in virtually all climatic zones, such as high as well as low rainfall areas. Each year one or the other part of the world experiences drought and suffers from huge economic losses, and this has been the case ever since the birth of human civilization. The impacts produced by droughts are numerous. Historical droughts have affected large populations (and represent up to 35% of those affected by natural disasters), often resulting in significant fatalities (50% of the mortality due to natural disasters), whereas 7% of world economic losses have been attributed to their occurrence (Below et al. 2007; Núñez et al. 2011). Thus, droughts are of great importance in the planning and management of water resources (Mishra and Singh 2010).

The Han River, which is divided into three regions: the Danjiangkou Reservoir sub-basin (upper sub-basin), the middle sub-basin, and the lower sub-basin, is a tributary of the Yangtze River. This river is the source of water for the middle route of the well-known South-to-North Water Diversion Project (SNWDP) in China. The middle route, located between the Danjiangkou Reservoir in the Han River and Beijing, will transfer 14 billion m^3 of water annually from the Han River to Beijing by 2030. As some of its water is transferred via the SNWDP, the Han River has an impact on socioeconomic development both in the middle and lower sub-basins and northern China.

Various indices have been developed to detect and monitor droughts, and commonly, Palmer Drought Severity Index (PDSI) and the Standardized Precipitation Index (SPI) are more frequently used indices for drought characterization (Mishra and Singh 2010). Palmer (1965) proposed a moisture index (Palmer Drought Severity Index, PDSI) based on water budget accounting using

L. Chen and S. Guo, *Copulas and Its Application in Hydrology and Water Resources*, Springer Water, https://doi.org/10.1007/978-981-13-0574-0_5

precipitation and temperature data. McKee et al. (1993) proposed the concept of standardized precipitation index (SPI) based on the long-term precipitation record for the desired period. PDSI has several limitations (see Alley 1984; Guttman 1991, 1998). For instance, the soundness of proposed water balance model is questionable, the temporal scale of PDSI is not clear, and the values of PDSI possess neither a physical (such as required rainfall depth) nor statistical meaning (such as recurrence probability) (Kao and Govindaraju 2007). Due to the limitations of PDSI, Guttman (1998) recommended the use of SPI as a primary drought index because it is simple, spatially invariant in its interpretation, and probabilistic. Therefore, SPI series is used for this book.

Drought properties are usually investigated separately by univariate frequency analysis (e.g., Tallaksen et al. 1997; Fernández and Salas 1999; Cancelliere and Salas 2004; Serinaldi et al. 2009). Since droughts are complex phenomena, one variable cannot provide a comprehensive evaluation of droughts (Shiau et al. 2007). A separate analysis of drought duration distribution and drought severity distributions cannot reveal the significant correlation between them. Instead of using traditional univariate analysis for drought assessment, a better approach for describing drought characteristics is to derive the joint distribution of drought variables (Mishra and Singh 2010). For example, Shiau and Shen (2001), Bonaccorso et al. (2003), Kim et al. (2003), González and Valdés (2003), Salas et al. (2005) and Cancelliere and Salas (2010) proposed different methods to investigate the joint distribution of drought duration and drought severity or intensity. These bivariate distributions have either complex mathematical derivations or their parameters are obtained by fitting the observed or generated data (Shiau 2006).

Until now most of the work has focused on bivariate cases. Investigators have used many different ways to build bivariate distributions of drought duration and severity. Actually, drought events have some other characteristics, such the minimum SPI (values) in one drought event, and drought interval time, which are mutually correlated. The studies mentioned above have only included some of the drought characteristics. However, it is important for design engineers and water resources planners to know not only the frequency of droughts but also the risk of having droughts of differing duration, severity, interval time and the minimum SPI value within a drought period. For this purpose, Chen et al. (2013) established a multivariate distribution for analyzing the probabilities and return periods of drought events with more variables. In order to simplify inference procedures and to derive flexible multivariate distributions, copulas can be efficiently employed.

The content of this chapter is therefore to employ the Archimedean and meta-elliptical copulas to construct four-dimensional joint distributions for droughts. The dry events are divided into four states, and copula functions are built in each state to investigate their applicability in drought analysis. The drought risk is defined and analyzed based on the return period (recurrence interval) of drought events, which has become standard practice for the risk-based design of hydraulic structures.

5.2 Definition of Drought and Univariate Variable

5.2.1 *Definition of Drought Events*

Drought identification based on an SPI series can be carried out by assuming a drought period as a consecutive number of time intervals where SPI values are less than 0 (Shiau 2006). Figure 5.1 illustrates the time series of SPI and drought events. Each drought event is characterized by four main properties, drought duration D_d, drought severity S_d, minimum SPI(MSPI) I_d, and drought interval time L_d. The definitions of these variables are discussed below:

Drought duration D_d is defined as the number of consecutive intervals (months) where SPI remains below the threshold value 0 (Shiau 2006).

Drought severity S_d is defined as a cumulative SPI value during a drought period, $S_d = \sum_{i=1}^{D_d} SPI_i$ where SPI_i means the SPI value in the *i*th month (Mishra and Singh 2010).

The Minimum SPI (MSPI) value, defined as the minimum SPI value within a drought period (Serinaldi et al. 2009), is used in this chapter to describe the peak of drought events.

The drought interval time L_d is defined as the period elapsing from the initiation of drought to the beginning of the next drought (Song and Singh 2010).

Based on the SPI values, the dry states can be divided into four states shown in Table 5.1 (Mishra et al. 2009).

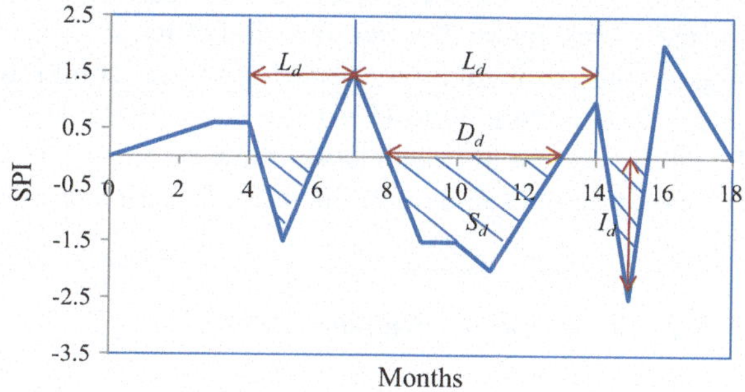

Fig. 5.1 Definition sketch of drought events

Table 5.1 Drought classification based on SPI

SPI	Classes
2.00 and above	Extremely wet
1.50 to 1.99	Very wet
1.00 to 1.49	Moderately wet
−0.99 to 0.99	Near normal
−1 to −1.49	Moderately dry
−1.5 to −1.99	Severely dry
−2.00 and less	Extremely dry

5.2.2 *Distributions of Univariate Drought Variables*

Generally, drought duration is fitted as a geometric distribution (Kendall and Dracup 1992; Mathier et al. 1992) by treating it as a discrete random variable. Similarly, Shiau and Shen (2001) computed drought interval time as equal to the sum of drought duration and non-drought duration, on the assumption that drought and non-drought durations follow a geometric distribution. As Sklar's theorem requires the continuity of marginal distributions, the continuous distribution is needed in this study. Two continuous distributions used mostly in drought analysis are exponential and gamma distributions. For example, the exponential distribution was selected for fitting drought duration (Zelenhastic and Salvai 1987). The gamma distribution has generally been used to describe drought severity (Shiau 2006). According to five drought states considered, twenty marginal distributions will be determined. Exponential or gamma distributions cannot fit every case well. Thus, the P-III distribution, the generalized Pareto distribution and the generalized extreme value distribution, which have also been used to describe hydrologic variables, are considered. Among these distributions, the one with the smallest root mean square error (*RMSE*) value between observed and theoretical probabilities will be selected.

5.3 Return Period for Drought Events

A common approach to hydraulic and hydrologic design is frequency analysis or the recurrence interval or return period of hydrologic events (Shiau and Shen 2001). In particular, drought return periods provide useful information for proper water use under drought conditions (Serinaldi et al. 2009). The return period of a drought can be defined as the average elapsed time or mean interval time between occurrences of two droughts (Shiau and Shen 2001; Serinaldi et al. 2009).

5.3.1 *Univariate Return Period*

Shiau and Shen (2001) calculated the return period of a drought event with severity equal to or greater than a certain value S_d. Shiau (2006) calculated the return period of a drought event with a duration equal to or greater than a certain value D_d. Similarly, the return period of drought intensity can be obtained using the same formula expressed as

$$T_d = \frac{E(L_d)}{1 - F_{D_d}(x)}; \quad T_d = \frac{E(L_d)}{1 - F_{S_d}(x)}; \quad T_d = \frac{E(L_d)}{1 - F_{I_d}(x)} \tag{5.1}$$

where $E(L_d)$ is the expected drought interval time.

5.3.2 Multivariate Return Period

Salas et al. (2005) extend Eq. 5.1 to a more general case of drought events defined regarding either severity or MSPI and duration. The interval time between two drought events E is $T_E = \sum_{j=1}^{N_d} L_{d_j}$, where L_{d_j} is the interval time between any two droughts in general (i.e., droughts not necessarily characterized by E); and N_d is the number of droughts until the next drought event E occurs. Then, the return period T is the expected value of T_E, and can be expressed as

$$T = E(T_E) = E(N_d)E(L_d) \tag{5.2}$$

where $E(N_d) = 1/P(E)$. The multivariate return period can be calculated based on Eq. 5.2.

Shiau (2006) defined the bivariate joint return period T_{and} and T_{or} as

$$T_{and} = \frac{E(L_d)}{1 - P(X_i \geq x_i, X_j \geq x_j)} = \frac{E(L_d)}{1 - F(x_i) - F(x_j) + C(F(x_i), F(x_j))} \tag{5.3}$$

$$T_{or} = \frac{E(L_d)}{1 - P(X_i \geq x_i \, or \, X_j \geq x_j)} = \frac{E(L_d)}{1 - C(F(x_i), F(x_j))} \tag{5.4}$$

According to Eq. 5.2, the trivariate return period can be defined as

$$\begin{aligned} T_{and} &= \frac{E(L_d)}{1 - P(X_i \geq x_i, X_j \geq x_j, X_k \geq x_k)} \\ &= E(L_d)/(1 - F(x_i) - F(x_j) - F(x_k) + C(F(x_i), F(x_j)) \\ &\quad + C(F(x_i), F(x_k)) + C(F(x_j), F(x_k)) - C(F(x_i), F(x_j), F(x_k))) \end{aligned} \tag{5.5}$$

$$T_{or} = \frac{E(L_d)}{1 - P(X_i \geq x_i or X_j \geq x_j or X_k \geq x_k)} = \frac{E(L_d)}{1 - C(F(x_i), F(x_j), F(x_k))} \tag{5.6}$$

5.3.3 Conditional Return Period

Shiau (2006) defined the bivariate conditional return period as

$$T_{x_i|x_j} = \frac{E(L_d)}{(1 - F_{x_j}(x_j))(1 - F_{x_i}(x_i) - F_{x_j}(x_j) + C(x_i, x_j))} \tag{5.7}$$

where $T_{x_i|x_j}$ denotes the conditional return period for X_i given $X_j \geq x_j$.

The bivariate conditional return period of drought duration, severity and MSPI are calculated.

5.4 Data Set

5.4.1 Historical Data

The daily rainfall data from 1961 to 2007 in the upper Han River basin is used to evaluate drought characteristics. The Han River is a left tributary of the Yangtze River with a length of 1532 km. Daily rainfall data from nine gauged stations, including the Ankang, Foping, Hanzhong, Lueyang, Shangzhou, Shiquan, Wanyuan, Xixia, and Zhenan are used in this study. The locations of these stations are shown in Fig. 5.2. The average areal rainfall of this basin is calculated based on these nine stations.

5.4.2 Rainfall Data Generation

Since a drought may last for several months or even years, the recorded series are usually short for evaluating drought characteristics (Mishra et al. 2009). In order to avoid this disadvantage, 500-year daily rainfall data are generated based on historical data characteristics. The Markov model is applied to produce precipitation occurrence, and the two-parameter gamma distribution is used to generate the precipitation quantity (Chen et al. 2010).

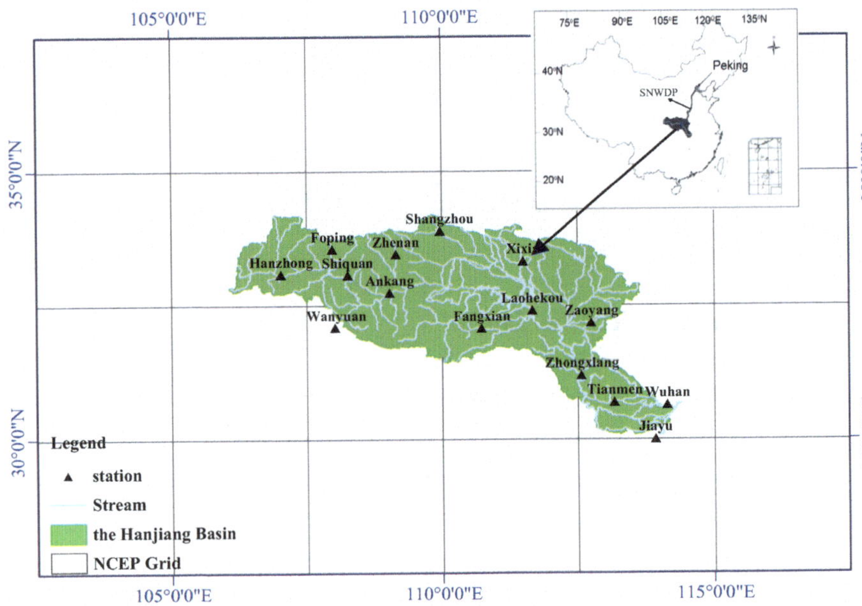

Fig. 5.2 Location of stations in the Hanjiang basin and the middle route of the SNWDP in China

5.5 Application

5.5.1 *Comparisons Between Historical and Synthetic Precipitation Series*

The three-order Markov model is applied to produce precipitation occurrence, and the two-parameter gamma distribution is used to generate the precipitation amount. In order to test the rationality of the rainfall simulation model, synthetic data with the same length of the historical data is generated. Their mean and standard variation values are calculated and given in Table 5.2. The absolute and relative errors between historical and synthetic data are also shown in this table. Results demonstrate that the maximum absolute error of mean values is 0.1 mm, corresponding to the maximum relative error of 4.65%, and the relative error for the standard deviation is less than 10%. Therefore, the synthetic data can be applied to the calculation hereafter. This model is used to generate daily rainfall data from the nine stations with a length of 500 years. The observed and theoretical cumulative monthly rainfall is shown in Fig. 5.3, in which the shape and the cumulative monthly values of these two series are nearly the same. The SPI series for different time scales are obtained and shown in Fig. 5.4. The monthly SPI values are used for analysis hereafter.

5.5.2 *Correlation Analysis*

The Pearson and Kendall correlation coefficients for all drought variables are given in Tables 5.3. Results confirm that for drought events, normal dry state, and moderate dry states, all variables show positive association and a highly correlated relationship between monthly rainfall data is observed. For the left dry states, some drought variables show a small negative correlation, but the negative correlations are small, close to zero. This means that the association between the two variables can be negligible and the fully nested copulas can be used in these cases.

5.5.3 *Estimation of Marginal Distributions*

The Ex, gamma, P-III, GP and GEV distributions are applied to fit every drought variable. The CDF and PDF of these distributions are given in Table 1.1 of Chap. 1. Parameters of marginal distributions are estimated by L-moments (Hosking 1990). The distribution with the smallest *RMSE* values between observed and theoretical probabilities is selected. The chosen distributions and their estimated parameters are listed in Table 5.4. Figure 5.5 compares computed and

Table 5.2 Comparisons of statistical variables between historical and synthetic data

Variables	Mean				Standard deviation			
Stations	Historical	Synthetic	Absolute error	Relative error (%)	Historical	Synthetic	Absolute error	Relative error (%)
Ankang	2.20	2.20	0.00	0.00	5.42	5.13	0.29	5.35
Foping	1.88	1.89	0.01	0.53	4.63	4.53	0.10	2.16
Hanzhong	2.35	2.35	0.00	0.00	5.63	5.38	0.25	4.44
Lueyang	2.17	2.16	0.01	0.46	5.14	4.76	0.38	7.39
Shangzhou	1.88	1.80	0.08	4.26	4.63	4.32	0.31	6.70
Shiquan	2.42	2.42	0.00	0.00	5.88	5.51	0.37	6.29
Wanyuan	3.37	3.31	0.06	1.78	8.09	7.36	0.73	9.02
Xixia	2.36	2.35	0.01	0.42	6.02	5.65	0.37	6.15
Zhenan	2.15	2.05	0.10	4.65	5.05	4.69	0.36	7.13

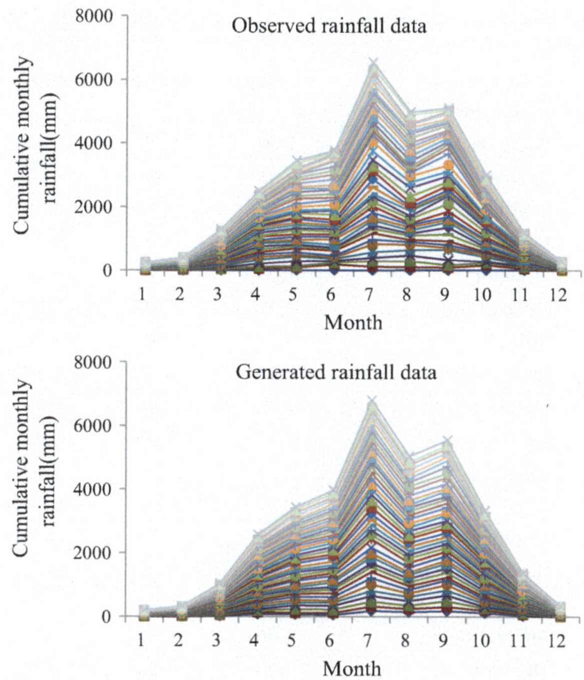

Fig. 5.3 Observed and simulated cumulative monthly rainfall

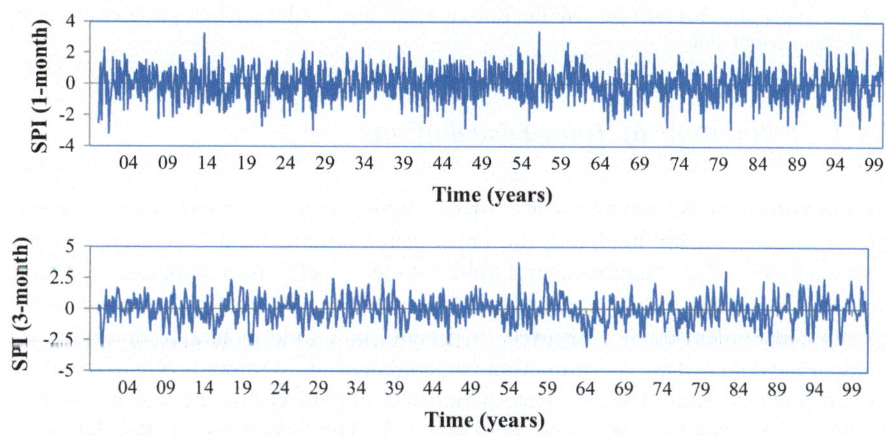

Fig. 5.4 SPI values with different time scales

empirical marginal distributions of observed drought duration, interval, severity, and MSPI in the case of SPI < 0 and normal dry events, which demonstrate that the theoretical and empirical values fit well for all marginal distributions.

Table 5.3 Values of Pearson and Kendall correlation coefficients for all drought variables in different drought states

Drought states	Correlation coefficient	Duration	Severity	MSPI	Interval time
Dry event	Duration	1.00	0.59	0.34	0.60
	Severity	0.82	1.00	0.77	0.37
	MSPI	0.42	0.78	1.00	0.21
	Interval time	0.71	0.59	0.27	1.00
Near normal dry	Duration	1.00	0.49	0.26	0.35
	Severity	0.77	1.00	0.82	0.18
	MSPI	0.33	0.78	1.00	0.10
	Interval time	0.37	0.29	0.13	1.00
Moderately dry	Duration	1.00	0.41	0.13	0.02
	Severity	0.93	1.00	0.88	0.02
	MSPI	0.15	0.49	1.00	0.01
	Interval time	0.01	0.00	−0.04	1.00
Severe dry	Duration	1.00	0.23	0.09	−0.05
	Severity	0.90	1.00	0.97	−0.03
	MSPI	0.11	0.53	1.00	−0.02
	Interval time	−0.06	−0.06	−0.01	1.00
Extreme dry	Duration	1.00	0.24	0.12	−0.03
	Severity	0.78	1.00	0.97	−0.03
	MSPI	0.17	0.74	1.00	−0.03
	Interval time	−0.02	−0.04	−0.04	1.00

Note the super-diagonal elements are the Kendall correlation, and the sub-diagonal elements are the Pearson correlation

5.5.4 *Estimation of Joint Distributions*

Four-dimensional Archimedean and meta-elliptic copulas are tested to determine the best-fit copula for modeling the dependence amongst the four drought characteristics. For the Archimedean family, three widely used copulas, including Gumbel-Hougaard, Frank, and Clayton, are used; for the meta-elliptical copulas, normal and Student copula are used. A pseudo-likelihood technique involving the ranks of the data is used for estimating parameters of the symmetric and asymmetric Archimedean copulas. The estimated parameters of both symmetric and asymmetric Archimedean copulas are given in Table 5.5. The inversion of Kendall's tau method (Genest et al. 2007) is used to estimate the parameters of the Normal copula, while the maximum pseudo-likelihood method is used to estimate the parameters of Student copula. The estimated parameter values of both Normal and Student copulas are given in Table 5.6.

Table 5.4 Selected marginal distributions and their estimated parameters

Events	Severity	MSPI	Duration	Interval time
Dry event	P-III	GP	GAM	GAM
	1.63	0.08	2.73	2.17
	1.47	1.50	0.75	1.43
	1.83			
Near normal dry	GAM	P-III	GAM	GAM
	1.22	1.05	4.33	2.31
	1.34	0.53	0.35	1.47
		1.52		
Moderately dry	GP	GP	GAM	GP
	1.01	0.99	18.66	−0.03
	1.25	0.44	0.06	11.37
Severe dry	GEV	GP	GAM	GAM
	1.65	1.51	61.79	1.01
	0.13	0.40	0.02	22.35
	−0.26	0.82		
Extreme dry	GP	GP	GAM	GAM
	2.02	2.01	60.33	1.18
	0.37	0.46	0.02	34.94
	−0.16	0.16		

Note the characters above the numbers in each cell mean the selected distributions

In order to select an appropriate copula, the root mean square error (*RMSE*) and the Akaike information criterion (*AIC*) are used (Zhang and Singh 2006). The formulas for calculating these indexes are given in Chap. 2. The *RMSE* and *AIC* values of the Archimedean and meta-elliptical copulas are shown in Table 5.7, which indicates that the asymmetric copulas given a better fit than the symmetric copula for the Archimedean family. Generally meta-elliptical copulas fitted better than the Archimedean copulas, except for the Clayton copula. The *RMSE* and *AIC* values of the Normal copula exhibited a better fit than the Student copula.

Figure 5.6 compares observed and theoretical joint probabilities, which indicates that the observed and theoretical values fitted each other well. For drought events with SPI values less than 0 and near normal dry state, the Normal copula gives a better fit than others. For the remaining three states, the Clayton copula is better. The theoretical probabilities of moderately, severely and extremely dry states calculated by both the Normal and Clayton copulas are shown in Fig. 5.6. It is indicated from the *RMSE* and *AIC* values in Table 5.7 and the fitting results in Fig. 5.6 that the difference between Normal and Clayton copulas is small. Therefore, the Normal copula is an appropriate copula for all of the dry events in the upper Han River basin.

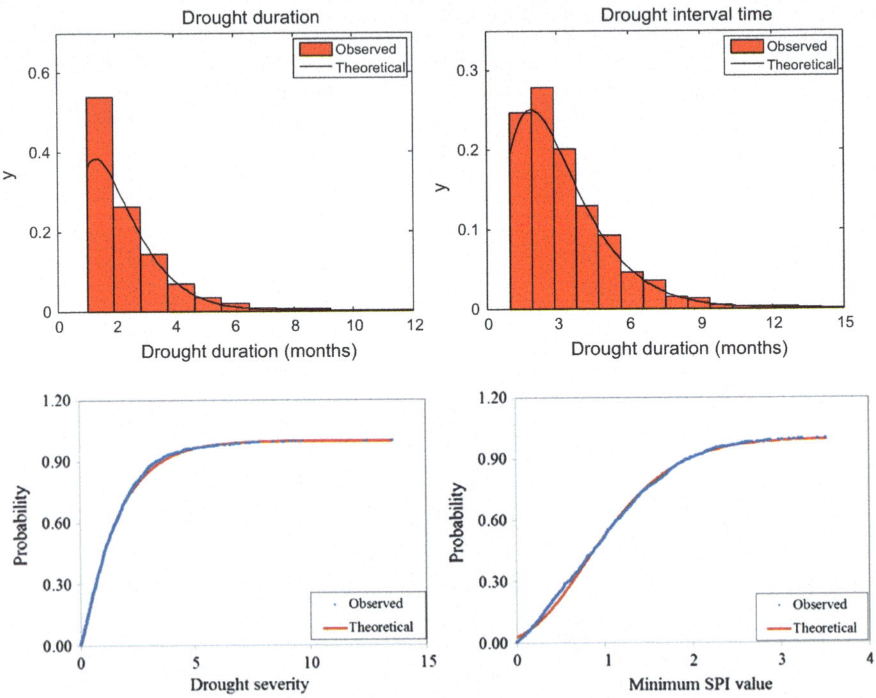

(a) SPI<0

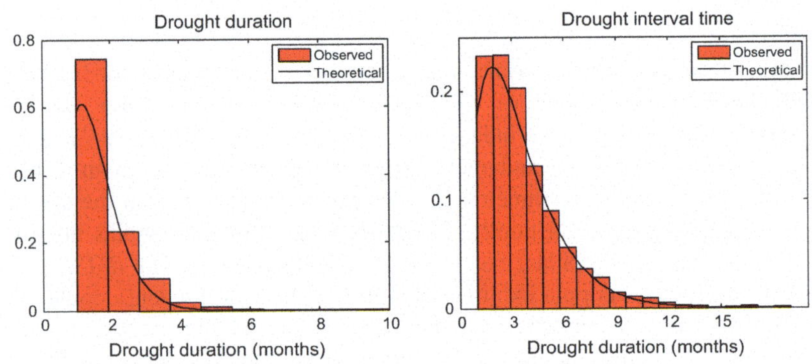

Fig. 5.5 Frequency curves of marginal distributions

5.5.5 *Return Period Analysis*

The average drought interval time estimated from both observed data and theoretical distributions is 3.09 months. Therefore, the calculated value of 3.09 months is used hereafter. First, the univariate return period is analyzed based on Eq. 5.1. Given the return values T of 5, 10, 20, 50 and 100 years, the marginal probabilities

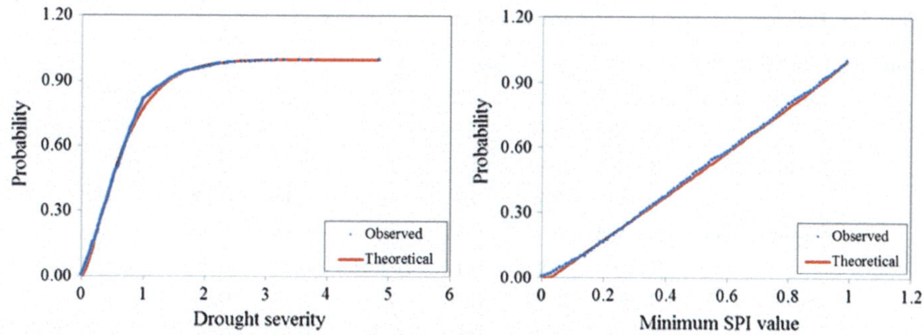

(b) Normal dry event

Fig. 5.5 (continued)

Table 5.5 Estimated parameters of symmetric and asymmetric Archimedean copulas

Dry events	Family	Symmetric	Asymmetric		
		θ	θ_1	θ_2	θ_3
Drought (SPI < 0)	Gumbel	1.66	1.40	1.40	2.80
	Frank	4.62	3.38	3.38	12.50
	Clayton	1.27	0.54	0.55	6.78
Near normal	Gumbel	1.32	1.18	1.26	2.53
	Frank	2.70	1.38	1.39	14.95
	Clayton	0.79	0.28	0.28	8.51
Moderately dry	Gumbel	1.07	1.00	1.02	2.83
	Frank	2.69	0.17	0.17	21.52
	Clayton	0.39	0.11	0.11	14.97
Severe dry	Gumbel	1.03	1.00	1.00	6.26
	Frank	0.66	0.001	0.10	35
	Clayton	0.27	0.04	0.06	59.53
Extreme dry	Gumbel	1.03	1.00	1.00	8.81
	Frank	0.66	0.001	0.26	30.0
	Clayton	0.28	0.04	0.06	66.26

$F(x)$ are derived. The magnitudes corresponding to the return periods defined by separate values of drought duration, severity and MSPI are calculated by solving the inverse function F and given in Table 5.8.

The joint return period is related to the marginal probabilities and joint probability. The marginal probabilities have been calculated before for analyzing the univariate return period. The joint probabilities of drought variables are also given in Table 5.8. The bivariate and trivariate return periods T_{and} and T_{or} are then calculated and listed in Table 5.9, which shows that the univariate return period is larger than the joint return period T_{or} and less than the joint return period T_{and}. The

Table 5.6 Estimated parameters of normal and student copulas

Number	Classifications	Copulas	ρ_1	ρ_2	ρ_3	ρ_4	ρ_5	ρ_6	υ
1	Drought event	Normal	0.92	0.73	0.50	0.45	0.30	0.70	
		Stuent	0.88	0.59	0.51	0.32	0.26	0.79	18.95
2	Normal dry	Normal	0.93	0.62	0.27	0.34	0.15	0.42	
		Stuent	0.90	0.33	0.25	0.16	0.12	0.50	8.67
3	Moderately dry	Normal	0.92	0.51	0.03	0.16	0.008	0.03	
		Stuent	0.97	−0.17	0.003	−0.12	0.01	0.05	3.26
4	Severe dry	Normal	0.98	0.30	−0.04	0.12	−0.03	−0.06	
		Stuent	0.85	0.60	0.28	0.14	0.06	0.47	8.19
5	Extreme dry	Normal	0.98	0.30	−0.05	0.15	−0.05	−0.03	
		Stuent	0.99	−0.09	−0.31	−0.07	−0.30	0.01	10.95

trivariate joint return period T_{and} is larger than bivariate one in the same row in Table 5.9, but the trivariate joint return period T_{or} is less than the bivariate one. This is because that adding one variable in the model makes the exceedance probabilities $P\ (X_1 > x_1,\ X_2 > x_2,\ X_3 > x_3)$ smaller than the two bivariate one $P\ (X_1 > x_1,\ X_2 > x_2)$.

According to Eq. 5.7, the values of bivariate conditional return period of drought duration, severity and MSPI are calculated and plotted in Fig. 5.7, which shows that the conditional return period increases when the values of drought variables increase. The derived conditional return periods of drought duration, severity and MSPI can be used to evaluate the risk of a specific water resources system. For instance, based on the derived conditional distribution, water resources managers are informed that the probabilities for a drought severity greater than 1.0 and 2.0 given drought duration exceeding one month are equal to 6.65 and 12.85 years, respectively.

5.5.6 *Drought Probability Analysis*

In this chapter, drought events are defined by drought duration, severity, interval time and MSPI. It is necessary to know the occurrence probabilities of arbitrary drought events. Table 5.8 gives D_d, S_d, I_d and L_d values corresponding to different values of $F(x)$. The joint probabilities of some drought events $E = \{D_d \leq d_d, S_d \leq s_d, I_d \leq i_d, L_d \leq l_d\}$ are also given in Table 5.8. Taking $F(x)$ equal to 0.1, for example, the corresponding magnitudes in Table 5.8 for drought duration, severity, MSPI and time interval are 0.25, 0.24, 0.65 and 0.95, respectively. The joint probability of this drought event is 0.01. Based on this model, the joint probability of any drought event can be obtained. The joint probability increases when marginal probabilities increase. From this point of view, the calculated results seem justified.

Table 5.7 RMSE and AIC values of different copulas

Events	Family	Archimedean						Meta-elliptical	
		Gumbel		Frank		Clayton		Normal	Student
		A	B	A	B	A	B		
Drought event	RMSE	0.048	0.028	0.031	0.022	0.052	0.049	0.013	0.022
	AIC	−8897	−10,482	−10,179	−11,189	−8662	−8841	−12,721	−11,177
Normal year	RMSE	0.058	0.05	0.055	0.037	0.055	0.045	0.018	0.036
	AIC	−7321	−8154	−7899	−8975	−7899	−8441	−10,931	−9043
Moderately dry	RMSE	0.053	0.099	0.056	0.034	0.079	0.028	0.032	0.041
	AIC	−2906	−2284	−2852	−3342	−2511	−3534	−3396	−3148
Severe dry	RMSE	0.103	0.024	0.095	0.019	0.086	0.013	0.025	0.034
	AIC	−1144	−1874	−1184	−1992	−1235	−2183	−1847	−1690
Extreme dry	RMSE	0.102	0.02	0.092	0.021	0.083	0.014	0.022	0.043
	AIC	−637	−1089	−666	−1076	−695	−1189	−1057	−867

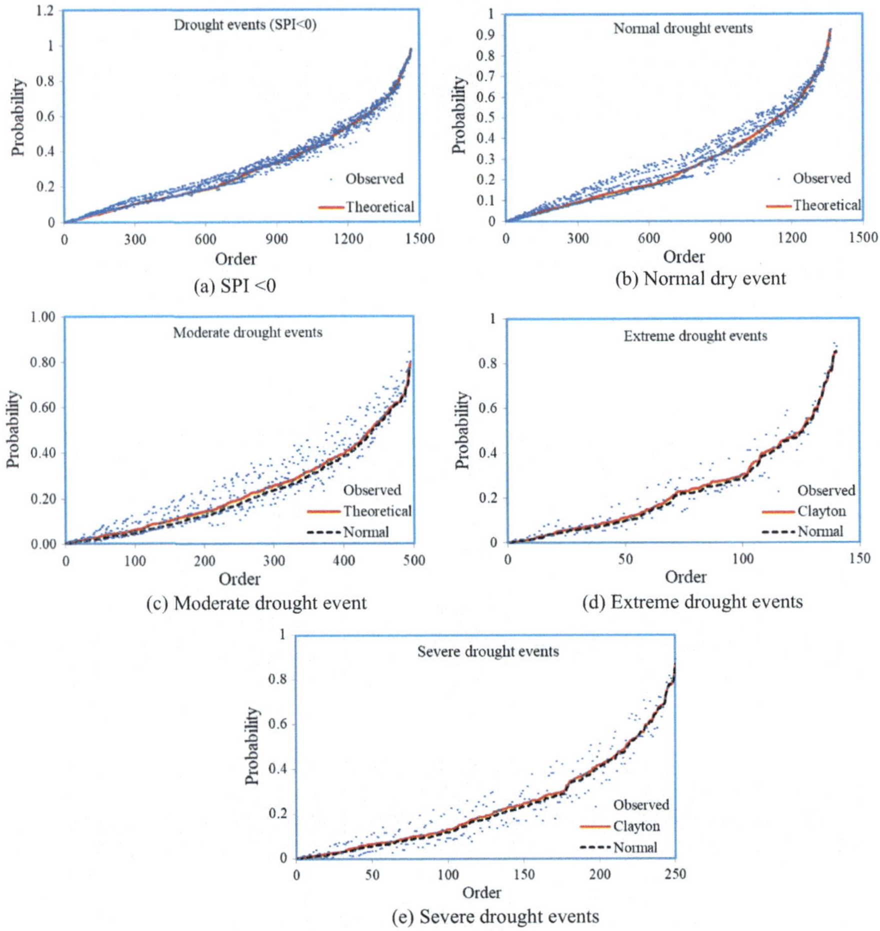

Fig. 5.6 Observed and theoretical joint probabilities for the four drought characteristics described in the text

Table 5.8 Results of joint and conditional distributions

$F(x)$	D_d	S_d	I_d	L_d	Joint probabilities	Conditional probabilities
0.1	0.25	0.24	0.65	0.95	0.01	0.14
0.3	0.68	0.57	1.22	1.82	0.10	0.33
0.5	1.21	0.95	1.78	2.67	0.25	0.49
0.7	1.98	1.40	2.49	3.76	0.46	0.66
0.9	3.58	2.03	3.81	5.78	0.78	0.86
0.99	6.84	2.57	6.24	9.55	0.97	0.98

Table 5.9 Joint return periods (years) of drought events E

T	$F(x)$	D_d	S_d	Id	$S_d > s_d, I_d > i_d$			$D_d > d_d, S_d > s_d$		
					$F(x,y)$	T_{and}	T_{or}	$F(x,y)$	T_{and}	T_{or}
5	0.38	1.44	0.88	0.72	0.32	5.55	4.55	0.27	6.13	4.22
10	0.69	2.45	1.94	1.38	0.63	12.24	8.45	0.59	15.15	7.46
20	0.85	3.31	2.95	1.83	0.81	26.51	16.06	0.78	36.39	13.79
50	0.94	4.34	4.27	2.21	0.92	72.66	38.11	0.90	113.35	32.07
100	0.97	5.08	5.25	2.39	0.96	154.81	73.85	0.95	265.09	61.62
T	$F(x)$	D_d	S_d	I_d	$D_d > d_d, I_d > i_d$			$D_d > d_d, S_d > sd, Id > i_d$		
					$F(x,y)$	T_{and}	T_{or}	$F(x,y,z)$	T_{and}	T_{or}
5	0.38	1.44	0.88	0.72	0.21	6.85	3.94	0.21	6.95	3.92
10	0.69	2.45	1.94	1.38	0.54	19.80	6.69	0.53	19.99	6.59
20	0.85	3.31	2.95	1.83	0.75	55.47	12.20	0.74	55.84	11.81
50	0.94	4.34	4.27	2.21	0.89	211.55	28.35	0.88	216.57	26.81
100	0.97	5.08	5.25	2.39	0.94	576.19	54.75	0.94	594.73	50.73

The probability of events $E = \{D_d \leq d_d, S_d \leq s_d, I_d \leq i_d\}$ under the condition $L_d \leq l_d$ can be defined as:

$$P(D_d \leq d_d, S_d \leq s_d, I_d \leq i_d | L_d \leq l_d) = \frac{P(D_d \leq d_d, S_d \leq s_d, I_d \leq i_d, L_d \leq l_d)}{P(L_d \leq l_d)} \tag{5.8}$$

The conditional probability defined in Eq. 5.8 is calculated, as given in Table 5.8. When $F(x)$ equals 0.99, the corresponding magnitudes for drought duration, severity, MSPI and interval time in Table 5.8 are 6.84, 2.57, 6.24 and 9.55, respectively. The conditional probability of event $E = \{D_d \leq d_d, S_d \leq s_d, I_d \leq i_d\}$ under the condition $L_d \leq l_d$ is 0.98. Thus, the conditional probability of any drought event can be obtained from Eq. 5.8.

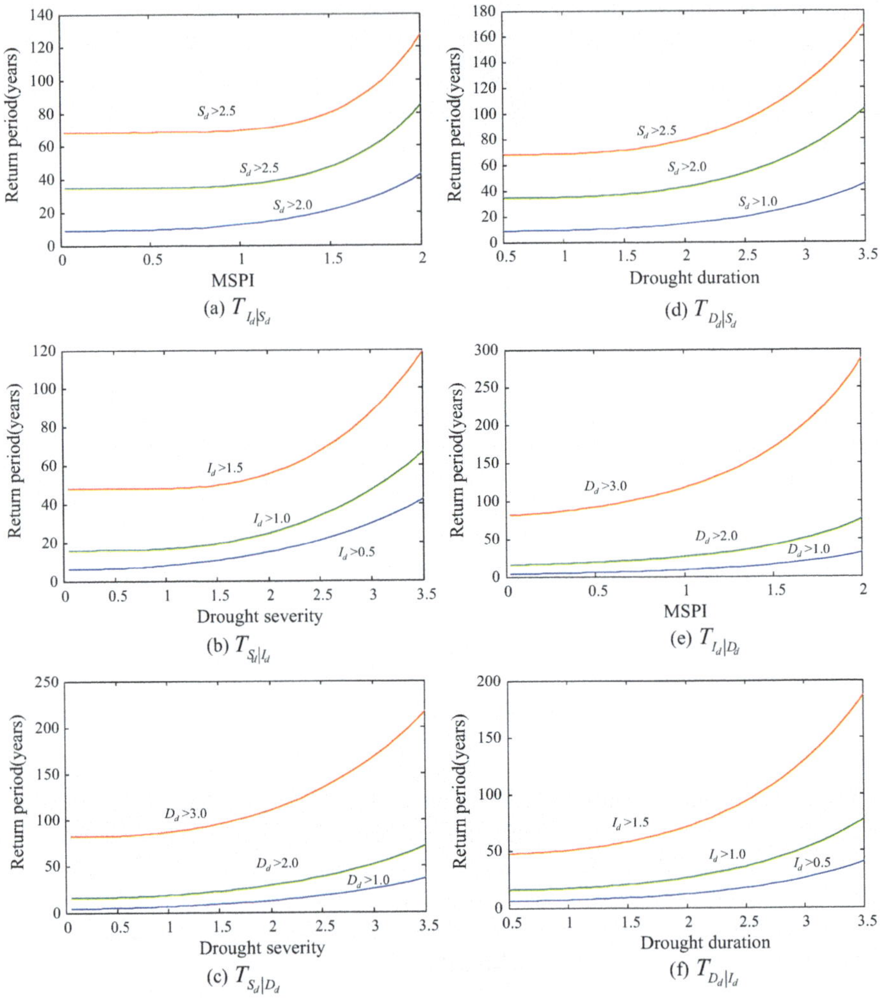

Fig. 5.7 Conditional return periods of drought events

5.6 Conclusion

In this Chapter, drought is defined by drought duration, severity, MSPI and interval time. The upstream Han River is selected as a case study. The exponential, gamma, GP, P-III and GEV distributions are applied to fit univariate data series. The Archimedean and meta-elliptical copulas are used to establish the joint multivariate distributions. The joint probabilities and return period are then estimated. The main conclusions of this chapter are following:

(1) The established marginal distributions of the four drought variables fit the empirical data well and can be used for drought analysis.
(2) Five copula functions are constructed for different drought states. The *RMSE* and *AIC* values are used to select the appropriate copula. Results of fitting indicate that it is applicable to use copulas for multivariate drought analysis. The Normal copula basically fits data series well and it is suggested for computing probabilities and return period analysis.
(3) The drought risk is estimated based on joint probabilities and return periods, which give important information for water management and planning.

References

Alley WM (1984) The palmer drought severity index: limitations and assumptions. J Clim Appl Meteor 23:1100–1109

Below R, Grover-Kopec E, Dilley M (2007) Documenting drought-related disasters: a global reassessment. J Environ Develop 16(3):328–344

Bonaccorso B, Cancelliere A, Rossi G (2003) An analytical formulation of return period of drought severity. Stoch Env Res Risk A 17(3):157–174

Cancelliere A, Salas JD (2004) Drought length properties for periodic-stochastic hydrological data. Water Resour Res 40:W02503. https://doi.org/10.1029/2002WR001750

Cancelliere A, Salas JD (2010) Drought probabilities and return period for annual streamflows series. J Hydrol 391(1–2):77–89

Chen J, Brissette PF, Leconte R (2010) A daily stochastic weather generator for preserving low-frequency of climate variability. J Hydrol 388:480–490

Chen L, Singh VP, Guo S, Mishra AK, Guo J (2013) Drought analysis based on copulas. J Hydrol Eng 18(7):797–808

Estrela T, Vargas E (2012) Drought management plans in the European Union. The case of Spain. Water Resour Manag 26(6):1537–1553

Fernández B, Salas JD (1999) Return period and risk of hydrologic events I: mathematical formulation. J Hydrol Eng 4(4):297–307

Genest C, Favre AC, Béliveau J, Jacques C (2007) Metaelliptical copulas and their use in frequency analysis of multivariate hydrological data. Water Resour Res 43(9):W09401

González J, Valdés JB (2003) Bivariate drought recurrence analysis using tree ring reconstructions. J Hydrol Eng 8(5):247–258

Guttman NB (1991) A sensitivity analysis of the Palmer hydrologic drought index. J Am Water Resour Assoc 27(5):797–807

Guttman NB (1998) Comparing the Palmer drought index and the standardized precipitation index. J Am Water Resour Assoc 34:113–121

Hosking JRM (1990) L-moments: analysis and estimation of distributions using linear combinations of order statistics. J R Stat Soc B 52:105–124

Kao SC, Govindaraju RS (2007) A bivariate frequency analysis of extreme rainfall with implications for design. J Geophys Res 112:D13119. https://doi.org/10.1029/2007JD008522

Kendall DR, Dracup JA (1992) On the generation of drought events using an alternating renewal-reward model. Stoch Hydrol Hydraul 6(1):55–68

Kim TW, Valdés JB, Yoo C (2003) Nonparametric approach for estimating return periods of droughts in arid regions. J Hydrol Eng 8(5):237–246

Mathier L, Perreault L, Bobe B, Ashkar F (1992) The use of geometric and gamma-related distributions for frequency analysis of water deficit. Stoch Hydrol Hydraul 6(4):239–254

McKee TB, Doesken NJ, Kliest J (1993) The relationship of drought frequency and duration to time scales. In: Proceedings of the 8th conference of applied climatology, 17–22 Jan, Anaheim, CA. American Meteorological Society, Boston, MA. pp 179–184

Mishra AK, Singh VP (2010) A review of drought concepts. J Hydrol 391(1–2):202–216

Mishra A, Singh VP, Desai V (2009) Drought characterization: a probabilistic approach. Stoch Env Res Risk A 23(1):41–55

Núñez JH, Verbist K, Wallis JR, Schaefer MG, Morales L, Cornelis WM (2011) Regional frequency analysis for mapping drought events in north-central Chile. J Hydrol 405(3–4): 352–366

Palmer WC (1965) Meteorological drought. Research paper no. 45, US Department of Commerce, Weather Bureau, Washington, DC

Salas JD, Fu C, Cancelliere A, Dustin D, Bode D, Pineda A, Vincent E (2005) Characterizing the severity and risk of drought in the Poudre River, Colorado. J Water Res Plan Man 131(5): 383–393

Serinaldi F, Bonaccorso B, Cancelliere A, Grimaldi S (2009) Probabilistic characterization of drought properties through copulas. Phys Chem Earth, Parts A/B/C 34(10–12):596–605

Shiau J (2006) Fitting drought duration and severity with two-dimensional copulas. Water Resour Manag 20(5):795–815

Shiau JT, Shen HW (2001) Recurrence analysis of hydrologic droughts of differing severity. J Water Resour Plan Man 127(1):30–40

Shiau JT, Feng S, Nadarajah S (2007) Assessment of hydrological droughts for the Yellow River, China, using copulas. Hydrol Process 21(16):2157–2163

Song S, Singh VP (2010) Frequency analysis of droughts using the Plackett copula and parameter estimation by genetic algorithm. Stoch Env Res Risk A 24(5):783–805

Tallaksen LM, Madsen H, Clausen B (1997) On the definition and modeling of stream drought duration and deficit volume. Hydrol Sci J 42(1):15–33

Zelenhastic E, Salvai A (1987) A method of streamflow drought analysis. Water Resour Res 23(1):156–168

Zhang L, Singh VP (2006) Bivariate flood frequency analysis using the copula method. J Hydrol Eng 11(2):150–164

Chapter 6
Flood Coincidence Risk Analysis Using Multivariate Copula Functions

6.1 Introduction

Disastrous floods can be caused by unusual combinations of hydrometeorological factors and river basin conditions. Topography, land cover, and temporal and spatial distribution of rainfall play a dominant role in the generation of floods, which can be reflected in the contributions that major tributaries make to the mainstream flow. The coincidence of flood flows of mainstream and its tributaries may determine the peak flow. Therefore, the risk of flooding due to the combination of flood flows from different rivers is important for hydraulic design. The combination risk arises when large floods occur simultaneously in the mainstream as well as in its tributaries, and this risk is characterized regarding flood magnitude and occurrence date. Traditional methods focus only on the flood magnitudes, and a more realistic approach is therefore needed.

The traditional approach to the risk assessment entails determining the probability that a pre-selected value of the flood characteristic will be exceeded or equivalently determining the return period (Prohaska et al. 2008). This approach is based on univariate frequency analysis or regional frequency analysis. However, this approach does not consider the correlation of flows from different regions. The risk of combining floods involves at least two sites in the mainstream and its tributaries or two tributaries. This suggests that a multivariate hydrological analysis, which considers the dependence between flood variables, is needed.

Prohaska et al. (2008) used a two-dimensional probability distribution to evaluate the coincidence of floods on two adjacent streams, on the assumption that floods followed log-normal distribution. The use of the log-normal distribution for representing the frequency distribution of peak flow is not supported by hydrologic practices in many countries. For example, the Pearson three (P-III) distribution is assumed for frequency analysis of flood peaks in China (MWR 1993), log-Pearson

L. Chen and S. Guo, *Copulas and Its Application in Hydrology and Water Resources*, Springer Water, https://doi.org/10.1007/978-981-13-0574-0_6

three in the U.S. (IACWD 1982) and the generalized logistic (GL) distribution in the UK (Robson and Reed 1999). Further, Prohaska's study is limited to only two variables. It is usual that there is more than one tributary of the mainstream.

For these reasons, a new multivariate model, based on the copula function, is applied in this study. Most of these studies involve bivariate copulas (Kao and Govindaraju 2010; De Michele and Salvadori 2003; Favre et al. 2004; Shiau et al. 2006; Dupuis 2007; Zhang and Singh 2006, 2007b). Trivariate copula functions also have been used. Grimaldi and Serinaldi (2006) applied the Archimedean copula to model the trivariate joint distribution of floods. Serinaldi and Grimaldi (2007) described an inference procedure to carry out a trivariate frequency analysis via asymmetric Archimedean copulas. Zhang and Singh (2007a, c) applied the Archimedean copulas to trivariate frequency analysis of floods as well as rainfall events. Kao and Govindaraju (2008) applied the Plackett copulas to trivariate statistical analysis of extreme rainfall events (e.g., Song and Singh 2010b); Song and Singh (2010b) modeled the joint probability distribution of drought duration, severity and inter-arrival time using a trivariate Plackett copula. Applications of four-dimensional copula functions in hydrological fields have also been reported recently. De Michele et al. (2007) introduced a method for constructing multivariate distributions, given 2-copulas for each bivariate marginal law and applied the method to provide a four-dimensional characterization of sea state statistics. Serinaldi et al. (2009) used a four-dimensional student copula to analyze drought probabilistic characteristics. Since more variables are involved, four-dimensional copulas will be used in this study.

The content of this chapter is to apply a multivariate copula to analyze the coincidence flood risk of rivers. The upper Yangtze and Colorado River are selected as case studies. Daily flow data from four sites at the upper Yangtze and Colorado River is chosen. Four-dimensional copula functions are applied to construct the joint distribution of flood occurrence dates and magnitudes. The von Mises distribution is used to describe the flood occurrence dates, while the Pearson type three (P-III) and log Pearson type three distributions are selected as the marginal distribution of annual maximum flood peaks. The coincidence probabilities of flood magnitudes and occurrence dates are analyzed. The conditional probabilities for the Three Gorges Reservoir (TGR) are calculated.

6.2 Methodology

In this section, copula functions are selected to construct the joint distribution. The detailed information of copula theory can be found in Chap. 2. The von Mises distribution is selected as a marginal distribution function for flood occurrence dates, and the characteristic and expression of the von Mises distribution are described in Chap. 3.

The Flood Estimation Handbook (Reed 1999) states the flood risk assessment is to estimate the risk of a flood occurrence. The Environment Agency's Strategy for Flood Risk Management 2003/4-2007/8 (EA 2003) states that one task of flood risk estimation is to estimate the chance of a probability of a certain flood event. A methodology is presented herein for the estimation of a kind of special flood event, namely the coincidence of flood flows in the main river and its tributary. The term coincidence is used to denote the simultaneous occurrence of floods at two (or more) rivers. The degree of coincidence is measured by the probability of flood events. The theoretical background draws from the practical application of a multivariate probability distribution function, or its conditional probabilities (Prohaska et al. 2008). As flood events are characterized by flood occurrence dates and magnitudes, both of the two factors should be considered. This study considered the quantitative characteristics of simultaneous floods on the main river and its tributaries, and the flood dates of simultaneous floods.

First, flood magnitude is selected as a reference variable for analysis (Favre et al. 2004). The P-III and log P-III distributions are selected as marginal distribution functions for flood magnitude. The copula function is used to establish the joint distribution. The exceedance probability of coinciding flood volumes considered in flow profiles is defined as:

$$P^T_{Q_n} = P(Q_1 > q^T_1, Q_i > q^T_i, \ldots, Q_n > q^T_n) \tag{6.1}$$

where $P^T_{Q_n}$ is the exceedance probability of coinciding flood magnitudes; i is the ith gauge station; n is the number of variables and can be equal to two, three, and four in this study; Q_1... Q_i ... Q_n are flow magnitudes; $q^T_1 \ldots q^T_i \ldots q^T_n$ mean the design flood volume for the return period T.

Second, flood date is selected as a reference variable for analysis. In this study, if annual maximum floods occur within dt days, the floods were defined as contemporary temporal occurrences. The coincidence probability of flood dates at two or more considered inflow profiles is defined as:

$$P^t_n = P_t(t_k < T_i \le t_{k+1}, t_k - dt_{ij} < T_j \le t_{k+1} + dt_{ij}, \ldots, t_k - dt_{in} < T_n \le t_{k+1} + dt_{in}) \tag{6.2}$$

where i, j represent any river in the data set, and gauge station j is located downstream of the catchment; T_i means the random variable of flood occurrence dates, and dt is the time interval and equals one day in this study. The flood travel time between the two sites also should be considered. Equation 6.2 can compute the probabilities of simultaneous floods for two, three rivers in the basin. To calculate P^t_n, the marginal distribution for T_i is needed to build first. The von Mises distribution was selected as a marginal distribution function for flood occurrence dates. The detailed information for deriving the distribution of flood occurrence dates is given in Chaps. 3 and 4. Then, the joint distribution is built for evaluating the coincidence probability of flood dates. The detailed information for establishing copulas is given in Chap. 2.

Third, both the flood magnitudes and flood dates are selected as reference variables for risk analysis. Assuming that the flood occurrence dates are independent of flood magnitudes and flood peaks occur simultaneously at two or more rivers in the same basin, the flood coincidence probabilities of rivers for given flood magnitudes were estimated as

$$P_n^T = \sum_{t=1}^{N} P_n^t \cdot P(Q_1 > q_1^T, Q_i > q_i^T, \ldots, Q_n > q_n^T) \tag{6.3}$$

6.3 Data

The upper Yangtze River, which is the longest river in China and third longest in the world, is selected as a case study. The Three Gorges Project (TGP) is located on the Yangtze River. Floods in the middle and lower reaches of the Yangtze River mainly stem from the upper region of Yichang site, which is also the control site for TGP. Usually, the flood volume of upper Yichang site is about 50% of the total flow volume of the Yangtze River, about 90% of the Jingjiang River reach, which is regarded as the most key area for flood prevention. Hence, studying flood characteristics in upper Yangtze River is an important task for flood prevention.

The upper Yangtze River comprises a complex of tributaries, principally Yalong River, Min River, Jialing River on the left bank, and Wu River on the right bank. A schematic of the regional main tributary rivers and gauging stations is shown in Fig. 6.1. Some basic features of the available data are given in Table 6.1. Yalong

Fig. 6.1 Locations of regional tributary rivers and gaging stations

Table 6.1 Major tributaries to the upstream Yangtze River

Major tributary	Catchment area (km^2) >80,000 km^2	Record of length	Yangtze River	Major tributary	Catchment area (km^2) >80,000 km^2	Record of length
				Yalong River	144,200	1951–2007
	485,099	1951–2007	Jinsha River			
Min River	135,400	1951–2007				
Jialing River	157,900	1951–2007				
				Wu River	87,920	1951–2007
	1,005,501	1951–2007	Yichang (TGR)		1,005,501	1951–2007

River joins Jinsha River which is also recognized as part of the Yangtze River. Therefore, the Jinsha River, instead of the Yalong River, is used in this study. Relative frequencies of annual maximum floods in these rivers are calculated, as graphed in Fig. 6.1. It is shown that large floods always occur in July and August, except in the Wu River, in which the highest RF occurs in the middle of June. Therefore, it is more likely that floods in Jinsha River, Min River, and Jialing River occur simultaneously. Therefore, Jinsha, Min and Jialing rivers are selected in this study. Yichang site at the location of TGP is an important site on the Yangtze River and is also selected.

The Colorado River, in the Southwestern United States and northwestern Mexico, approximately 1450 miles (2330 km) long (Munro 1992), is also selected as a case study. The Colorado River above Lees Ferry is defined as upper Colorado River basin with about 17,800 square miles. The Colorado River originates in the mountains of central Colorado and flows about 230 miles southwest into Utah. There are some tributaries in the upper Colorado River basin, principally Green River, Gunnison River and San Juan River. A schematic of the regional main tributary rivers and gauging stations is shown in Table 6.2. The Green River, located in the western United States, is the chief tributary of the Colorado River. The watershed of the river, the Green River basin, covers parts of Wyoming, Utah, and Colorado. It is only slightly smaller than that of the Colorado when the two rivers merge. The average yearly mean flow of the river at Green River, Utah, is 173.3 m^3/s (6121 cubic feet) (Enright et al. 2008). The Gunnison River is a significant tributary of the Colorado River, 264 km (164 miles) long, in the southwest state of Colorado (U.S. Geological Survey 2011). It is the fifth largest tributary of

Table 6.2 Major tributaries to the upper Colorado River

Major tributary	Catchment area (km^2)	Record of length	Colorado River	Major tributary	Catchment area (km^2)	Record of length
	20,800	1933–2011	Above Cameo		20,800	1933–2011
Green River	44,850	1894–2011		Gunnison River	20,533	1896–2011
				San Juan River	12,000	1914–2011
	111,800	1921–2011	Lees Ferry		111,800	1921–2011

the Colorado River, with a mean flow of 122 m^3/s (4320 ft^3/s). The San Juan River is a tributary of the Colorado River in the southwestern United States, about 616 km (383 miles) long, the mean flow of which is about 62.4 m^3/s (2205 cubic feet per second) at its mouth (U.S. Geological Survey 2011). Comparing with the other two major tributaries, the mean flow of San Juan River is relatively smaller. Therefore, only Green and Gunnison River are considered in this study. As Lees Ferry is the division site between upper and lower Colorado River, this site is considered. The site near Grand Junction (named upper Cor. hereafter) is selected for analyzing the flow above Cameo of Colorado River. Therefore, four sites in Colorado River basin are considered in this study.

Pairwise dependence structures of the four stations in the two river basins are estimated. Empirical estimates of bivariate Kendall's τ of flood magnitudes and occurrence dates for all the pairs of interest here are given in Tables 6.3 and 6.4. The correlation coefficient between the Beibei and Yichang stations in upper

Table 6.3 Values of Kendall's τ of flood magnitudes and occurrence dates for all pairs of the four stations in the upper Colorado River

Stations	Above Cameo	Green River	Gunnison River	Lees Ferry
Above Cameo	1.00	0.68	0.66	0.49
Green River	0.37	1.00	0.58	0.42
Gunnison River	0.19	0.32	1.00	0.50
Lees Ferry	0.19	0.19	0.13	1.00

Note Upper triangular matrix is Kendall's τ of flood magnitude, and the lower triangular matrix is Kendall's τ of flood dates. The meaning is the same hereafter

Table 6.4 Values of Kendall's τ of flood magnitudes and occurrence dates for all pairs of the four stations in the upper Yangtze River

Stations	Pingsha	Gaochang	Beibei	Yichang
Pingsha	1.00	0.11	−0.08	0.28
Gaochang	0.07	1.00	0.03	0.21
Beibei	0.08	0.08	1.00	0.32
Yichang	0.19	0.18	0.34	1.00

Yangtze River is negative, but it is very small and close to 0. This means that the association between the two variables can be negligible and the Gumbel copula is therefore used.

6.4 Application

6.4.1 Estimation of Marginal Distributions

In order to show the validity of the mixed von Mises distribution, other distributions, such as Gumbel, normal, and Pearson III distributions, are selected as possible marginal distributions for the upper Yangtze River. Parameters of the mixed von Mises distribution are estimated by the maximum likelihood method. Parameters of other distributions are estimated by the L-moment method. Then these distributions are fitted to the data and compared with the mixed von Mises distribution. The best-fitted distributions are selected using the root mean square error (RMSE) values shown in Table 6.5 (Zhang and Singh 2007b). It is found that the mixed von Mises distribution has the smallest RMSE values for the flood dates

Table 6.5 RMSE Values of different probability distributions of flood occurrence dates in the upper Yangtze River (%)

Distribution	Pingshan	Gaochang	Beibei	Yichang
Mixed von Mises	1.898	1.413	1.725	2.067
Generalized logistic (GLO)	4.028	3.291	3.646	7.148
Generalized Pareto (GP)	3.688	3.911	2.574	3.465
Pearson type 3 (P-III)	3.184	2.773	2.725	5.591
Generalized extreme-value (GEV)	2.927	2.688	2.654	5.988
Gamma distribution	4.539	2.806	4.450	6.280
Normal distribution	3.560	2.887	3.021	7.393
Gumbel distribution	6.641	4.525	4.255	5.985
Wakeby distribution	2.796	2.566	1.848	3.465
Kappa distribution	2.734	2.664	2.102	2.195
Exponential distribution	10.432	8.735	8.426	6.340

Note The bold characters mean the minimum value of each column at all four stations in upper Yangtze River. The values of estimated parameters of the von Mises distribution of both river basins are listed in Table 6.6. The Kolmogorov-Smirnov (KS) test is selected as the goodness-of-fit test to evaluate the validity of the assumption that the flood occurrence dates followed the mixed von Mises distribution. Results shown in Table 6.6 indicate that this assumption cannot be rejected at the 5% significance level. The frequency histograms of the flood occurrence dates fitted by the mixed von Mises distribution for AM sample series in upper Colorado River are shown in Fig. 6.2a–d. The marginal distribution curves of flood occurrence dates in upper Yangtze River are shown in Fig. 6.3, in which the line represents the theoretical distribution, and the crosses the empirical frequencies of observations. Figures 6.2 and 6.3 indicate that all the theoretical distributions fitted the observed data reasonably well.

The values of estimated parameters of the P-III and log P-III distributions are given in Table 6.6. A chi-square goodness-of-fit test is performed to test the assumption, H_0, that the flood magnitude followed the P-III or LP-III distribution. It is shown that P-III or LP-III distribution is valid for flood magnitudes at four sites

Table 6.6 Parameters and hypothesis test results of margin distributions

Stations	Mixed von Mises distribution				P-III distribution			
	u_i	K_i	p_i	K-S	α	β	δ	$\chi 2$
Pingshan	4.33 5.19 3.46	67.99 8.89 3.70	0.17 0.44 0.39	0.042 (0.176)	10.41	0.0008	3279.29	0.36 (3.84)
Gaochang	4.29 3.65 2.88	3.04 300.00 7.90	0.60 0.16 0.24	0.039 (0.176)	4.726	0.0005	6660.33	0.33 (3.84)
Beibei	5.44 2.96 5.29	0.00 7.20 4.25	0.21 0.46 0.33	0.035 (0.176)	330.6	0.002	−110580	0.57 (5.99)
Yichang	2.99 4.20 5.35	13.48 6.63 11.21	0.50 0.28 0.23	0.045 (0.176)	156.25	0.0016	−49825	0.92 (3.84)
Above Cameo	3.32	3.69	1.0	0.064 (0.155)	4.919	0.0896	3.783	0.13 (3.84)
Green River	3.02	2.38	1.00	0.057 (0.155)	3.493	0.1050	3.966	0.003 (3.84)
Gunnison River	2.57 4.29	2.25 4.97	0.96 0.04	0.031 (0.155)	30.451	0.0503	2.418	0.54 (3.84)
Lees ferry	2.29 3.52 5.84	198.29 1.00 260.56	0.09 0.74 0.17	0.144 (0.155)	10.058	0.0952	3.584	0.79 (3.84)

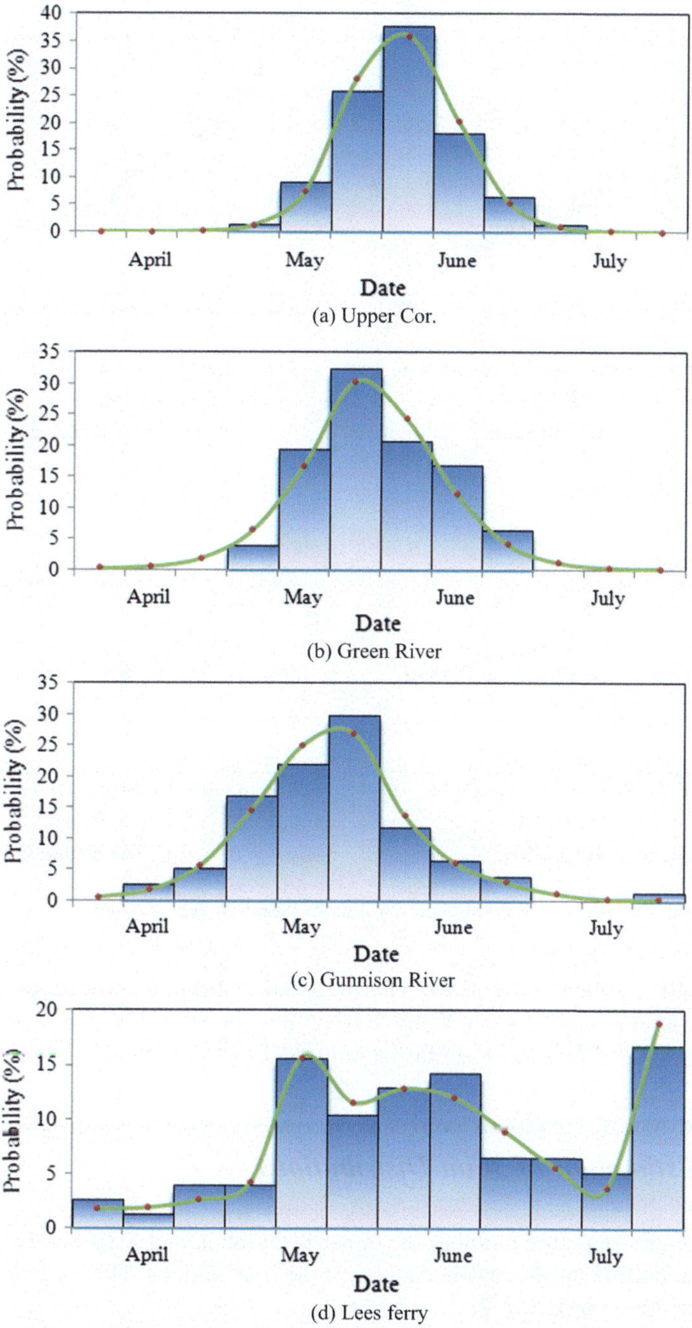

Fig. 6.2 Frequency histograms of flood occurrence dates fitted by the mixed von Mises distribution for the four stations in upper Colorado River

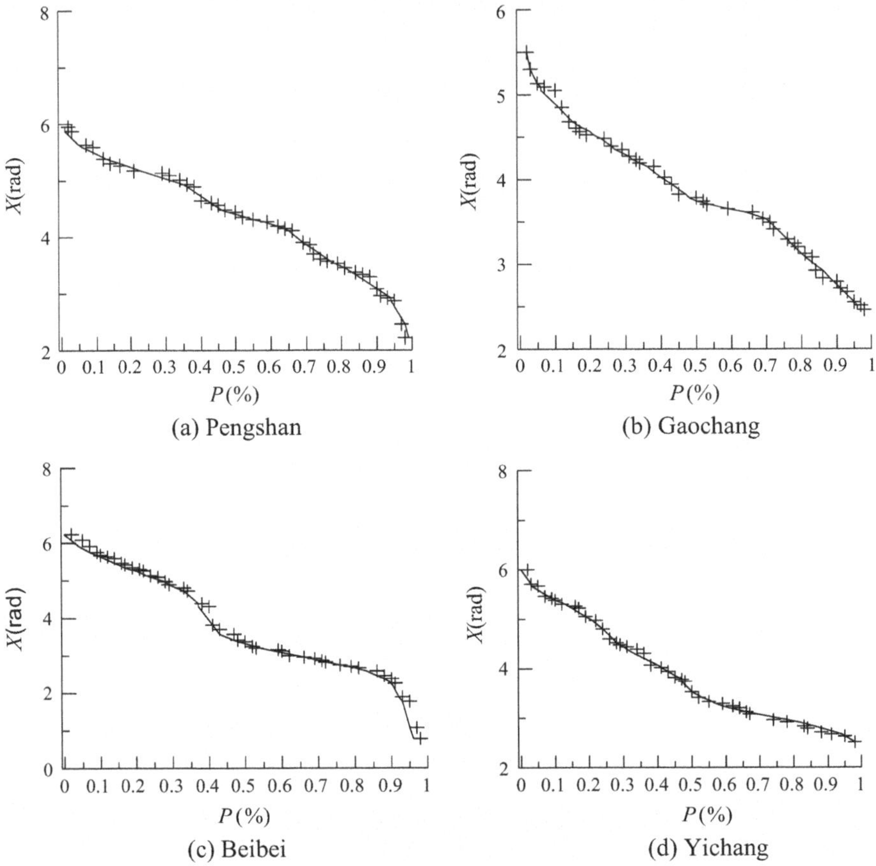

Fig. 6.3 Frequency curves of flood occurrence dates based on AM samples

studied with a critical value 0.05. The marginal distribution frequency curves of flood magnitudes in the Upper Yangtze River are shown in Fig. 6.4. It is seen that graphically the P-III distribution fit the empirical distribution.

6.4.2 Estimation of Joint Distributions

A four-variate symmetric Gumbel, asymmetric Gumbel, and X-Gumbel copulas are used for modelling the dependence amongst the four stations. The formulas of these copulas are given in Chap. 2.

A pseudo-likelihood technique involving the ranks of the data is used for estimating parameters of the four-variate symmetric Gumbel and asymmetric Gumbel copulas. For the Yangtze River, the value of estimated parameter of symmetric

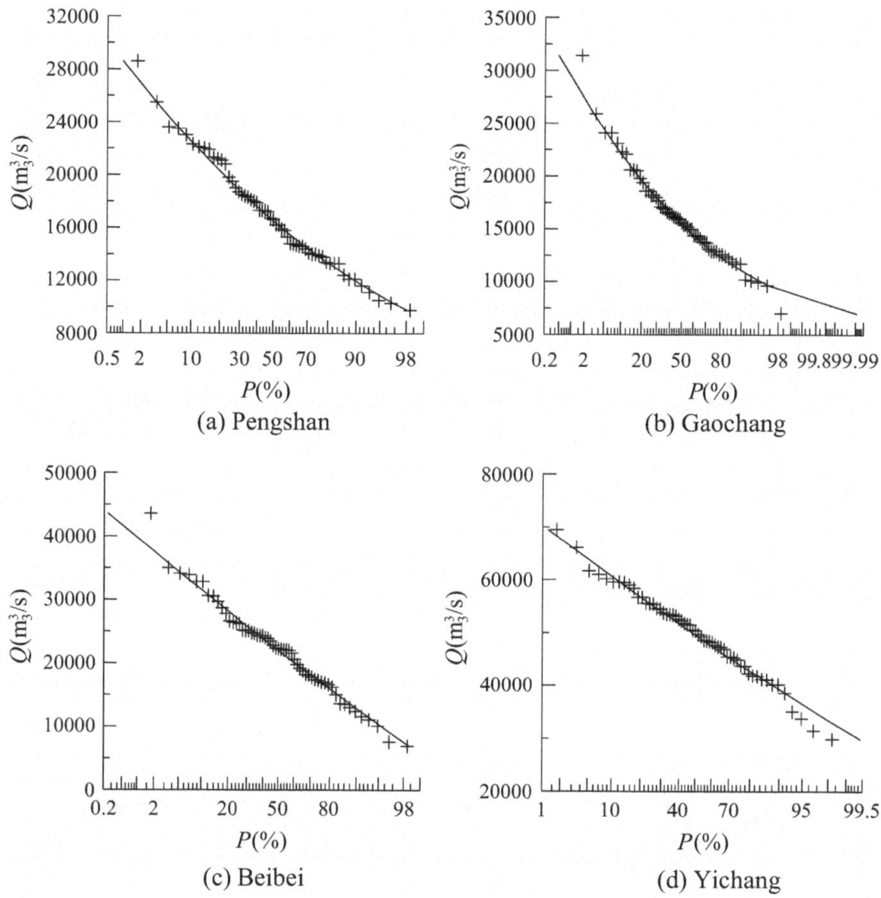

Fig. 6.4 Frequency curves of flood magnitudes based on AM samples

Gumbel is $\hat{\theta} = 1.14$ for flood magnitudes, and $\hat{\theta} = 1.20$ for flood occurrence dates. Estimates of parameters of the asymmetric Gumbel are $\hat{\theta}_1 = 1.06$, $\hat{\theta}_2 = 1.16$, and $\hat{\theta}_3 = 1.46$ for flood magnitudes; and $\hat{\theta}_1 = 1.18$, $\hat{\theta}_2 = 1.20$, and $\hat{\theta}_3 = 1.38$ for the flood occurrence dates. For the Colorado River, the value of estimated parameter of symmetric Gumbel is $\hat{\theta} = 1.99$ for flood magnitudes, and $\hat{\theta} = 1.22$ for flood occurrence dates. Estimates of parameters of the asymmetric Gumbel are $\hat{\theta}_1 = 1.82$, $\hat{\theta}_2 = 2.35$, and $\hat{\theta}_3 = 2.72$ for flood magnitudes; and $\hat{\theta}_1 = 1.13$, $\hat{\theta}_2 = 1.30$, and $\hat{\theta}_3 = 1.54$ for the flood occurrence dates. Pickand's dependence function, which was recommended by Salvadori and De Michele (2010), is used for estimating parameters of the X-Gumbel copula. The values of parameters of X-Gumbel for flood magnitudes and flood dates in Upper Colorado River are given in Table 6.7.

Table 6.7 Parameters of X-Gumbel joint distributions

Rivers	Parameters	a_1	a_2	a_3	a_4	x	s
Upper Yangtze River	Magnitude	0.039	1.0	0.92	0.63	2.99	1.46
	Dates	0.999	0.132	0.137	0.576	1.09	2.19
Upper Colorado River	Magnitudes	0.707	0.773	0.725	0.268	3.267	2.763
	dates	0.221	0.396	1.000	0.193	1.181	3.372

The empirical and fitted Pickands' function for all the pairs of stations and three copula models of flood magnitudes in the two basins are plotted in Figs. 6.5 and 6.6, respectively. The symmetric Gumbel dependence functions are the same in all the plots. The asymmetric Gumbel dependence functions are different corresponding to different pairs. The asymmetric Gumbel copula fits better than the symmetric one. The X-Gumbel provides a better fit than the other two models. The empirical joint probabilities of flood occurrence dates and flood peak magnitudes are plotted against theoretical probabilities, as shown in Fig. 6.7, in which the theoretical joint probabilities, F, of the real occurrence combinations of x and y are estimated. Figure 6.7 shows that no significant difference between empirical and theoretical joint probabilities can be detected.

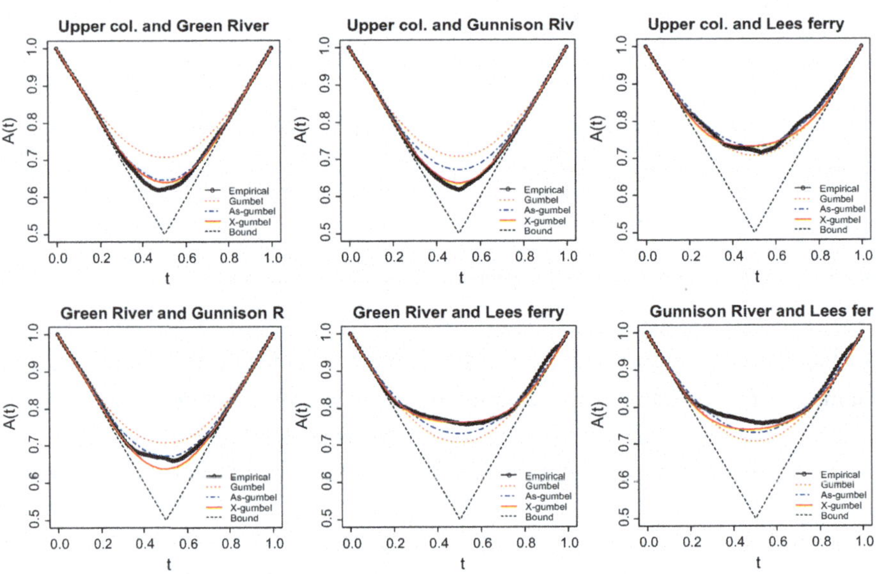

Fig. 6.5 Plots of empirical and fitted Pickand's dependence functions of flood magnitude for all pairs of stations and the three models

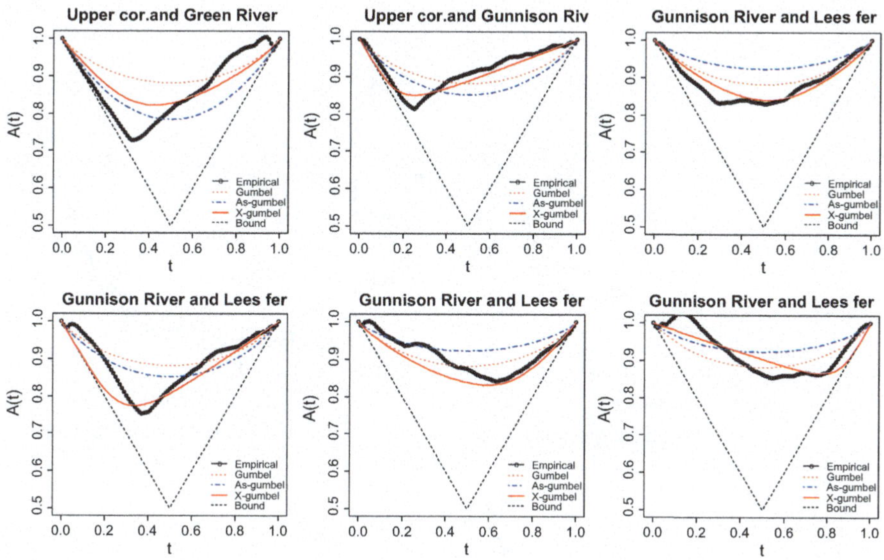

Fig. 6.6 Plots of empirical and fitted Pickand's dependence functions of flood occurrence dates for all pairs of stations and the three models

6.4.3 Analysis of Flood Coincidence Risk

6.4.3.1 Coincidence Probabilities Analysis

According to the analysis above, the X-Gumbel copula is used for the flood coincidence risk analysis hereafter. The exceedance probabilities of coinciding *T*-year flood volumes at two and three considered inflow profiles are calculated as shown in Tables 6.8 and 6.9. The average exceedance probabilities of 100, 50, 10, 5, and 2-year for the four sites are 0.0075, 0.015, 0.0763, 0.1561 and 0.4196, respectively.

The coincidence probabilities of flood dates in two, three and four rivers, P_2^t, P_3^t and P_4^t, are evaluated as shown in Fig. 6.8a–e, respectively. For the Jinsha and Min Rivers, the higher coincidence probabilities occur in late July and middle August. According to the observed data, there are seven times that the flood occurred simultaneously in the two rivers, five of which is within this period. For the Jinsha and Jialing Rivers, the curve demonstrates the multi-modal characteristic, and the higher coincidence probabilities occur in the middle July and early September, which indicates that the flood control water level of the Three Gorges Reservoir (TGR) should not be raised too high and certain flood control storage is needed for TGR. For the Jialing and Min tributaries, the highest probability occurs in July. Six of eight flood events that occurring simultaneously in the two rivers, are within this

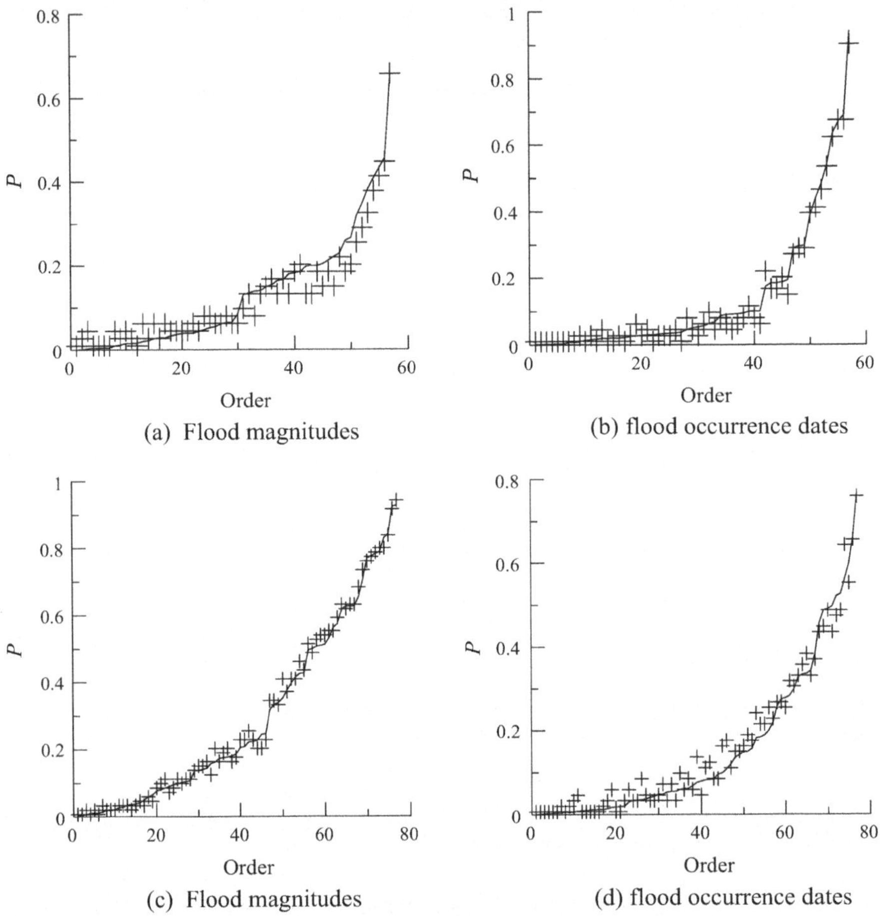

Fig. 6.7 Joint distribution and empirical probabilities of observed combinations based on **a** and **b** are for flood magnitudes, and flood occurrence dates in upper Yangtze River; **c** and **d** are for flood magnitudes and flood occurrence dates in upper Colorado River

period. For the three rivers in the upper Yangtze River, July has the highest coincidence probabilities. For the four stations, the higher probabilities occur in July. It is indicated that in May and June, the coincidence probabilities are very small, which means the low coincidence risk. Therefore, it is possible to raise the flood control water level of TGR in the two months. All the analysis mentioned above demonstrates that the calculated results are in accordance with historical data.

The coincidence probabilities of T-year design flood for two and three tributaries are calculated based on Eq. 6.3, and results are listed in Tables 6.10 and 6.11. Results are reasonable from the point of view that the coincidence probabilities

Table 6.8 The exceedance probability of coinciding *T*-year flood volumes at two considered inflow profiles

Tributaries	T	100	50	10	5	2
Upper Col. and Green Rivers	100	0.00746	0.00921	0.00997	0.00999	0.01000
	50	0.00929	0.01495	0.01976	0.01995	0.019994
	10	0.00998	0.01982	0.07607	0.09390	0.099553
	5	0.00999	0.01997	0.09458	0.15562	0.196065
	2	0.01000	0.02000	0.09969	0.19677	0.418765
Upper Col. and Gunnison Rivers	100	0.00751	0.00927	0.00997	0.00999	0.01000
	50	0.00929	0.01505	0.01979	0.01996	0.02000
	10	0.00998	0.01981	0.07654	0.09441	0.09962
	5	0.00999	0.01996	0.09458	0.15647	0.19651
	2	0.01000	0.02000	0.09966	0.19669	0.42025
Green and Gunnison Rivers	100	0.00749	0.00930	0.00998	0.00999	0.01000
	50	0.00925	0.01502	0.01982	0.01997	0.02000
	10	0.00997	0.01978	0.07639	0.09467	0.09969
	5	0.00999	0.01995	0.09420	0.15621	0.19682
	2	0.01000	0.01999	0.09959	0.19632	0.41979

increase when the return period T is decreasing. The average coincidence probabilities of 100 and 10-year design flood in two tributaries are 0.000143 and 0.001467, respectively. The coincidence probabilities of 1000 and 500-year design flood in three tributaries are 3.63×10^{-6} and 3.71×10^{-5}. From Tables 6.10 and 6.11, the coincidence probabilities of any other return period can be obtained directly or by interpolation.

6.4.3.2 Conditional Probabilities Analysis

The flood control standard of TGR is 1000 years. To analyze the effect of the upper tributaries on TGR, the conditional probabilities are calculated. The conditional probabilities of the occurrence of the T-year flood at the TGR, given the occurrence of flood in the upper tributaries can be defined as:

$$\begin{aligned} &P(Q_n > q_n^T | Q_1 > q_1^T, \ldots, Q_{n-1} > q_{n-1}^T) \\ &= P(Q_1 > q_1^T, \ldots, Q_n > q_n^T) / P(Q_1 > q_1^T, \ldots, Q_{n-1} > q_{n-1}^T) \end{aligned} \tag{6.4}$$

where n is the number of random variables and is from two to four; $Q_1, \ldots, \ldots Q_n$ are flow magnitudes in any of the two rivers; $q_1^T, \ldots, q_n^T$ mean the T-year design flood. For the case n equal to 2, the conditional probabilities of T-year design flood for the Yangtze River at TGR, given the flood volume in one of the upper tributary by

Table 6.9 The exceedance probability of coinciding *T*-year flood volumes at three considered inflow profiles

Upper Col.	Gunnison / Green	100	50	10	5	2
100	100	0.00671	0.00744	0.00751	0.00751	0.00751
	50	0.00744	0.00899	0.00929	0.00929	0.00929
	10	0.00749	0.00925	0.00996	0.00997	0.00998
	5	0.00749	0.00925	0.00997	0.00999	0.00999
	2	0.00749	0.00925	0.00997	0.00999	0.01000
50	100	0.00741	0.00896	0.00927	0.00927	0.00927
	50	0.00904	0.01346	0.01505	0.01505	0.01505
	10	0.00930	0.01502	0.01971	0.01980	0.01981
	5	0.00930	0.01502	0.01978	0.01994	0.01996
	2	0.00930	0.01502	0.01978	0.01995	0.01999
10	100	0.00746	0.00921	0.00996	0.00997	0.00997
	50	0.00929	0.01495	0.01969	0.01979	0.01979
	10	0.00997	0.01975	0.06874	0.07602	0.07654
	5	0.00998	0.01982	0.07605	0.09207	0.09456
	2	0.00998	0.01982	0.07639	0.09419	0.09944
5	100	0.00746	0.00921	0.00997	0.00999	0.00999
	50	0.00929	0.01495	0.01976	0.01993	0.01996
	10	0.00998	0.01982	0.07573	0.09180	0.09439
	5	0.00999	0.01995	0.09247	0.14128	0.15632
	2	0.00999	0.01997	0.09466	0.15613	0.19480
2	100	0.00746	0.00921	0.00997	0.00999	0.01000
	50	0.00929	0.01495	0.01976	0.01995	0.01999
	10	0.00998	0.01982	0.07607	0.09389	0.09939
	5	0.00999	0.01997	0.09457	0.15553	0.19452
	2	0.01000	0.02000	0.09954	0.19528	0.38732

specifying $Q_1 > q_1^T$, is obtained by Eq. 6.4. In a similar manner, the conditional probabilities of Yichang Station given the flood volume of two or three upper rivers are obtained. The calculated conditional probabilities are listed in the Table 6.12.

Table 6.12 shows that for a fixed return period in the upper rivers, the conditional probabilities show an increasing trend when the return period of TGR decreases. For example, given the occurrence of 1000-year design flood in Jinsha River, the conditional probabilities of 1000 and 10-year design flood in TGR are 0.35 and 0.71, respectively. The conditional probabilities of TGR given *T*-year design floods in three rivers are greater than those given *T*-year design floods in two rivers, and the conditional probabilities of TGR given *T*-year design floods in two rivers is greater than those only given *T*-year flood in one river. From these points of view, results of the calculation are reasonable. It is shown in Table 6.12, the

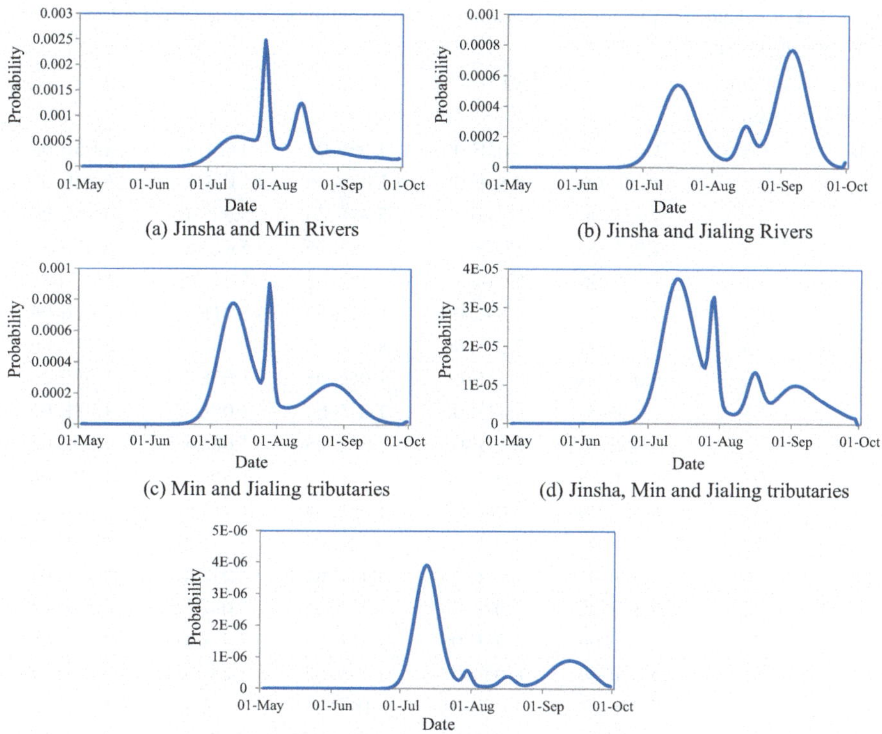

(a) Jinsha and Min Rivers

(b) Jinsha and Jialing Rivers

(c) Min and Jialing tributaries

(d) Jinsha, Min and Jialing tributaries

(e) Jinsha, Min, Jialing, and Yangtze (Yichang Station) Rivers

Fig. 6.8 The coincidence probabilities of flood dates on each day in the upper Yangtze River and its tributaries

Table 6.10 Coincidence probabilities considering flood magnitudes and occurrence dates in two of the tributaries in the upper Yangtze River

Rivers	*T*	100	50	10	5	2
Upper Col. and Green Rivers	100	0.00023	0.00029	0.00031	0.00031	0.00031
	50	0.00029	0.00047	0.00062	0.00062	0.00062
	10	0.00031	0.00062	0.00237	0.00293	0.00311
	5	0.00031	0.00062	0.00295	0.00486	0.00612
	2	0.00031	0.00062	0.00311	0.00614	0.01307
Upper Col. and Gunnison Rivers	100	0.00006	0.00007	0.00007	0.00007	0.00007
	50	0.00007	0.00011	0.00015	0.00015	0.00015
	10	0.00007	0.00015	0.00057	0.00070	0.00074
	5	0.00007	0.00015	0.00070	0.00116	0.00144
	2	0.00007	0.00015	0.00074	0.00144	0.00287
Green and Gunnison Rivers	100	0.00014	0.00018	0.00019	0.00019	0.00019
	50	0.00018	0.00029	0.00038	0.00038	0.00038
	10	0.00019	0.00038	0.00146	0.00181	0.00191
	5	0.00019	0.00038	0.00180	0.00299	0.00377
	2	0.00019	0.00038	0.00191	0.00376	0.00804

Table 6.11 Coincidence probabilities considering flood magnitudes and occurrence dates in three tributaries of the upper Yangtze River

Jinsha	Jialing	100	50	10	5	2
	Min					
100	100	3.63E-06	4.02E-06	4.06E-06	4.06E-06	4.06E-06
	50	4.02E-06	4.86E-06	5.02E-06	5.02E-06	5.02E-06
	10	4.05E-06	5E-06	5.38E-06	5.39E-06	5.39E-06
	5	4.05E-06	5E-06	5.39E-06	5.4E-06	5.4E-06
	2	4.05E-06	5E-06	5.39E-06	5.4E-06	5.4E-06
50	100	4.00E-06	4.84E-06	5.01E-06	5.01E-06	5.01E-06
	50	4.88E-06	7.27E-06	8.13E-06	8.13E-06	8.13E-06
	10	5.03E-06	8.11E-06	1.06E-05	1.07E-05	1.07E-05
	5	5.03E-06	8.11E-06	1.07E-05	1.08E-05	1.08E-05
	2	5.03E-06	8.11E-06	1.07E-05	1.08E-05	1.08E-05
10	100	4.03E-06	4.98E-06	5.38E-06	5.39E-06	5.39E-06
	50	5.02E-06	8.08E-06	1.06E-05	1.07E-05	1.07E-05
	10	5.39E-06	1.07E-05	3.71E-05	4.11E-05	4.13E-05
	5	5.39E-06	1.07E-05	4.11E-05	4.97E-05	5.11E-05
	2	5.39E-06	1.07E-05	4.13E-05	5.09E-05	5.37E-05
5	100	4.03E-06	4.98E-06	5.38E-06	5.40E-06	5.40E-06
	50	5.02E-06	8.08E-06	1.07E-05	1.08E-05	1.08E-05
	10	5.39E-06	1.07E-05	4.09E-05	4.96E-05	5.10E-05
	5	5.40E-06	1.08E-05	5.00E-05	7.63E-05	8.45E-05
	2	5.40E-06	1.08E-05	5.11E-05	8.44E-05	1.05E-04
2	100	4.03E-06	4.98E-06	5.38E-06	5.40E-06	5.40E-06
	50	5.02E-06	8.08E-06	1.07E-05	1.08E-05	1.08E-05
	10	5.39E-06	1.07E-05	4.11E-05	5.07E-05	5.37E-05
	5	5.40E-06	1.08E-05	5.11E-05	8.40E-05	1.05E-04
	2	5.40E-06	1.08E-05	5.38E-05	1.06E-04	2.09E-04

Jialing River has the most significant impact on the flow of TGR. The coefficient of correlation between Jialing River (at Beibei Station) and Yangtze River (at Yichang Station) is 0.318, the largest value in Table 6.4, which shows the close relationship between the two rivers. It is demonstrated that from Table 6.13, the higher conditional probabilities of TGR are generally obtained when the flows of Jinsha and Jialing Rivers are known. Table 6.14 gives the conditional probabilities of TGR when *T*-year design flood in the upper three rivers are known. It can be seen that when the three rivers upper have a 1000-year flood, the conditional probabilities of TGR is 1.0.

Table 6.12 Conditional probabilities $P(Q_Y > q_Y^T | Q_1 > q_1^T)$ of TGR, under the condition of the flood occurring in one of upper Yangtze River

Yichang	Return Period	1000	500	100	50	10
Jinsha River	1000	0.35	0.47	0.71	0.78	0.89
	500	0.24	0.35	0.62	0.71	0.86
	100	0.07	0.12	0.36	0.48	0.74
	50	0.04	0.07	0.24	0.36	0.66
	10	0.01	0.02	0.07	0.13	0.41
Min River	1000	0.27	0.37	0.58	0.65	0.79
	500	0.18	0.27	0.50	0.58	0.75
	100	0.06	0.10	0.28	0.38	0.62
	50	0.03	0.06	0.19	0.28	0.55
	10	0.01	0.02	0.06	0.11	0.34
Jialing River	1000	0.39	0.53	0.77	0.83	0.93
	500	0.26	0.39	0.68	0.77	0.90
	100	0.08	0.14	0.40	0.54	0.79
	50	0.04	0.08	0.27	0.40	0.72
	10	0.01	0.02	0.08	0.14	0.44

Table 6.13 Conditional probabilities $P(Q_Y > q_Y^T | Q_1 > q_1^T, Q_2 > q_2^T)$ of TGR, under the condition of the flood occurring in two of upper Yangtze River

Yichang	Return period	1000	500	100	50	10
Jinsha and Jialing River	1000	0.99	1.00	1.00	1.00	1.00
	500	0.97	0.98	1.00	1.00	1.00
	100	0.72	0.81	0.93	0.98	1.00
	50	0.45	0.62	0.80	0.88	1.00
	10	0.05	0.10	0.31	0.40	0.76
Jinsha and Min River	1000	0.73	0.96	1.00	1.00	1.00
	500	0.42	0.73	1.00	1.00	1.00
	100	0.09	0.18	0.73	0.96	1.00
	50	0.04	0.09	0.42	0.73	1.00
	10	0.01	0.02	0.09	0.18	0.75
Jialing and Min River	1000	0.90	0.95	1.00	1.00	1.00
	500	0.81	0.89	0.99	1.00	1.00
	100	0.50	0.63	0.88	0.95	1.00
	50	0.32	0.47	0.75	0.86	0.99
	10	0.05	0.10	0.32	0.43	0.80

Table 6.14 Conditional probabilities $P(Q_Y > q_Y^T | Q_1 > q_1^T, Q_2 > q_2^T, Q_3 > q_3^T)$ of TGR, under the condition of the flood occurring in three of upper Yangtze River

Return period	1000	500	100	50	10
1000	1.00	1.00	1.00	1.00	1.00
500	0.98	0.99	1.00	1.00	1.00
100	0.76	0.86	0.97	1.00	1.00
50	0.50	0.68	0.86	0.94	1.00
10	0.06	0.12	0.38	0.48	0.87

6.5 Conclusions

The flood combination risk, which reflects the probability of coincidence of multi-dimensional flood peaks, is important for reservoir operation and flood management. The copula function is used to establish the joint distribution of flood magnitudes and flood occurrence dates. The coincidence probabilities of flood magnitudes and dates are calculated. The conditional probabilities of TGR for different return periods are analyzed. The main conclusions of this study are summarized as follows:

(1) Symmetric Gumble, asymmetric Gumble and X-Gumble copula function are used. The X-Gumble copula provides the best fit. Therefore, the X-Gumble copula is used for the combination risk analysis in this chapter.
(2) By analyzing the coincidence probabilities of flood magnitudes and flood dates, this Chapter contributes to better practical knowledge in the area of engineering hydrology, particularly about the assessment of flood events and the performance of comprehensive flood-risk analyses. According to the analysis results, it is possible to raise the flood control water level of TGR in May and June. To the contrary, in September, the flood control water level of the TGR should not be raised too high, and certain flood control storage is needed for TGR. The flow in Jialing River has the most significant impact on the inflow in TGR. If the three upper rivers have a 1000-year design flood, the TGR also experiences a 1000-year flood. The coincidence probabilities or conditional probabilities of any other return period can be obtained directly from Tables 6.8, 6.9, 6.10, 6.11, 6.12, 6.13 and 6.14 or by interpolation.

References

De Michele C, Salvadori G (2003) A generalized Pareto intensity duration model of storm rainfall exploiting 2-copulas. J Geophys Res 108(D2):4067. https://doi.org/10.1029/2002JD002534
De Michele C, Salvadori G, Passoni G, Vezzoli R (2007) A multivariate model of sea storms using copulas. Coastal Eng 54(10):734–751

Dupuis DJ (2007) Using copulas in hydrology: benefits, cautions, and issues. J Hydrol Eng 12 (4):381–393

Enright M, Wilberg DE, Tibbetts JR (2008) Water Resources Data, Utah, Water Year 2004. U.S. Geological Survey: 120

Favre AC, Adlouni S, Perreault L, Thiémonge N, Bobée B (2004) Multivariate hydrological frequency analysis using copulas. Water Resour Res 40(W01101):12

Grimaldi S, Serinaldi F (2006) Design hyetographs analysis with 3-copula function. Hydrol Sci J 51(2):223–238

Interagency Advisory Committee on Water Data (IACWD) (1982) Guidelines for determining flood flow frequency: bulletin 17B of the Hydrology Subcommittee. Office of Water Data Coordination, U.S. Geological Survey, Reston, Virginia

Kao SC, Govindaraju RS (2008) Trivariate statistical analysis of extreme rainfall events via the Plackett family of copulas. Water Resour Res 44(2):W02415. https://doi.org/10.1029/2007WR006261

Kao SC, Govindaraju RS (2010) A copula-based joint deficit index for droughts. J Hydrol 380 (1–2):121–134

Ministry of Water Resources (MWR) (1993) Regulation for calculating design flood of water resources and hydropower projects. Chinese Water Resource Hydro Press, Beijing, China (in Chinese)

Munro P (1992) A Mojave dictionary. UCLA, Los Angeles

Prohaska S, Ilic A, Majkic B (2008) Multiple-coincidence of flood waves on the main river and its tributaries. IOP Conf Ser Earth Environ Sci 4:012013

Reed D (1999) The flood estimation handbook-1: overview. Institute of Hydrology, Wallingford

Robson A, Reed D (1999) Flood estimation handbook, vol.3: statistical procedure for flood frequency estimation. Institute of Hydrology, Wallingford, UK

Salvadori G, Michele CD (2010) Multivariate multiparameter extreme value models and return periods: A copula approach. Water Resour Res 46, W10501, https://doi.org/10.1029/2009WR009040

Serinaldi F, Grimaldi S (2007) Fully nested 3-copula: procedure and application on hydrological data. J Hydrol Eng 12(4):420–430

Serinaldi F, Bonaccorso B, Cancelliere A, Grimaldi S (2009) Probabilistic characterization of drought properties through copulas. Phys Chem Earth 34(10–12):596–605

Shiau JT, Wang HY, Chang TT (2006) Bivariate frequency analysis of floods using copulas. J Am Water Resour Assoc 42(6):1549–1564

Song S, Singh VP (2010) Frequency analysis of droughts using the Plackett copula and parameter estimation by genetic algorithm. Stoch Environ Res Risk Assess 24(5):783–805. https://doi.org/10.1007/s00477-010-0364-5

U.S. Geological Survey (2011) National hydrography dataset high-resolution flowline data. The National Map

Zhang L, Singh VP (2006) Bivariate flood frequency analysis using the copula method. J Hydrol Eng 11(2):150–164

Zhang L, Singh VP (2007a) Gumbel-Hougaard copula for trivariate rainfall frequency analysis. J Hydrol Eng 12(4):409–419

Zhang L, Singh VP (2007b) Bivariate rainfall frequency distributions using Archimedean copulas. J Hydrol 332:93–109

Zhang L, Singh VP (2007c) Trivariate flood frequency analysis using the Gumbel-Hougaard copula. J Hydrol Eng 12(4):431–439

Chapter 7
Copula-Based Method for Multisite Monthly and Daily Streamflow Simulation

7.1 Introduction

Stochastic simulation of flow discharge sequences is needed for water resources planning and management. It may help prepare for events that have not yet been observed in the past but nonetheless can be expected in the future (Szilagyi et al. 2006). Since multiple reservoirs and river sections are often considered in a system's operation plan, and more information is needed accompanying the rapidly increasing construction of reservoirs, there is a need to generate concurrent multisite streamflow series. Generated multisite flow data can serve as useful input for the design of reservoirs, evaluation of alternative operating policies for a system of reservoirs, and assessment of risk and reliability of water resources system operation (Srinivas and Srinivasan 2005).

A variety of methods have been proposed for stochastic multisite streamflow simulation. The first type is autoregressive moving average (ARMA) model and its variants, which assume that the current flow is linearly related to the previous observations. In these models, the actual flow is transformed to an alternative variable that satisfies the assumptions of linearity and normal probability distribution in the model structure (Sharma and O'Neill 2002). The disadvantages of ARMA are the limitation in representing nonlinear dependences and nonstandard probability distribution forms (Sharma and O'Neill 2002; Hao and Singh 2013). These limitations may result in less than an accurate representation of flows that are likely to occur and may lead to biased reservoir operating policies. In addition, these models are usually incapable of representing sudden bursts or jumps, which constitute a feature which often observed in short-period streamflow (Sharma et al. 1997).

The second type entails disaggregation models, which divide annual flows into seasonal or monthly flows or divide the aggregate basin flows into flows at individual sites (Kumar et al. 2000). A linear stochastic framework for streamflow disaggregation was proposed first by Valencia and Schaake (1973) and

L. Chen and S. Guo, *Copulas and Its Application in Hydrology and Water Resources*, Springer Water, https://doi.org/10.1007/978-981-13-0574-0_7

subsequently modified and improved by several researchers (Mejia and Rousselle 1976; Lane 1979; Salas et al. 1980; Stedinger and Vogel 1984; Stedinger et al. 1985). However, the parametric assumption of the probability distribution of streamflow is usually invoked (Kumar et al. 2000). To overcome this assumption, a nonparametric approach for space or time disaggregation based on the kernel density estimation was proposed by Lall et al. (1996) and Tarboton et al. (1998). Since then, nonparametric models, such as moving block bootstrapping method (Srinivas and Srinivasan 2005) or K-nearest neighbor resampling method (Prairie et al. 2007), have been developed for multisite streamflow simulation. However, these methods, such as kernel method, are known to be inefficient and cumbersome to implement in higher dimensions. This limits their extension to space and time disaggregation (Prairie et al. 2007). The traditional bootstrapping or k-nearest neighbor model generates only observed values that are in the sample data (Lee and Salas 2008).

Though a variety of methods have been proposed in the hydrologic literature for multisite simulation of streamflow, none of the methods seem to have gained universal acceptability among practicing engineers for various water resources applications (Srinivas and Srinivasan 2005). Recently, copulas have been used for stochastic simulation of hydrological data, because they are flexible in choosing arbitrary marginal distributions, representing the dependence structure, extending to more than two variables, and permitting separate analysis of marginal distributions and dependence structure. The copula method is simple in the sense that it does not have to deal with the uncertainties associated with the identification of normalizing transformation. Lee and Salas (2008) used the copula method for modeling and simulation of annual streamflows. Bárdossy and Pegram (2009) used the copula method for multisite precipitation simulation. Hao and Singh (2013) proposed an entropy-copula method for single-site monthly streamflow simulation, in which the marginal distribution was built using the entropy method and the joint distribution of adjacent monthly streamflow was built using the copula method. These studies indicate that the copula method can be an effective tool for the stochastic simulation of hydrological data.

Srinivas and Srinivasan (2005) indicated that an ideal model for stochastic simulation of multisite multi-seasonal streamflow should ensure the preservation of marginal distributions at various temporal levels at each site and cross-correlations among sites. The copula method can use any marginal distributions. Furthermore, since copulas are capable of exhibiting the dependence between two or more random variables and modeling the general dependence in multivariate data (e.g., Joe 1997; Nelsen 2006), cross-correlations among sites can be realized by building the multivariate joint distributions of hydrological series at different locations.

However, the ability of commonly used parametric copulas to model dependences in higher dimensions is rather restricted, e.g., for the Archimedean copulas. The conflicts between multiple dependences required in multisite streamflow stochastic simulation and the difficulty of simulating the dependences in higher dimensions have limited the application of copulas to multisite streamflow simulation. To overcome these difficulties, Hao and Singh (2013) proposed the maximum entropy copula method for multisite monthly streamflow simulation, in which the copula function was built by maximizing entropy. This method can extend copulas to higher dimensions. However, it involves too many parameters that need to be estimated. To illustrate, consider bivariate copulas as an example. The method needs to estimate $2m + 2$ (m is the maximum order of moments, $m = 3$ in that study) parameters. Therefore, a simple multisite stochastic simulation method based on copulas with less parameter needs to be developed.

Very often a hydrologic stochastic process must be studied at different time scales (Koutsoyiannis 2001). In comparison with monthly or yearly applications, less research has been reported on multisite daily flow simulation which may be desirable for practical use. For example, a daily record of multisite streamflow is usually required for simulating the behavior of multi-reservoir operating policies. The traditional ARMA model has been shown to produce an adequate synthetic record for periods as short as five days (Xu et al. 2003). Kumar et al. (2000) indicated that disaggregating monthly streamflow to daily flows involves a higher dimensional problem that cannot always be well represented by traditional parametric disaggregation techniques. They adopted K-NN bootstrap techniques in conjunction with an optimization scheme for spatial and temporal disaggregation of monthly streamflow to daily flows. However, until now, simulation of multisite daily streamflow has remained a challenge.

The content of this chapter is therefore to introduce an efficient, reliable and parsimonious method for generating monthly and daily multisite streamflows, which can capture the features exhibited by observed data (Chen et al. 2015). To this end, we establish multivariate copulas based on bivariate and conditional copulas to characterize the temporal and spatial dependences between multiple sites. This makes it easier to estimate the parameters of copulas. The monthly and daily flows are simulated based on the multivariate conditional distribution given the previous flow and the concurrent flow of the other site. Monthly data from three tributaries in the Colorado River basin, daily data from two gauging stations on mainstream and four gauging stations on tributaries in the upper Yangtze River basin, are used for case studies. Basic statistics of observed and simulated data are calculated and compared. Finally, the performance of the copula-based method is comprehensively evaluated.

7.2 Methodology

In this section, we develop a new method for multisite stochastic simulation. This method seeks to generate monthly or daily flow data at multiple sites for a period of N years. The sketch of the locations of these sites is shown in Fig. 7.1. There are two cases. Figure 7.1a shows that all the sites (A, B and C) are located on the tributaries. Figure 7.1b shows that some of the sites are located on the mainstream (A and D) and some on the tributaries (B and C). First, streamflow of single site A is simulated using the copula method proposed by Lee and Salas (2008, 2011).

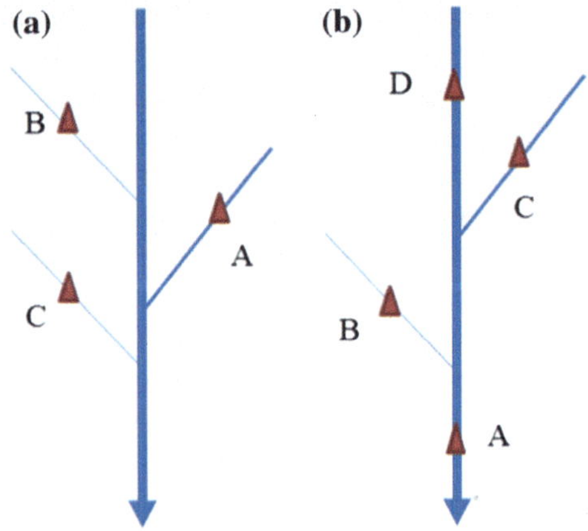

Fig. 7.1 Sketch showing locations of gauged stations

Streamflow of site B, which has spatial dependence with site A, is generated using the method described below. Streamflow of sites C, D and so on, which also have the spatial dependence with site A, are generated using the same method. For case (a), site A can be any of the stations in the tributaries. For case (b), we select the site located on the downstream of the mainstream as site A, because there is a strong dependence between site A and other sites upstream. Therefore, multisite streamflow simulation entails two steps. First, the streamflow of site A is generated using the copula method. Then, the streamflow at sites B, C and so on are simulated, based on the spatial dependence between site A and other sites.

7.2.1 Single-Site Streamflow Simulation Based on Bivariate Copulas

The most important issue for single-site streamflow simulation is to preserve the temporal dependences of site A. Bivariate copula functions are used to describe the correlation between consecutive months or days. Thus, the method for establishing the bivariate copulas is described first, and then the single-site stochastic simulation method proposed by Lee and Salas (2011) is described.

Supposes Y_t is thc simulated streamflow of site A at time t. The streamflow at time t is related to the previous streamflow at time $t-1$. To describe the correlation between them, the copulas are used to establish the joint distributions. Since the streamflow at time t is conditioned on the previous streamflow at $t-1$, the conditional distribution is employed to generate the streamflow data. The generating procedures with the conditional copula are summarized as follows

(1) Fit the marginal distributions $F(y_t)$ and $F(y_{t-1})$ using the P-III distribution, $u_1^t = F(y_t)$ and $u_1^{t-1} = F(y_{t-1})$.

(2) Establish the joint distribution $F(y_{t-1}, y_t)$ using copulas, $F(y_{t-1}, y_t) = C(u_1^{t-1}, u_1^t)$, and estimate their parameters based on the Kendall tau method.
(3) As the streamflow at the previous time $t-1$ is known, $u_1^{t-1} = F(y_{t-1})$ can be calculated using the fitted P-III distribution. Generate a uniform random number ε, $C(u_1^t | u_1^{t-1}) = \varepsilon$. Since u_1^{t-1} and ε are known, the value of u_1^t can be obtained from the inverse function $C^{-1}(u_1^t | u_1^{t-1})$.
(4) Derive the monthly or daily streamflow y_t of site A from the inverse function $F^{-1}(u_1^t)$.

Note that the first value y_1 is generated from the marginal distribution $F(y_1)$. Other detailed information can be found in Lee and Salas (2008, 2011).

7.2.2 *Multi-Site Streamflow Simulation*

The most important issue for multisite streamflow simulation is to preserve the temporal and spatial dependences. Multivariate copula functions are used to describe the correlation between hydrological variables. Thus, the method for establishing the multivariate copulas is described first, and then the multisite stochastic simulation method is proposed.

Suppose that X_t and X_{t-1} are the simulated streamflow of site B at time t and $t-1$, respectively. Y_t is the simulated streamflow of site A at time t. The value of X_t is related to both X_{t-1} and Y_t. There exists a temporal dependence between X_t and X_{t-1}, and a spatial dependence between X_t and Y_t. To describe the temporal and spatial dependences among sites, we can build three-dimensional copulas using the method mentioned above.

The trivariate copula function is given by:

$$\begin{aligned} C(u_1, u_2, u_3) &= F(F(y_t), F(x_{t-1}), F(x_t)) \\ &= \int_{\infty}^{x_{t-1}} C_{XY}(F_{Y_t|X_{t-1}}(y_t|x_{t-1}), F_{X_t|X_{t-1}}(x_t|x_{t-1}))F(\mathrm{d}x_{t-1}) \end{aligned} \tag{7.1}$$

where $u_1 = F(y_t)$, $u_2 = F(x_{t-1})$ and $u_3 = F(x_t)$ are the marginal distributions. $F_{X_t|X_{t-1}}(x_t|x_{t-1})$ and $F_{Y_t|X_{t-1}}(y_t|x_{t-1})$ are the conditional probability distributions given variable X_{t-1}, which can be obtained using Eq. 2.6 of Chap. 2.

Since the method mentioned above uses the bivariate Archimedean copulas to build multivariate copulas, we only need to estimate the parameters of bivariate copulas. There are several methods for parameter estimation of bivariate copulas. They include the Kendall tau method, maximum likelihood (ML) method, the method of inference function for margins (IFM), semi-parametric pseudo-maximum-likelihood method, and so on, which have been discussed in Chap. 2. When simulating daily streamflow data, there are 365 trivariate copulas that need to be established. It is difficult to estimate the parameters of these copulas using IFM or the semi-parametric pseudo-maximum-likelihood method, since usually some optimization algorithms are often used to obtain the maximum likelihood value and its corresponding parameters. On the contrary, the advantage of the proposed method is to use the bivariate copulas to derive the trivariate one. Therefore, the Kendall tau method can be used to estimate the parameters of trivariate copulas instead of the IFM or semi-parametric pseudo-maximum-likelihood method, which makes it possible for the practical application.

The Kendall tau correlation coefficients have a relationship with the copula function, and can be used to estimate the parameters of bivariate copulas, the relationship between Kendall tau and copula parameters are summarized in Table 2.3 of Chap. 2.

To obtain the values of $F_{X_t|X_{t-1}}(x_t|x_{t-1})$ and $F_{Y_t|X_{t-1}}(y_t|x_{t-1})$, the bivariate joint distributions of $C(u_1,u_2)$ and $C(u_3,u_2)$ are needed to be built. The parameters of these copulas can be estimated using the equations given in Table 2.3. In addition, the parameter of the joint distribution C_{XY} of Eq. 7.1 can also be estimated based on the Kendall tau. Since variables X_t and Y_t have a dependence with variable X_{t-1}, the partial correlation coefficient, which is a measure of a pair of random variables after removing the effects of other variables, needs to be involved. The partial correlation coefficient corresponding to the Kendall tau can be calculated as (Kendall 1948; Sidney 1956; Ebuh and Oyeka 2012)

$$\tau_{X_tY_t|X_{t-1}} = \frac{\tau_{X_tY_t} - \tau_{X_tX_{t-1}}\tau_{Y_tX_{t-1}}}{\sqrt{(1-\tau^2_{X_tX_{t-1}})(1-\tau^2_{Y_tX_{t-1}})}} \tag{7.2}$$

where $\tau_{X_tY_t|X_{t-1}}$ is the partial correlation given the controlling variable X_{t-1}; $\tau_{X_tY_t}$ is the Kendall tau rank correlation coefficient between X_t and Y_t; is the Kendall tau rank correlation coefficient between and X_t and X_{t-1}, and is the Kendall tau rank correlation coefficient between Y_t and X_{t-1}.

In order to sample from the continuous joint CDF, $F(F(y_t), F(x_{t-1}), F(x_t))$, for obtaining the random values of (Y_t, X_{t-1}, X_t), the conditional distribution method is employed in this book. For generation of multisite streamflow at time t while preserving the temporal correlation between consecutive months or days and spatial correlation between sites A and B, the streamflow values at time t at site B can be generated from the conditional distribution, given the streamflow value of its previous time at site B and of current time at site A. The multivariate probability distribution of three variables X_{t-1}, X_t and Y_t is thus built. When the method is applied to generate monthly streamflow of each year, 12 conditional distributions have to be used sequentially. When the method is applied to generate daily streamflow of each year, 365 conditional distributions have to be used sequentially. For time t, when the current streamflow y_t at site A and the previous flow x_{t-1} at site B are known, u_1 and u_2 can be calculated. The current flow x_t at site B can be obtained from the conditional distribution of u_3, given the values u_1 and u_2, which is expressed as:

$$G(u_3|u_1,u_2) = \frac{\partial_{u_1,u_2}C(u_1,u_2,u_3)}{\partial_{u_1,u_2}C_{u_1,u_2}(u_1,u_2)} = \frac{\partial_{u_1}C_{u_1u_3}(\partial_{u_2}C_{u_1u_2}(u_1,u_2),\partial_{u_2}C_{u_2u_3}(u_2,u_3))}{\partial_{u_1,u_2}C_{u_1,u_2}(u_1,u_2)} \tag{7.3}$$

where the variables u_1 and u_2 next to the partial derivation operator means that the derivative of a function C with respect to variables u_1 and u_2:

$$\partial_{u_1,u_2}C(u_1,u_2,u_3) = \frac{\partial C(u_1,u_2,u_3)}{\partial u_1 \partial u_2} \tag{7.4a}$$

$$\partial_{u_1,u_2}C_{u_1u_2}(u_1,u_2) = \frac{\partial C_{u_1u_2}(u_1,u_2)}{\partial u_1 \partial u_2} \tag{7.4b}$$

$$\partial_{u_2}C_{u_1u_2}(u_1,u_2) = \frac{\partial C_{u_1u_2}(u_1,u_2)}{\partial u_2} \tag{7.4c}$$

$$\partial_{u_2}C_{u_2u_3}(u_2,u_3) = \frac{\partial C_{u_2u_3}(u_2,u_3)}{\partial u_2} \tag{7.4d}$$

From Eq. 7.4a, 7.4b, 7.4c, 7.4d, $\partial_{u_2}C_{u_1u_2}(u_1,u_2)$ and $\partial_{u_2}C_{u_2u_3}(u_2,u_3)$ can be defined as Q_1 and Q_2,

$$Q_1(u_1,u_2) = \frac{\partial C_{u_1u_2}(u_1,u_2)}{\partial u_2}; \quad Q_2(u_2,u_3) = \frac{\partial C_{u_2u_3}(u_2,u_3)}{\partial u_2} \tag{7.5}$$

Then,

$$\begin{aligned} G(u_3|u_1,u_2) &= \frac{\partial_{u_1}C_{u_1u_3}(Q_1,Q_2)}{\partial_{u_1,u_2}C_{u_1,u_2}(u_1,u_2)} \\ &= \frac{\frac{\partial C_{u_1u_3}(Q_1,Q_2)}{\partial Q_1}\frac{\partial Q_1}{\partial u_1} + \frac{\partial C_{u_1u_3}(Q_1,Q_2)}{\partial Q_2}\frac{\partial Q_2}{\partial u_1}}{\frac{\partial C_{u_1,u_2}(u_1,u_2)}{\partial u_1 u_2}} \\ &= \frac{\frac{\partial C_{u_1u_3}(Q_1,Q_2)}{\partial Q_1}\frac{\partial Q_1}{\partial u_1} + 0}{\frac{\partial Q_1}{\partial u_1}} \\ &= \frac{\partial C_{u_1u_3}(Q_1,Q_2)}{\partial Q_1} = H(Q_2|Q_1) \end{aligned} \tag{7.6}$$

where H represents the conditional distribution. Q_1 and Q_2 can be calculated by Eq. 7.5. Knowing the values of u_1, u_2 and u_3, the intermediate variables Q_1 and Q_2 can be calculated. Substituting Q_1 and Q_2 into Eq. 7.6, the value of $G(u_3|u_1,u_2)$ is finally determined. It can be seen that $G(u_3|u_1,u_2)$ is actually the conditional distribution of Q_2 given Q_1.

This indicates that at time t the current streamflow X_t at site B is conditioned on the previous flow X_{t-1} at the same site and the current flow Y_t at the other site A, which has the spatial correlation with site B. Since u_1 and u_2 are known, Q_1 can be calculated using Eq. 7.5. A uniform random number ε is generated and $G(u_3|u_1,u_2)$ equals ε. Substituting the value of Q_1 and ε into Eq. 7.6, the value of Q_2 can be obtained by solving Eq. 7.6. Since $u_2 = F(x_{t-1})$ is already known, the subsequent u_3 is determined. The generated data in the real domain is obtained from the inverse function $x_t = F^{-1}(u_3)$.

The step by step generation procedure with the conditional probability function of Eq. 7.6 can be summarized as follows.

(1) Establish the marginal distribution based on the P-III distribution and estimate its parameters using the L-moment method.
(2) Calculate the Kendall tau rank correlation coefficients $\tau_{X_tY_t}$, $\tau_{X_tX_{t-1}}$ and $\tau_{Y_tX_{t-1}}$. Then the partial correlation coefficient $\tau_{X_tY_t|X_{t-1}}$ is obtained using Eq. 7.2.
(3) Simulate the single-site streamflow data using the bivariate copulas.
(4) To carry out the multisite streamflow simulation, three bivariate joint distributions, $C(u_1,u_2)$, $C(u_2,u_3)$ and $C_{XY}(F_{Y_t|X_{t-1}}(y_t|x_{t-1}), F_{X_t|X_{t-1}}(x_t|x_{t-1}))$, need to be built. The parameters of the first two copulas are estimated using the equations listed in Table 2.3. For the third copula, the partial correlation coefficient is calculated using Eq. 7.2. Then the parameter of the third copula C_{XY} is also derived based on the relationship between Kendall tau and copulas parameters.
(5) Calculate the conditional distribution $Q_1(u_1,u_2)$ based on the established joint distribution $C(u_1,u_2)$. A uniform random number ε is generated so that $G(u_3|u_1,u_2) = \varepsilon$. Substitute the value of Q_1 and ε into Eq. 7.6, the value of $Q_2(u_2,u_3)$ is obtained from the inverse function $H^{-1}(Q_2|Q_1)$.
(6) The subsequent u_3 is derived by solving the inverse function of Q_2. Finally, the generated streamflow data at gauging station B is obtained from the inverse function $x_t = F^{-1}(u_3)$.

Note that for the first data of site B, it is only related to the first data of site A. Therefore, the simulated streamflow data of the first day is directly generated from the conditional copula function $C(u_3|\, u_1)$.

7.3 Multisite Monthly Streamflow Simulation

The coupla-based method is applied for simulating multisite monthly streamflow in the Colorado River Basin. A schematic of the rivers and gauging stations is shown in Fig. 7.2. Monthly flows of three tributaries of Colorado River, namely Paria River, Little Colorado River and Virgin River, are generated using the proposed method. The data sets with a length of 103 years (1906–2008) from the sites at Lees Ferry located on Paria River (denoted as site A), near Cameron on Little Colorado

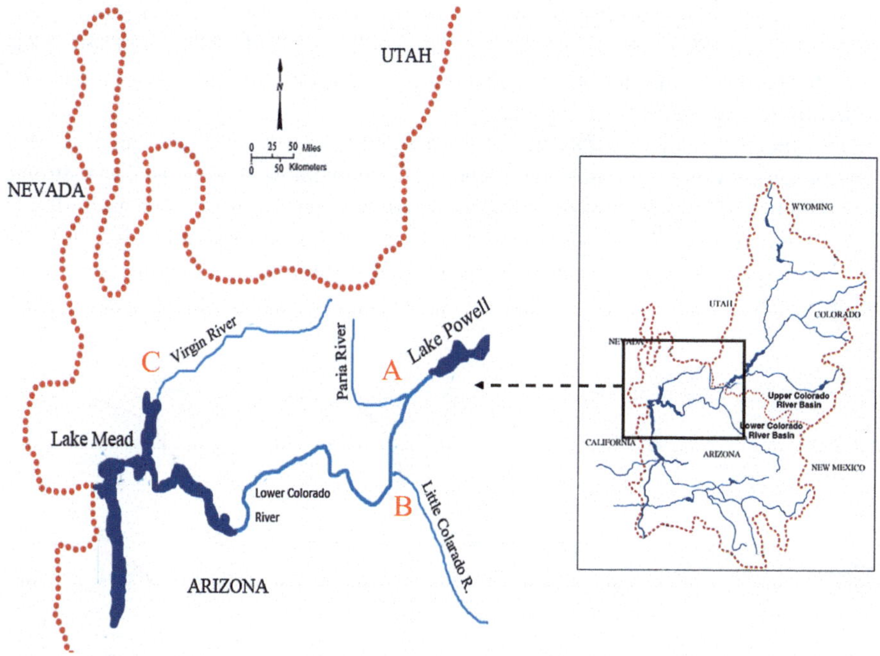

Fig. 7.2 Locations of rivers and gauging stations in the Colorado River basin

River (denoted as site B), and at Littlefield on Virgin River (denoted as site C) are used. More details about the datasets are given by Hao and Singh (2013).

The bivariate Kendall tau and partial correlation coefficient corresponding to the Kendall tau are estimated. These correlations include the temporal dependence of single site and the spatial dependence between sites. Results are given in Table 7.1. Several coefficients are negative, but are very small, almost close to zero, which means that the association between the two variables can be negligible, and the Gumbel copula therefore can be used in this study. The Kendall tau of streamflow

Table 7.1 Values of Kendall tau representing temporal and spatial dependences of sites A, B and C

Sites	τ	1	2	3	4	5	6	7	8	9	10	11	12
A	$\tau_{Y_{t-1}Y_t}$	0.38	0.24	0.50	0.55	0.49	0.35	0.04	0.26	0.11	0.11	0.37	0.17
B	$\tau_{X_{t-1}X_t}$	0.28	0.42	0.47	0.45	0.47	0.05	−0.05	0.27	0.15	0.10	0.31	0.29
C	$\tau_{X_{t-1}X_t}$	0.41	0.44	0.61	0.66	0.70	0.69	0.37	0.31	0.15	0.26	0.45	0.41
A–B	$\tau_{X_tY_t}$	0.21	0.44	0.36	0.39	0.37	0.39	0.36	0.39	0.32	0.27	0.31	0.13
	$\tau_{X_{t-1}Y_t}$	0.03	0.20	0.25	0.27	0.21	0.13	−0.01	0.12	0.03	0.02	0.08	−0.01
	$\tau_{X_tY_t\|X_{t-1}}$	0.21	0.40	0.28	0.32	0.31	0.39	0.36	0.37	0.32	0.27	0.30	0.14
A–C	$\tau_{X_tY_t}$	0.42	0.55	0.62	0.57	0.44	0.33	0.39	0.46	0.51	0.35	0.38	0.41
	$\tau_{X_{t-1}Y_t}$	0.25	0.27	0.42	0.46	0.31	0.21	0.21	0.17	0.06	0.17	0.25	0.24
	$\tau_{X_tY_t\|X_{t-1}}$	0.36	0.50	0.50	0.40	0.32	0.26	0.34	0.43	0.51	0.33	0.31	0.35

for two consecutive months at site A, $\tau_{Y_{t-1}Y_t}$, is used for single-site stochastic simulation. The other dependencies, such as $\tau_{X_tY_t}$, $\tau_{X_tX_{t-1}}$ and $\tau_{X_{t-1}Y_t}$, are used in the multisite stochastic simulation for calculating the partial correlation coefficient and estimating the parameters of copulas.

First, the marginal distributions of each month are constructed. The P-III distribution is fitted to the monthly data, and the L-moment method is used to estimate its parameters. The marginal distribution frequency curves of flood magnitudes of the Paria River, Little Colorado River, and Virgin River are shown in Figs. 7.3, 7.4 and 7.5, respectively. It is seen that the P-III distribution can fit empirical distribution well.

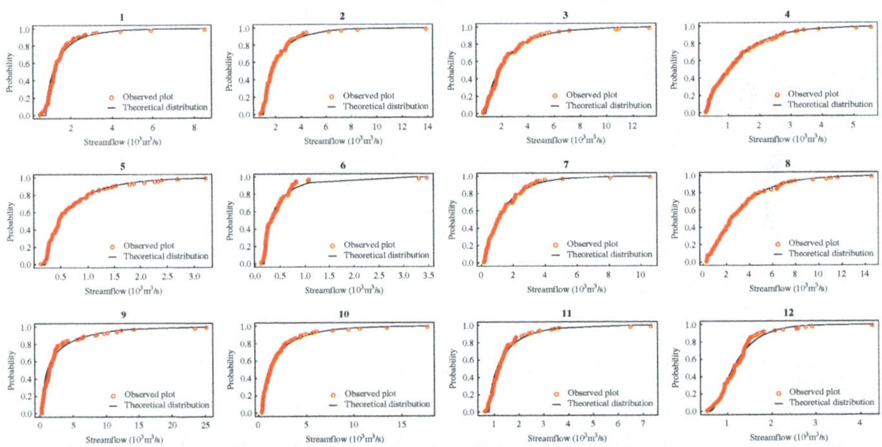

Fig. 7.3 Comparison of empirical and theoretical probabilities for streamflow of Paria River

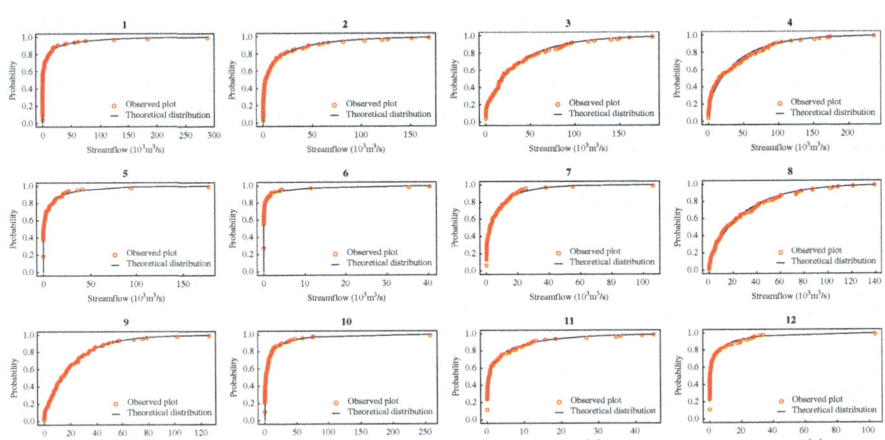

Fig. 7.4 Comparison of empirical and theoretical probabilities for streamflow of Little Colorado River

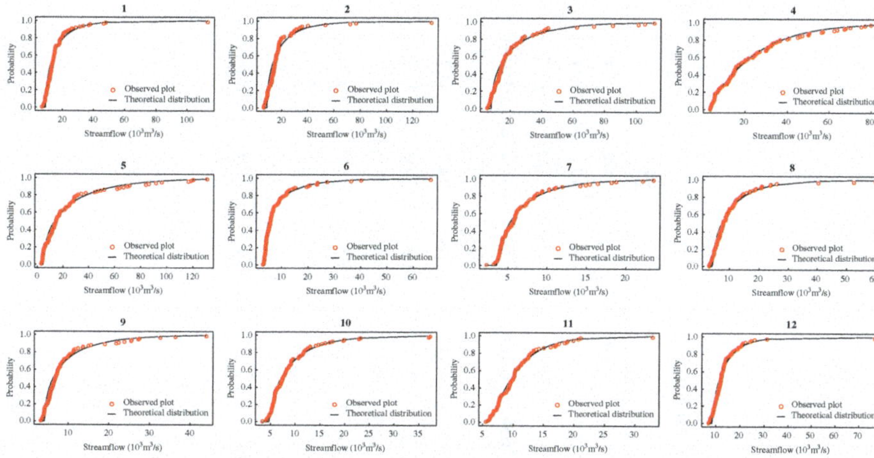

Fig. 7.5 Comparison of empirical and theoretical probabilities for streamflow of Little Colorado River

Second, the single-site monthly flow is simulated using the method described by Lee and Salas (2011) who concluded that the Gumbel copula shows a better fit than the Frank and Clayton copulas. Therefore, the Gumbel copula is used to generate the single-site monthly flow. To demonstrate the performance of single-site monthly streamflow simulation, basic statistics, including mean value, standard deviation (S_d), skewness (C_s), and lag-1 correlation (lag-1), are calculated and shown in Table 7.2. The relative error (RE) is calculated for each month, as given in Table 7.2, which shows that the maximum RE values corresponding to mean, standard deviation S_d, the coefficient of skewness C_s and lag-1 are 1.29, 4.68, 6.60 and 2.45%, respectively. In addition, boxplots are used to display the observed and simulated statistics, and the performance is judged to be good when a statistic falls within the boxplot (Salas and Lee 2011; Hao and Singh 2013). Boxplots of statistics (mean flow, standard deviation S_d and coefficient of skewness C_s) of the observed and simulated monthly streamflows for site A are shown in Fig. 7.6, which indicates that all these simulations show good results since those statistics fall within the boxplots for most of the months. Therefore, the copula-based stochastic simulation method performs well in preserving mean flow, S_d, C_s and lag-1 correlation. The generated single-site data can be used for multisite monthly streamflow simulation.

We illustrate the derivation of the joint distribution for monthly streamflow at sites A and B as an example. Denote monthly streamflow for month t at sites A and B as y_t and x_t, and for month $t-1$ at site B as x_{t-1}. Their corresponding marginal distributions are $u_1 = F(y_t)$, $u_2 = F(x_{t-1})$ and $u_3 = F(x_t)$. The trivariate joint distribution of monthly streamflow y_t, x_{t-1} and x_t are built using Eq. 7.1. To obtain the trivariate joint distribution, three bivariate distributions need to be established. They are $C(u_1, u_2)$, $C(u_2, u_3)$ and $C_{XY}(F_{Y_t|X_{t-1}}(y_t|x_{t-1}), F_{X_t|X_{t-1}}(x_t|x_{t-1}))$.

Table 7.2 Values of statistics and relative errors of monthly streamflow stochastic simulation at site A for each month

Items	Statistics	1	2	3	4	5	6	7	8	9	10	11	12
Observed	Mean (m^3/s)	1486	2225	2489	1292	694	449	1492	3214	2836	2146	1440	1249
	S_d	896	1554	2045	1015	611	385	1437	2753	4099	2576	877	492
	C_S	2.61	2.70	2.41	1.89	2.58	2.83	2.18	1.84	3.54	3.14	2.74	1.57
	Lag-1	0.3840	0.2364	0.4955	0.5507	0.4907	0.3539	0.0374	0.2638	0.1080	0.1097	0.3712	0.3840
Simulated	Mean (m^3/s)	1473	2206	2478	1291	688	445	1483	3172	2841	2089	1425	1246
	S_d	860	1488	1979	996	592	371	1399	2653	4037	2456	843	482
	C_S	2.52	2.59	2.31	1.83	2.50	2.73	2.09	1.72	3.44	3.08	2.68	1.48
	Lag-1	0.3842	0.2424	0.4997	0.5367	0.4887	0.3508	0.0367	0.2613	0.1067	0.1102	0.3655	0.3842
RE (%)	Mean(m^3/s)	0.87	0.83	0.45	0.04	0.95	0.71	0.63	1.29	0.16	2.65	1.05	0.24
	S_d	4.05	4.25	3.21	1.86	3.12	3.66	2.61	3.63	1.53	4.68	3.86	2.01
	C_S	3.56	4.03	4.13	3.55	2.98	3.51	4.02	6.60	2.83	1.95	2.17	5.74
	Lag-1	0.04	2.45	0.83	2.61	0.41	0.89	1.69	0.96	1.15	0.47	1.58	0.04

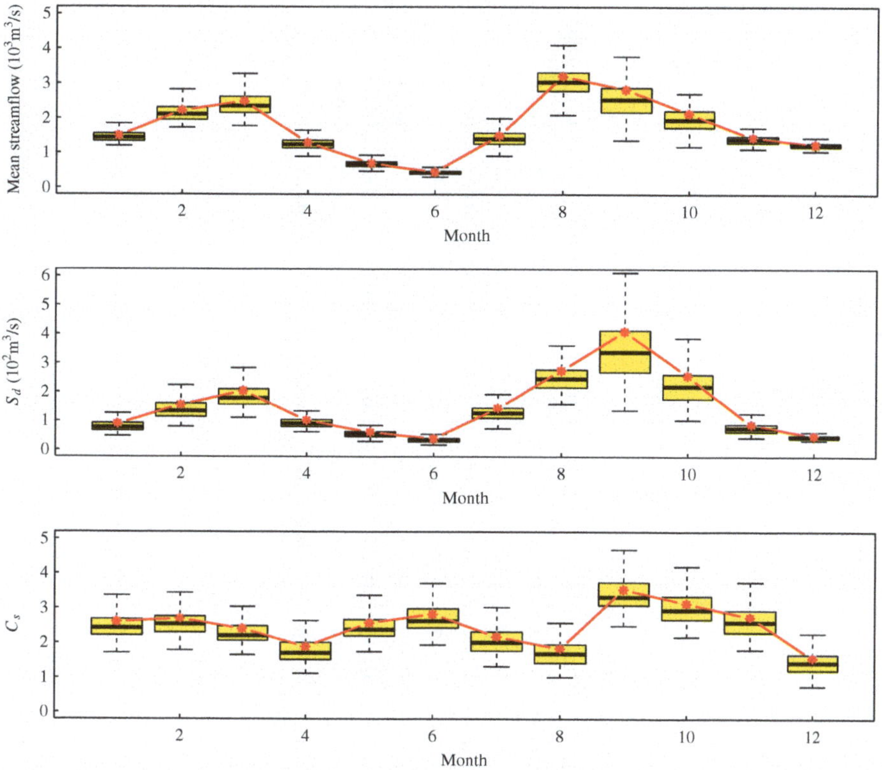

Fig. 7.6 Observed and simulated statistics of monthly streamflow at site A

The Gumbel, Frank and Clayton copula functions are employed to establish those bivariate distributions. And the same kind of copulas is used for each trivariate joint distribution. The Kendall tau method is used to estimate their parameters. For the third joint distribution, the partial correlation coefficient corresponding to the Kendall tau is applied instead of the regular one.

Both empirical and theoretical probabilities of trivariate joint distribution are computed. The theoretical copulas are calculated by the Gumbel, Frank and Clayton copulas, respectively. The root mean square error (*RMSE*) values between empirical and theoretical copulas of trivariate joint distribution are calculated and given in Table 7.3. It can be seen that there is no significant difference when using different copulas for calculation. The mean values of *RMSE* are calculated and listed in the last column of Table 7.3. The Gumbel copula has the smallest mean RMSE value for sites A–B and A–C. The empirical joint probabilities of the three variables are plotted against theoretical probabilities calculated by the Gumbel copulas for sites A–B and A–C, as shown in Figs. 7.7 and 7.8, which show that no significant difference between empirical and theoretical joint probabilities can be detected.

Generally, several copulas are used for establishing the joint distribution, and the one with the maximum likelihood value is usually selected. And the SP method is widely used for parameter estimation. In this study, the Gumbel copula coupled

Table 7.3 RMSE values of different trivariate copulas for monthly streamflow simulation

Sites	Copulas	1–2	2–3	3–4	4–5	5–6	6–7	7–8	8–9	9–10	10–11	11–12	12–1	Mean value
A–B	Gumbel	0.016	0.021	0.023	0.046	0.080	0.068	0.032	0.045	0.047	0.033	0.044	0.024	0.040
	Frank	0.035	0.042	0.045	0.062	0.086	0.073	0.024	0.022	0.027	0.032	0.039	0.032	0.043
	clayton	0.039	0.031	0.025	0.037	0.068	0.068	0.041	0.061	0.060	0.042	0.068	0.046	0.049
	Selected	0.015	0.017	0.018	0.041	0.085	0.090	0.017	0.016	0.025	0.024	0.049	0.028	0.035
A–C	Gumbel	0.026	0.021	0.038	0.053	0.039	0.036	0.021	0.019	0.032	0.026	0.020	0.023	0.029
	Frank	0.045	0.041	0.068	0.085	0.071	0.065	0.033	0.025	0.018	0.025	0.038	0.038	0.046
	clayton	0.032	0.030	0.037	0.053	0.041	0.038	0.029	0.033	0.049	0.029	0.029	0.026	0.035
	Selected	0.022	0.021	0.020	0.016	0.017	0.023	0.018	0.029	0.021	0.018	0.027	0.025	0.022

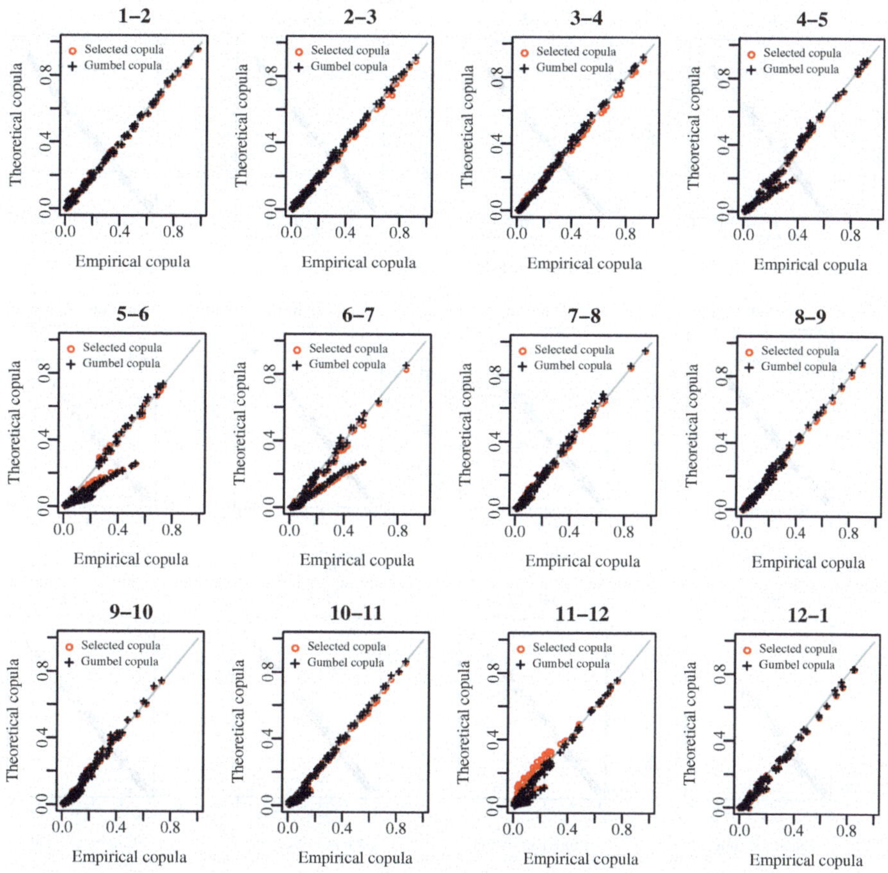

Fig. 7.7 Observed and theoretical joint probabilities of monthly streamflow x_{t-1}, x_t and y_t (y_t represents the monthly streamflow of Paria River at time t; and x_{t-1} and x_t represents the monthly streamflow of Little Colorado River at time $t-1$ and t, respectively)

with the Kendall tau parameter estimation method is used instead of the selected copula with the SP method. Before using the Gumbel copula with the Kendal tau parameter estimation method for multi-site stochastic simulation, further comparisons should be carried out. In order to further test the performance of the Gumbel copula, we use the Gumbel, Frank and Clayton copulas to construct each bivariate distribution and employed the SP method for parameter estimation. The copula with the maximum likelihood values is selected to construct the trivariate distribution. The performances of the selected and Gumbel copulas are compared. The *RMSE* values for those two copulas are computed and results are given in Table 7.3. Figures 7.7 and 7.8 show the empirical and theoretical probabilities of trivariate joint distribution calculated by the selected copula and the Gumbel copula. Table 7.3, and Figs. 7.7 and 7.8 indicate that both the selected and the Gumbel copula perform well, and sometimes the Gumbel copula fits better than the selected copulas. Generally there is no obvious difference between those two models.

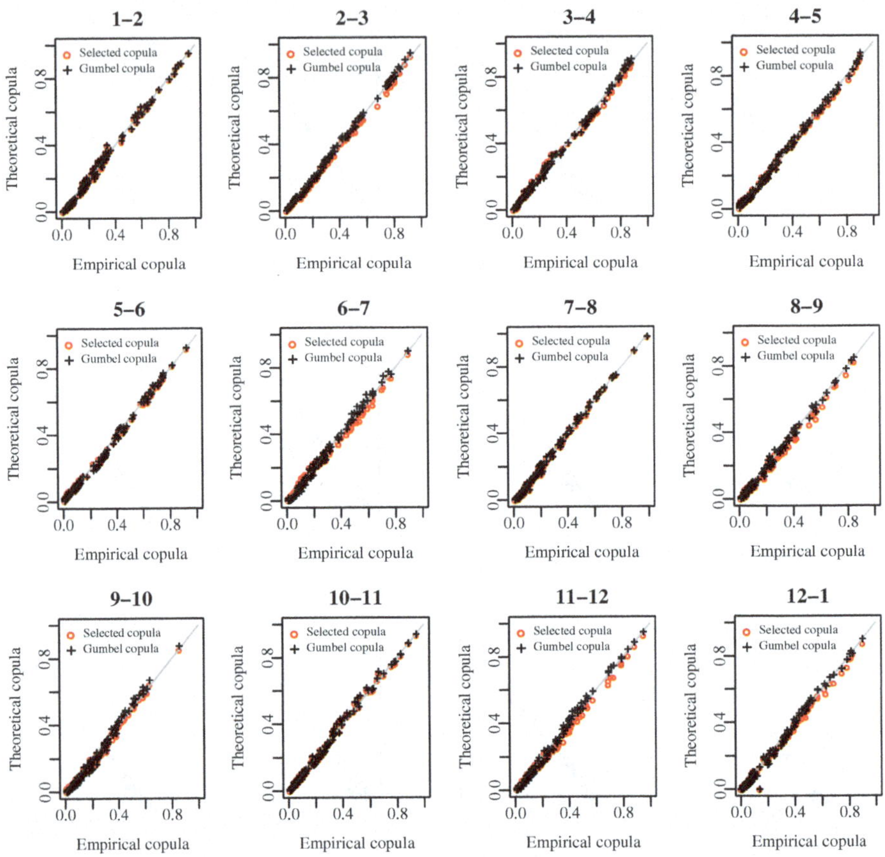

Fig. 7.8 Observed and theoretical joint probabilities of monthly streamflow x_{t-1}, x_t and y_t (y_t represents monthly streamflow of Paria River at time t; and x_{t-1} and x_t represents monthly streamflow of Virgin River at time $t-1$ and t, respectively)

Therefore, the model established by the Gumbel copula with the Kendal tau parameter estimation method is used hereafter to generate monthly streamflow at sites B and C.

Using the established trivariate joint distributions, multisite monthly flows are simulated. The monthly streamflow at sites B is generated based on the spatial dependences between sites A and B. Similarly, monthly streamflow at site C is generated based on the spatial dependence between sites A and C. Since the values of u_1 and u_2 are already known, the conditional probability distribution $Q_1(u_1, u_2)$ is obtained. A uniform random number ε is generated, and $G(u_3|u_1, u_2)$ equal ε. Substituting the value of Q_1 and ε into Eq. 7.6, subsequent u_3 is derived by solving the inverse function of $Q_2(u_2, u_3)$. Finally, the generated streamflow data at gauging stations B and C is obtained from the inverse function $x_t = F^{-1}(u_3)$, respectively. The performance of the Gumbel copula is evaluated by comparing mean flow, standard deviation (S_d) and skewness of the generated data with those of the observed data. Boxplots of statistics (mean flow, standard deviation S_d and coefficient of skewness C_s) of the observed and simulated monthly streamflows for sites B and C are shown in Figs. 7.9 and 7.10, respectively, which indicate that all the simulations show good results, since these statistics fall within the boxplots for most of the months. The mean values of the statistics and *RE* are given in Table 7.4 which shows that for site B, the mean value of *RE* is less than 5% except for June. The mean value of *RE* in June is higher and reaches 12.17%. Since the discharge in the Little Colorado River is generated from snowmelt and rainfall, there is no snowmelt and is less rainfall in June, which leads to very small discharge. The observed mean monthly streamflow for June is 1,151 m^3/s. From 1906 to 2008, there are 57 years in which the discharge in June is zero, which influences the simulated results. The standard deviation and skewness of *RE* at site B is less than 9%. For site C, the performance of the proposed method is satisfactory. The *RE* value of the mean is less than 1%. The *RE* values of the standard deviation and skewness are less than 7%.

Boxplots of the Kendall tau correlation of the observed and simulated monthly streamflows for three sites A, B and C are shown in Fig. 7.11, in which (a), (b) and (c) show the correlations for consecutive months at sites A, B and C, and (d), (e) and

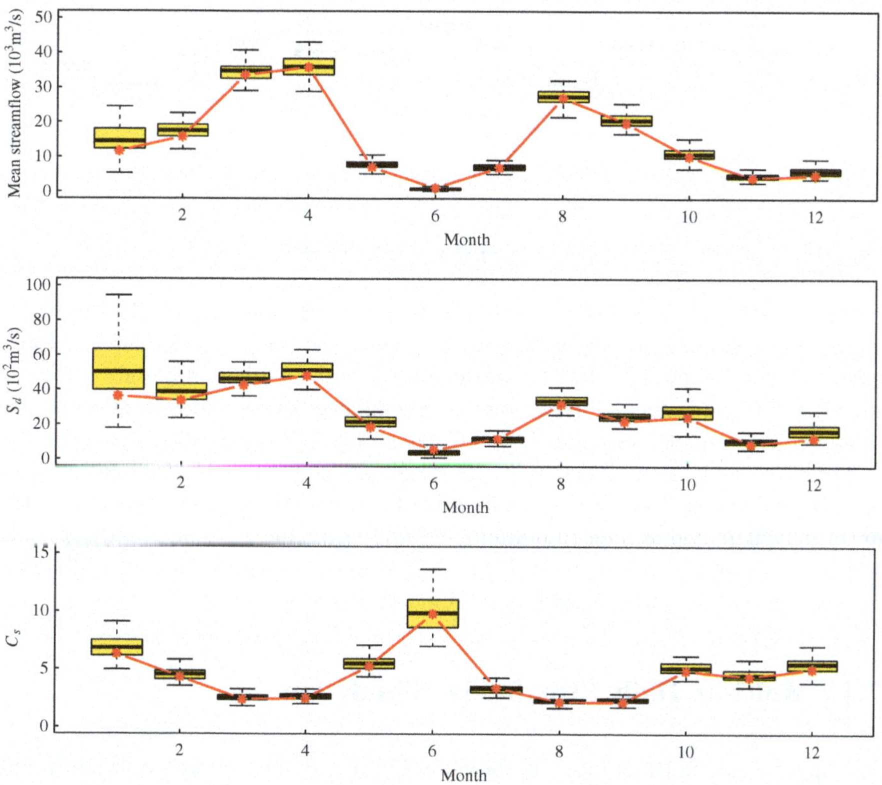

Fig. 7.9 Observed and simulated statistics of monthly streamflow at site B

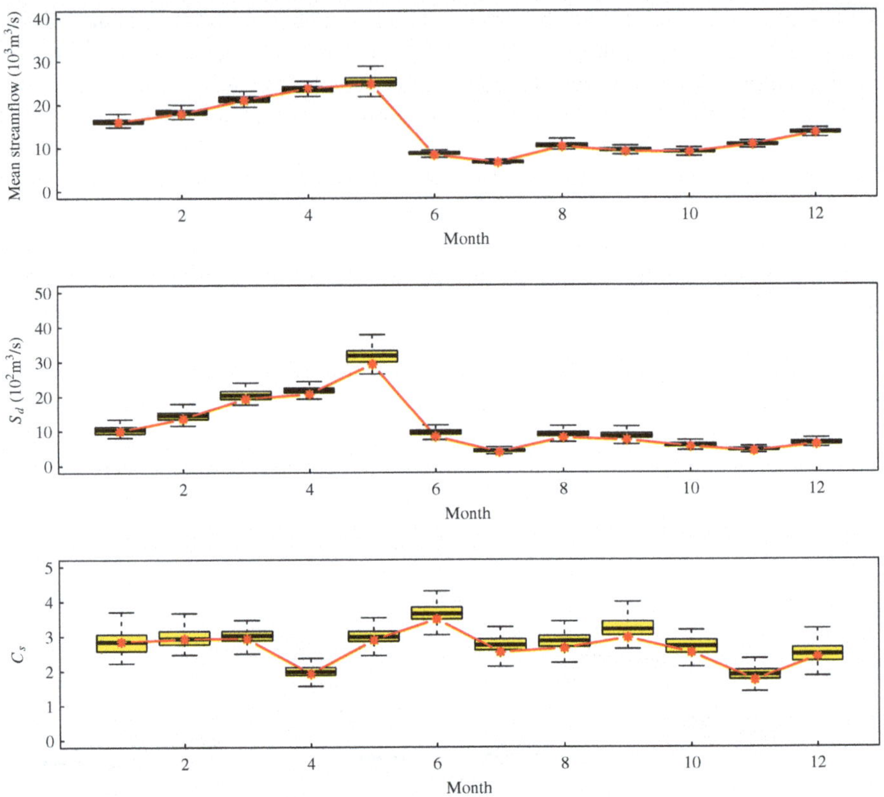

Fig. 7.10 Observed and simulated statistics of monthly streamflow at site C

(f) show the correlations for the same month at different sites. Figure 7.11 indicates that all these simulations show good results, since the observed Kendall tau correlation falls within the boxplots for most of the months. Although monthly flows at sites B and C are generated, based on the dependence between those and site A, the spatial correlation between sites B and C can still be preserved relatively well.

For flood and drought statistics, the minimum and maximum streamflows are usually considered. Monthly streamflow is often applied for drought analysis. The mean annual maximum and minimum streamflow of the generated and observed series are computed and shown in Table 7.5, which demonstrates that the difference between those two series regarding extreme events is not large.

7.4 Multisite Daily Streamflow Simulation

The upper Yangtze River basin as shown in Fig. 6.1 is selected as a case study. We denote the Yichang station on the Yangtze River as site A, the Pingshan station on the Jinsha River as site B, the Gaochang station on the Min River as site C, the Lijiawan station on the Tuo River as site D, the Beibei station on the Jialing River

Table 7.4 Values of statistics and relative errors (RE) of monthly streamflow at sites B and C generated by multisite stochastic simulation method based on the Gumbel copula

Sites	Items	Statistics	1	2	3	4	5	6	7	8	9	10	11	12
		Mean	11708	15804	33653	35892	7147	1151	7004	27064	19887	10316	4143	4939
	Observed	S_d	36401	33834	42621	48141	18748	5622	11926	31838	22182	24508	9098	12575
		C_S	6.29	4.33	2.40	2.48	5.28	9.81	3.38	2.19	2.16	4.89	4.38	5.07
		Mean	12109	15268	33119	35340	6931	1011	7041	27007	20072	10400	4262	5159
B	Simulated	S_d	38333	31593	41434	47556	18395	5250	11760	31747	22147	24878	9529	13587
		C_S	6.41	4.17	2.35	2.47	5.33	10.46	3.28	2.15	2.10	4.93	4.43	5.25
		Mean	3.43	3.40	1.58	1.54	3.03	12.17	0.53	0.21	0.93	0.82	2.88	4.46
	RE (%)	S_d	5.31	6.62	2.78	1.21	1.88	6.62	1.38	0.29	0.16	1.51	4.74	8.05
		C_S	1.81	3.69	2.13	0.43	1.02	6.57	2.87	1.77	2.75	0.75	1.33	3.51
		Mean	16121	18044	21157	23827	24881	8420	6667	10300	9140	8997	10719	13453
	Observed	S_d	10059	13591	19409	20755	29398	8396	3913	8053	7349	5413	4205	6087
		C_S	2.86	2.93	2.93	1.92	2.89	3.49	2.54	2.66	2.96	2.52	1.74	2.40
		Mean	16062	17898	21009	23911	24798	8424	6642	10285	9214	9016	10707	13393
C	Simulated	S_d	9814	12725	18418	20161	28892	8345	3676	7825	7372	5321	4109	5744
		C_S	2.78	2.75	2.79	1.86	2.85	3.46	2.36	2.55	2.89	2.42	1.63	2.23
		Mean	0.37	0.81	0.70	0.35	0.33	0.05	0.37	0.14	0.81	0.22	0.11	0.45
	RE (%)	S_d	2.43	6.37	5.11	2.86	1.72	0.61	6.05	2.83	0.31	1.69	2.29	5.64
		C_S	2.73	6.27	4.96	3.20	1.47	1.05	7.09	4.05	2.11	4.19	6.25	6.95

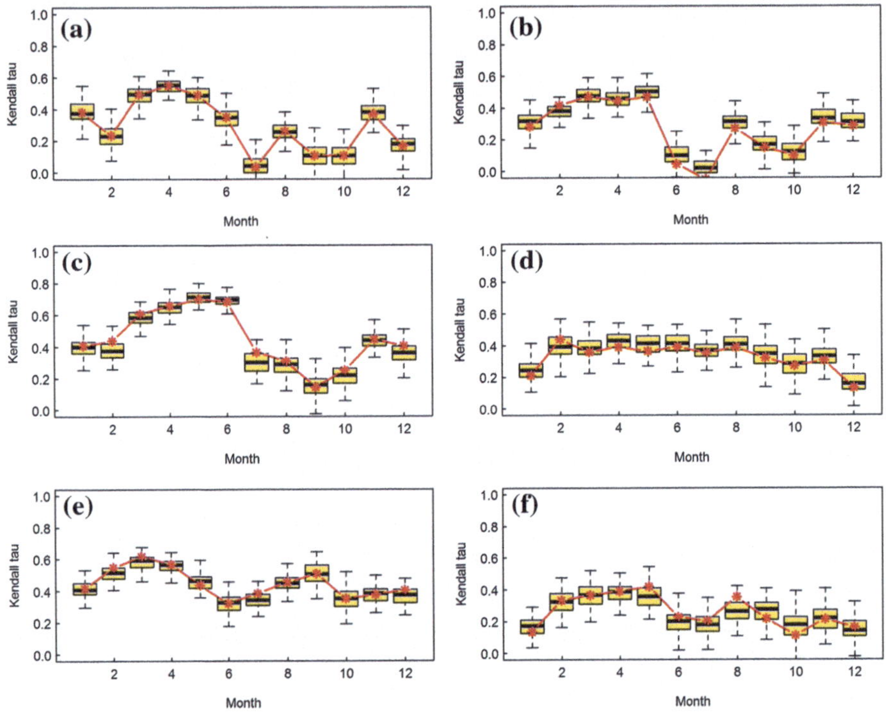

Fig. 7.11 Observed and simulated Kendall tau correlation of monthly streamflow for consecutive months at sites A (**a**), B (**b**) and C (**c**) and for the same month at different sites A–B (**d**), A–C (**e**) and B–C (**f**)

Table 7.5 Results of mean annual maximum and minimum monthly streamflow for observed and simulated series

Sites	Observed		Simulated	
	Maximum	Minimum	Maximum	Minimum
A	5948	326	5923	307
B	73,768	69	74,296	34
C	35,459	4730	34,970	4407

as site E, and the Wulong station on the Wu River as site F. The Yichang gauging station is on the mainstream and has a high dependence with the other gauging stations. In addition, since the flow generation process is very different among these tributaries, the dependences among sites B, C, D, E, and F are relative small, especially for the sites Beibei gauging station on the Jialing River and the Pingshan gauging station on the Jinsha River. The Kendall correlation coefficient between these two sites is only 0.05, which can be taken as independent. Therefore, daily steamflow at sites B, C, D, E, and F are generated based on the simulated daily flow data at sites A. In other words, the temporal dependences of single sites and spatial dependences between sites A and B (A–B), A and C (A–C), A and D (A–D), A and E (A–E) and A and F (A–F) are simulated using the trivariate copulas.

First, the marginal distributions of each day are constructed. The P-III distribution is fitted to the daily data. The L-moments method is used for estimating their parameters. Second, the daily streamflow at site A is simulated using the copula method proposed by Lee and Salas (2011) and applied by Hao and Singh (2013). Due to the satisfactory performance of the Gumbel copula for monthly streamflow simulation recommended by Lee and Salas (2011), the Gumbel copula is used to simulate the single-site daily flow data at site A. The basic statistics of observed and simulated data, including the mean value, standard deviation and skewness, are calculated and shown in Fig. 7.12. It can be seen that the mean value, standard deviation and skewness of the simulated data fit those of observed data well.

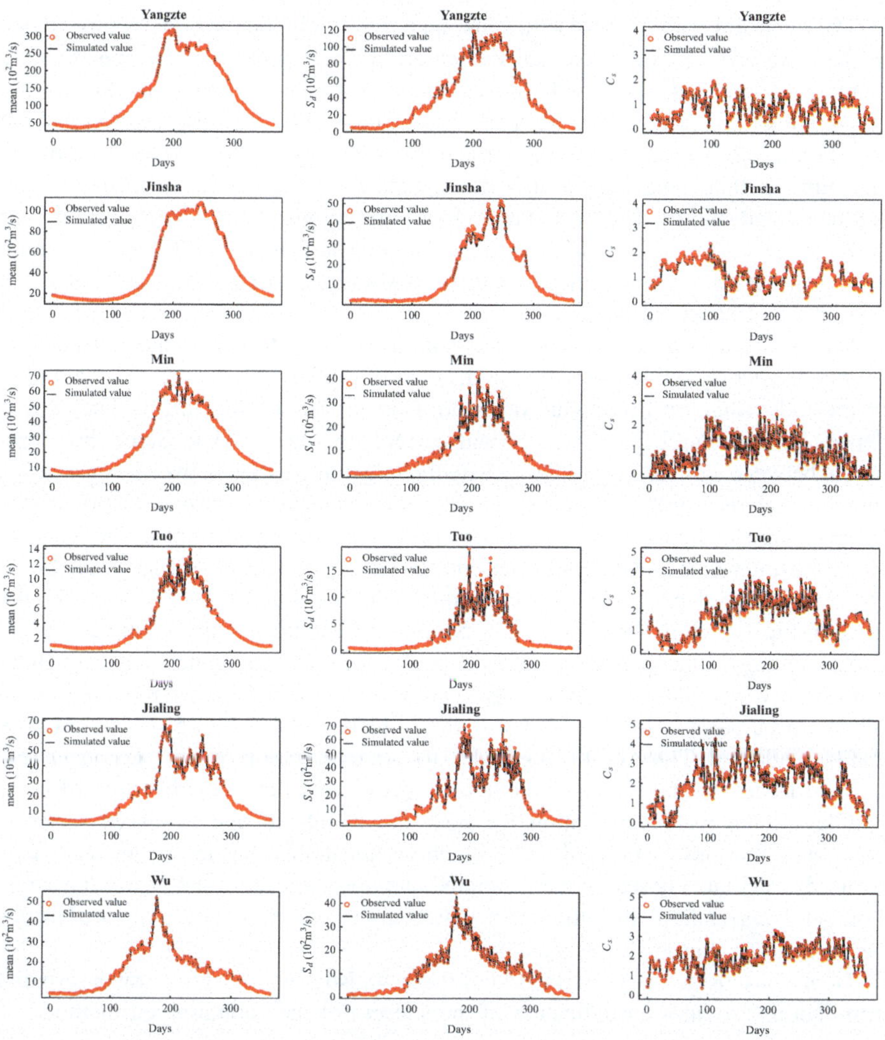

Fig. 7.12 Basic statistics of observed and generated data

For multi-site simulation, the same method is used to establish the trivairate joint distribution between sites A–B, A–C, A–D, A–E, and A–F. There are three bivariate copulas that need to be built for each day. If we use three kinds of copulas (Gumbel, Frank and Clayton) and select the best one among them, $3^3 \times 365$ copulas need to be built for simulating daily flows for the whole year. This will require a lot of calculation work, which makes this method impossible for practical applications. According to the calculated results of monthly streamflow simulation, the Gumbel copula with the Kendall parameter estimation method performs well. Therefore, the Gumbel copula model is used hereafter for multi-site daily streamflow simulation. The simulated results of basic statistics at sites B, C, D, E and F are shown in Fig. 7.12, which indicate that the mean value, standard deviation and skewness of the simulated data fit those of observed data well.

The temporal correlation coefficients, namely correlations for consecutive days, of the observed and simulated daily streamflow from June to September for five gauging stations B, C, D, E and F are calculated and drawn in boxplots. Results are given in Fig. 7.13, which indicate a good agreement, since the observed Kendall tau correlation falls within the boxplots for most of the days. The performance of the algorithm is then assessed through the spatial correlation between observed and simulated daily flows. The mean lag-0 cross correlations of the whole year for sites A–B, A–C, A–D, A–E, and A–F are calculated and listed in Table 7.6, which indicates that the cross correlation values of simulated streamflows are consistent with those of observed flows.

We have discussed the lag-1 correlation above. However, the correlations beyond lag-1 may influence modeling the overall stochastic structure of streamflow. In the following, we investigate other autocorrelations besides lag-1. The dependences between lag-t (t = 1, …, 5) and current flow are studied. Since the correlation between consecutive days is usually larger than that between consecutive months, daily streamflow data is employed. We calculate the mean autocorrelations between lag-t and current flow for each of the five gauging stations in the Yangtze River Basin, which indicate that the correlation coefficient is still large till lag-5. However, these dependencies do not remove the impact of lag-1. To describe the relationship between lag-t (t = 2, …, 5) and current flow while taking away the effects of lag-1, the partial correlation coefficients are calculated and shown in Table 7.7, which indicate that the partial correlation coefficients are relative low. In other words, compared with the lag-1, the influence of lag-2 to 5 on current flow can be neglected. Thus, in this method, only autocorrelation of lag-1 is considered.

In order to test the proposed method for preserving the autocorrelation of lag-1 and lag-2, the mean correlation for both the simulated and observed series is computed and shown in Table 7.8, which demonstrates that the mean lag-1 correlations calculated based on the generated and observed series are nearly the same. For lag-2 correlation, the difference between the observed and generated series becomes large, but is still acceptable.

Hence, a satisfactory reproduction of these statistics and the temporal and spatial correlations provides an indication of the success of the copula-based method.

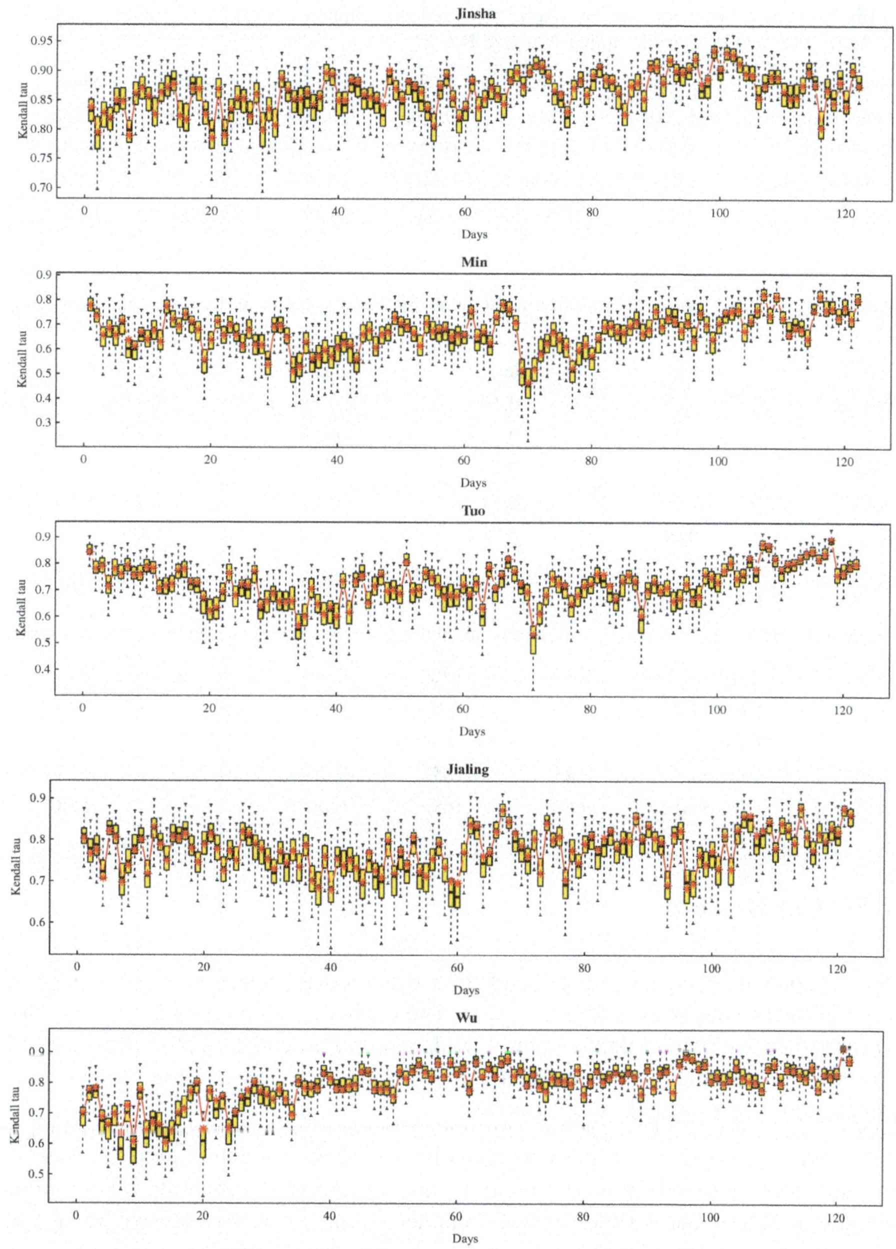

Fig. 7.13 Temporal dependences between consecutive days for gauging stations B, C, D, E and F in the flood season (from June to September)

Table 7.6 Lag-0 cross correlations among observed and simulated streamflows for sites A–B, A–C, A–D, A–E and A–F in the upper Yangtze River

Items	A–B	A–C	A–D	A–E	A–F	Mean
Observed	0.319	0.311	0.262	0.409	0.389	0.338
Simulated	0.315	0.317	0.258	0.403	0.378	0.334
Absolute error	0.004	0.006	0.004	0.006	0.011	0.004
RE (%)	1.25	1.93	1.53	1.47	2.83	1.18

Table 7.7 Results of partial correlation coefficients for the five rivers in the upper Yangtze River Basin

Rivers	lag-2	lag-3	lag-4	lag-5
Jinsha	0.08	0.10	0.11	0.11
Min	0.11	0.13	0.13	0.13
Tuo	0.12	0.14	0.14	0.14
Jialing	0.10	0.11	0.11	0.11
Wu	0.06	0.08	0.08	0.09

Table 7.8 Mean lag-1 and lag-2 correlation calculated based on the simulated and observed series

Lags	Types	Jinsha	Min	Tuo	Jialing	Wu
Lag-1	Generated	0.90	0.69	0.78	0.82	0.81
	Observed	0.90	0.69	0.79	0.82	0.81
Lag-2	Generated	0.85	0.57	0.70	0.75	0.73
	Observed	0.83	0.53	0.66	0.71	0.68

7.5 Conclusion

This chapter introduces a simple and robust method for space-time simulation of streamflow of both small and large rivers. The copula-based method can be used for generating monthly and daily streamflow data at multiple sites. Three tributaries of Colorado River and the upper Yangtze River are selected as case studies. The conclusions are summarized as follows:

(1) Trivariate copulas are used to describe the temporal and spatial correlation structure of streamflow. The bivariate and conditional probability distributions are used to construct multivariate copulas. Comparison between empirical and theoretical joint probabilities shows no significant differences for the case studies used in this paper. Therefore, the copulas generally can be used for multisite stochastic simulation. For simulating streamflow at multiple sites, trivariate copulas are constructed, although the method can be used for higher dimensions. Streamflow at sites B, C and so on are generated, based on the

streamflow at site A. This avoids errors caused by establishing the complicated models and reduces the amount of calculation.

(2) The copula-based method is applied for simulating multisite monthly and daily streamflow. Statistical attributes, such as mean, standard deviation, skewness, lag-1 correlation and lag-0 cross correlation, are effectively reproduced, which show that the generated data at both higher and lower time scales capture the distribution properties of the single site and preserve the spatial correlation of streamflow at different locations.

(3) The main advantage of the copula-based method is that parameters of the models can be easily estimated based on the Kendall tau method. This makes it possible to generate daily streamflow data because 365 trivariate copulas need to be built for multisite daily flow generation. Furthermore, this method can preserve linear and non-linear correlation structures and can be used for any marginal distribution. Compared with other multisite stochastic simulation methods, the algorithm is simple, which makes it possible for practical use.

References

Bárdossy A, Pegram GGS (2009) Copula based multisite model for daily precipitation. Hydrol Earth Syst Sci 13:2299–2314

Chen L, Singh VP, Guo S, Zhou J (2015) Copula-based method for multisite monthly and daily streamflow simulation. J Hydrol 528:369–384

Ebuh GU, Oyeka ICA (2012) A nonparametric method for estimating partial correlation coefficient. J Biom Biostat 3:156. https://doi.org/10.4172/2155-6180.1000156

Hao Z, Singh VP (2013) Modeling multisite streamflow dependence with maximum entropy copula. Water Resour Res 49:7139–7143. https://doi.org/10.1002/wrcr.20523

Joe H (1997) Multivariate models and dependence concepts. Chapman and Hall, London

Kendall M (1948) Rank correlation methods. Griffin, Oxford, England

Koutsoyiannis D (2001) Coupling stochastic models of different timescales. Water Resour Res 37 (2):379–391

Kumar D, Lall NU, Petersen MR (2000) Multisite disaggregation of monthly to daily streamflow. Water Resour Res 36(7):1823–1833. https://doi.org/10.1029/2000WR900049

Lall UB, Rajagopalan B, Tarboton DG (1996) A nonparametric wet/dry spell model for resampling daily precipitation. Water Resour Res 329:2803–2823

Lane WL (1979) Applied stochastic techniques, Users Manual Eng. and Res. Center, Bur of Reclam, Denver, Colorado, U.S

Lee T, Salas J (2011) Copula-based stochastic simulation of hydrological data applied to Nile River flows. Hydrol Res 42(4):318–330

Lee T, Salas J (2008) Using copulas for stochastic streamflow generation. World Environmental and Water Resources Congress 2008:1–10. https://doi.org/10.1061/40976(316)572

Mejia JM, Rousselle J (1976) Disaggregation models in hydrology revisited. Water Resour Res 122:185–186

Nelsen RB (2006) An introduction to copulas, 2nd edn. Springer-Verlag, New York

Prairie J, Rajagopalan B, Lall U, Fulp T (2007) A stochastic nonparametric technique for space-time disaggregation of streamflows. Water Resour Res 43:W03432. https://doi.org/10.1029/2005WR004721

Salas JD, Delleur JW, Yevjevich V, Lane WL (1980) Applied modeling of hydrologic time series. Water Resour Publ, Highlands Ranch, Colo

Sharma A, O'Neill R (2002) A nonparametric approach for representing interannual dependence in monthly streamflow sequences. Water Resour Res 38(7):1100. https://doi.org/10.1029/2001WR000953

Sharma A, Tarboton DG, Lall U (1997) Streamflow simulation: A nonparametric approach. Water Resour Res 33(2):291–308. https://doi.org/10.1029/1096WR02839

Sidney S (1956) Nonparametric statistics for the behavioral sciences. McGraw-Hill Series in Psychology, New York

Srinivas VV, Srinivasan K (2005) Hybrid moving block bootstrap for stochastic simulation of multi-site multi-season streamflows. J Hydrol 302:307–330

Stedinger JR, Vogel RM (1984) Disaggregation procedures for generating serially correlated flow vectors. Water Resour Res 201:47–56

Stedinger JR, Pei D, Cohn TA (1985) A condensed disaggregation model for incorporating parameter uncertainty into monthly reservoir simulations. Water Resour Res 215:665–675

Szilagyi J, Balint G, Csik A (2006) Hybrid, markov chain-based model for daily streamflow generation at multiple catchment sites. J Hydrol Eng 11(7):245–256

Tarboton DG, Sharma A, Lall U (1998) Disaggregation procedures for stochastic hydrology based on nonparametric density estimation. Water Resour Res 34(1):107–119. https://doi.org/10.1029/97WR02429

Valencia DR, Schaake JL Jr (1973) Disaggregation processes in stochastic hydrology. Water Resour Res 93:580–585

Xu Z, Schumann A, Li J (2003) Markov cross-correlation pulse model for daily streamflow generation at multiple sites. Adv Water Resour 26(3):325–335

Chapter 8
Uncertainty Analysis of Hydrologic Forecasts Based on Copulas

8.1 Introduction

Hydrologic forecasting is a crucial non-structural flood mitigation measure and provides an essential basis for flood warning, flood control and reservoir operation (Guo et al. 2004; Calvo and Savi 2009; Chen et al. 2014a; Zhang et al. 2015; Fan et al. 2016; Liu et al. 2017; Wu et al. 2017). Forecasting models that are widely used at present are typically deterministic, and model outputs are provided to users in the form of deterministic values (Chen and Yu 2007; Coccia and Todini 2011; Ma et al. 2013; Bergstrand et al. 2014; Li et al. 2014). However, a hydrological forecasting model is only a simulation of the real hydrological processes and is therefore imperfect and not precise (Ravines et al. 2008; Wetterhall et al. 2013). These models accept hydrological input, meteorological input, etc., and utilize conceptualized model parameters; and these complex factors cause inevitably uncertainties in the hydrologic forecasts (Freer et al. 1996; Montanari 2007; Montanari and Grossi 2008; Renard et al. 2010; Chen et al. 2014b). The principle of rational decision-making under uncertainty indicates that when a deterministic forecast turns out to be wrong, the consequences will probably be worse than a situation where no forecast is available (Krzysztofowicz 1999; Wetterhall et al. 2013; Ramos et al. 2013). A rational decision maker who wants to make optimal decisions should therefore take forecast uncertainty explicitly into account (Verkade and Werner 2011; Ramos et al. 2013). Therefore, quantitative assessment of inherent uncertainty is a critical issue. Hydrologic forecasting services are trending toward providing users with probabilistic forecasts, in place of traditional deterministic forecasts.

The transition from a deterministic forecast to a probabilistic forecast is based on quantification of the uncertainty inherent in the deterministic forecast. The Bayesian Forecasting System (BFS) proposed by Krzysztofowicz (1999) provides a general framework to produce probabilistic forecasts via any deterministic hydrological model. Various probabilistic forecasting systems suited to different purposes have

L. Chen and S. Guo, *Copulas and Its Application in Hydrology and Water Resources*, Springer Water, https://doi.org/10.1007/978-981-13-0574-0_8

been developed within this framework (Reggiani and Weerts 2008; Calvo and Savi 2009; Biondi et al. 2010; Weerts et al. 2011; Sikorska et al. 2012; Pokhrel et al. 2013).

In the BFS, the total uncertainty is decomposed into input uncertainty and hydrological uncertainty. The hydrological uncertainty processor (HUP) is a component of the BFS that quantifies the hydrological uncertainty and produces probabilistic forecast under the hypothesis that there is no input uncertainty (Krzysztofowicz and Kelly 2000). Through Bayes' theorem, the HUP combines a prior distribution, which describes the natural uncertainty about the realization of a hydrologic process, with a likelihood function which quantifies the uncertainty in model forecasts, and outputs a posterior distribution, conditional upon the deterministic forecasts. This posterior distribution provides a complete characterization of uncertainty, including quantiles, prediction intervals and probabilities of exceedance for specified thresholds which are needed by rational decision makers and information providers who want to extract forecast products for their customers.

The HUP can be implemented in many ways, as different mathematical models for prior distribution and likelihood function can be developed. Krzysztofowicz and Kelly (2000) introduced a meta-Gaussian HUP, which was developed by converting both original observations and model forecasts into a Gaussian space by using the Normal Quantile Transform (NQT). This meta-Gaussian HUP has been widely used by many researchers in the fields of hydrology and meteorology (Chen and Yu 2007; Biondi et al. 2010; Biondi and De Luca 2013; Chen et al. 2013a).

The prior density and likelihood function are conditional probability distributions. It is well known that copula function has an outstanding capability to model joint distributions and gives flexibility in choosing an arbitrary marginal distribution (e.g. non-Gaussian form), nonlinear and heteroscedastic dependence structure. The conditional probability distribution can be expressed in the explicit form using copula function (Favre et al. 2004; Nelsen 2006; Zhang and Singh 2006, 2007a, b, c; Genest and Favre 2007; Bárdossy and Li 2008; Chen et al. 2010; Zhang et al. 2011, 2012). These advantageous characteristics of the copula function motivate us to develop the prior distribution and likelihood function models in the original space directly without a data transformation procedure into Gaussian space. Liu et al. (2017) proposed a post-processor based on copula function for deterministic forecast model to produce probabilistic forecasts within the general framework of the HUP.

Despite the tremendous amount of resources invested in developing more hydrologic models, no one can convincingly claim that any particular model in existence today is superior to other models for all type of applications and under all conditions (Wu et al. 2015; Liu et al. 2016; Ba et al. 2017). Different models are capable of capturing different aspects of the hydrologic processes. The uncertainty of each model arises from parameters calibration, the design of the model structure, and input measurements, which partially brings underlying imprecise influence (Götzinger and 2008; Li et al. 2011; Hemri et al. 2015). One of the primary techniques to reflect different uncertainties in hydrological forecasts is to create an ensemble of forecast trajectories (Seo et al. 2006; Madadgar et al. 2014).

The Bayesian Model Average (BMA) method introduced by Raftery et al. (2005) follows a statistical technique to combine the advantages of different models. Different from other multi-model methods, the BMA method presents a more realistic description of predictive uncertainty, since the BMA predictive variance can be decomposed into two components: between-model variability and within–model variability (Ajami et al. 2007). The BMA method is a statistical procedure that infers consensus predictions by weighing individual predictions based on their probabilistic likelihood measures, with the better performing predictions receiving higher weights than, the worse performing ones. The method has been explored to improve both the accuracy and reliability of streamflow predictions (Vrugt and Robinson 2007; Liang et al. 2011). Duan et al. (2007) concluded that the combination of multi-model ensemble strategies using the BMA framework could quantify statements on prediction uncertainty and significantly improve verification performances. Zhou et al. (2016) compared the mean prediction of BMA with its individual parameter transfer method (physical similarity approach) and demonstrated that the probabilistic predictions of BMA could reduce the uncertainty with a significant degree. Nevertheless, the standard BMA method imposes lots of pseudo variation requirements, and this influences precise understanding of data variations, which gives rise to further development of this theoretical research (Madadgar and Moradkhani 2014).

Klein et al. (2016) used a mixture of marginal density distribution to estimate the predictive uncertainty of hydrologic multi-model ensembles by using pair-copula construction. Similar researches show that copula technique is an effective tool for reflecting the unclear and complex relationships because it can flexibly choose the arbitrary type of the marginal distributions instead of Gaussian distribution (Carreau and Bouvier 2016; Khajehei and Moradkhani 2017). According to the promising results of using copula functions in post-processing of hydrologic forecasts, Madadgar and Moradkhani (2014) firstly integrated copula functions with BMA to estimate the posterior distribution and found that Copula-BMA (CBMA) is an effective post-processor to relax any assumption on the distribution of conditional probability density function (PDF). The CBMA not only displayed better deterministic skill than BMA but also confirmed the impact of posterior distribution in calculating the weights of individual models by EM algorithm. Results indicated that the predictive distributions are more accurate and reliable. It is also shown that the post-processed forecasts have better correlation with observation after CBMA application. The CBMA method in the meteorological application does not need to assume the shape of the posterior distribution and leaves out the data-transformation procedure and demonstrates that predictive distributions are less bias and more confident with small uncertainty (Möller et al. 2013). Inspired by the ideas of Madadgar and Moradkhani (2014), a general framework of the combination of copula Bayesian processor with BMA (CBP-BMA) is proposed by He et al. (2018).

8.2 Hydrologic Uncertainty Processor Based on Copula Function

8.2.1 Hydrologic Uncertainty Processor

Let predict and H be the observed discharge whose realization h is being forecasted. Let estimator S be the output discharge generated by a corresponding deterministic forecast model whose realization s constitutes a point estimate of H. Let random variable H_0 represent the observed discharge at the time $n = 0$ when the forecast is prepared; then H_n ($n = 1, 2, \ldots, N$) is the observed discharge at lead time n; and S_n ($n = 1, 2, \ldots, N$) is the corresponding deterministic forecast discharge at lead time n. What the rational decision maker then needs is not a single number s_n, but the distribution function of predictand H_n, conditional on $H_0 = h_0$ and $S_n = s_n$. The purpose of the HUP is to supply such a conditional distribution function through Bayesian revision (Liu et al. 2017).

The Bayesian procedure for information revision of uncertainty involves two steps. First, the expected conditional density function of deterministic forecast discharge S_n given that $H_0 = h_0$ is derived via the total probability law:

$$\kappa_n(s_n|h_0) = \int_{-\infty}^{+\infty} f_n(s_n|h_0, h_n) \cdot g(h_n|h_0)dh_n \tag{8.1}$$

Second, the posterior density function of predictand H_n conditional on a deterministic forecast $S_n = s_n$ and observed discharge at the forecasting time $H_0 = h_0$, is derived via Bayes' theorem (Krzysztofowicz and Kelly 2000):

$$\phi_n(h_n|h_0, s_n) = \frac{f_n(s_n|h_0, h_n) \cdot g_n(h_n|h_0)}{\kappa_n(s_n|h_0)} \tag{8.2}$$

In concept, Bayes' theorem revises the prior density function $g_n(h_n/h_0)$, which characterizes the prior uncertainty about H_n, given observed discharge at the forecasting time $H_0 = h_0$ The extent of the revision is determined by the likelihood function $f_n(s_n/h_0, h_n)$, which characterizes the degree to which $S_n = s_n$ reduces the uncertainty about H_n. The result of this revision is the posterior density function $\Phi_n(h_n/h_0, s_n)$, which quantifies the uncertainty about H_n that remains after the deterministic forecast model generates forecast $S_n = s_n$.

8.2.2 Meta-Gaussian HUP

As we can see from Eq. 8.2, the posterior density depends on the prior density function and likelihood function. The most widely used technique to describe the prior density and likelihood functions is the meta Gaussian model. In this model, the NQT method (Bogner et al. 2012) is applied to convert both actual flow H_n and predicted flow S_n into the Gaussian space. Then the transformed $\{h_n/h_0\}$ and $\{s_n/h_0, h_n\}$ are assumed to be linear and normally distributed. Subsequently, linear regression method is then employed to determine the posterior density of H_n in the transformed Gaussian space, from which the posterior density function of H_n in the original space can be found. For the sake of the following comparison, the detailed procedure is presented as follows (Krzysztofowicz and Kelly 2000).

8.2.2.1 Normal Quantile Transform

Specifying and determining marginal distributions of the actual flow H_0, H_n and predicted flow $\{S_n: n = 1, \ldots, N\}$ is the first step. The actual flows $\{H_n: n = 0, 1, \ldots, N\}$ are considered as random variables. Given only such a record, there is usually no basis for assigning a probability distribution to flow H_n that differs from the distribution assigned to flow H_0, for any $n = 1, 2, \ldots, N$ within a few days. In other words, there is no a statistical difference between these $1 + N$ flow series (Koutsoyiannis and Montanari 2015). Therefore, we hold the opinion that the variables H_n follow the same marginal cumulative distribution functions (CDF) with H_0, and thus only the CDF of H_0 needed to be fitted. The predicted flows $\{S_n: n = 1, \ldots, N\}$ are considered as different random variables and different CDFs needed to be fitted for variable S_n.

Let Γ and $\overline{\Lambda}_n$ be the CDF of H_0 and S_n with corresponding densities γ and $\bar{\lambda}_n$, respectively. The NQT of a variate is defined as a composition of the inverse of the standard normal distribution Q, and the CDF of the variate is assumed to be strictly increasing. The transformed variates are

$$W_n = Q^{-1}[\Gamma(H_n)], \quad n = 0, 1, \ldots, N \tag{8.3}$$

$$X_n = Q^{-1}[\overline{\Lambda}_n(S_n)], \quad n = 0, 1, \ldots, N \tag{8.4}$$

where W_n and X_n are the normal quantiles of H_n and S_n, respectively. Q^{-1} is the inverse function of Q.

8.2.2.2 Modeling in the Transformed Space

(1) Prior density

The model for the prior density rests on the assumption that the actual river discharge process in the transformed space is governed by the normal-linear equation

$$W_n = cW_{n-1} + \Xi \tag{8.5}$$

where c is a parameter and Ξ is a variate stochastically independent of W_{n-1} and normally distributed with mean zero and variance $1 - c^2$. Consequently, the conditional mean and variance are

$$E(W_n | W_{n-1} = w_{n-1}) = cw_{n-1} \tag{8.6}$$

$$Var(W_n | W_{n-1} = w_{n-1}) = 1 - c^2 \tag{8.7}$$

The prior density for lead time n takes the form

$$g_{Q_n}(w_n | w_0) = \frac{1}{(1 - c^{2n})^{1/2}} \cdot q\left[\frac{w_n - c^n w_0}{(1 - c^{2n})^{1/2}}\right] \tag{8.8}$$

where q denotes the standard normal density and subscript Q_n denotes a density in the space of transformed variants.

(2) Likelihood function

The model for the likelihood function rests on the assumption that the stochastic dependence between the transformed variate is governed by the normal-linear equation

$$X_n = a_n W_n + d_n W_0 + b_n \Theta_n \tag{8.9}$$

where a_n, b_n and d_n are parameters and Θ_n is a stochastically independent variate of (W_n, W_0) and normally distributed with mean zero and variance δ_n^2. Consequently, the conditional mean and variance are

$$E(X_n | W_n = w_n, W_0 = w_0) = a_n w_n + d_n w_0 + b_n \tag{8.10}$$

$$Var(X_n | W_n = w_n, W_0 = w_0) = \delta_n^2 \tag{8.11}$$

The conditional density function is

$$f_{Q_n}(x_n | w_n, w_0) = \frac{1}{\delta_n} \cdot q\left(\frac{x_n - a_n w_n - d_n w_0 - b_n}{\delta_n}\right) \tag{8.12}$$

(3) Posterior density

The posterior density derived from the prior density and likelihood function takes the form as follows

$$\varphi_{Q_n}(w_n|x_n, w_0) = \frac{1}{T_n} q\left(\frac{w_n - A_n x_n - D_n w_0 - B_n}{T_n}\right) \tag{8.13}$$

In which $A_n = \frac{a_n t_n^2}{a_n^2 t_n^2 + \delta_n^2}$, $B_n = \frac{-a_n b_n t_n^2}{a_n^2 t_n^2 + \delta_n^2}$, $D_n = \frac{c_n \delta_n^2 - a_n d_n t_n^2}{a_n^2 t_n^2 + \delta_n^2}$, $T_n^2 = \frac{t_n^2 \delta_n^2}{a_n^2 t_n^2 + \delta_n^2}$, and $t_n^2 = 1 - c^{2n}$.

8.2.2.3 Posterior Density and Distribution in the Original Space

With all densities in the transformed space belonging to the Gaussian family, all densities in the original space belong to the meta-Gaussian family. The meta-Gaussian posterior density of actual river discharge conditional on model output discharge $S_0 = s_0$ and observed river discharge $H_0 = h_0$ takes the form

$$\phi_n(h_n|s_n, h_0) = \frac{\gamma(h_n)}{T_n \cdot q\{Q^{-1}[\Gamma(h_n)]\}} \cdot q\left\{\frac{Q^{-1}[\Gamma(h_n)] - A_n \cdot Q^{-1}[\overline{\Lambda}_n(s_n)] - D_n \cdot Q^{-1}[\Gamma(h_0)] - B_n}{T_n}\right\} \tag{8.14}$$

The corresponding meta-Gaussian posterior distribution takes the form

$$\Phi_n(h_n|s_n, h_0) = Q\left\{\frac{Q^{-1}[\Gamma(h_n)] - A_n \cdot Q^{-1}[\overline{\Lambda}_n(s_n)] - D_n \cdot Q^{-1}[\Gamma(h_0)] - B_n}{T_n}\right\} \tag{8.15}$$

8.2.3 Copula-Based HUP

Copula function is an effective tool used to develop prior distribution and likelihood function models, in which the predictand and the deterministic forecasts are allowed to have distribution functions of any form, along with nonlinear and heteroscedastic dependence structure. Therefore, it can be implemented in the original space directly without a data transformation procedure into Gaussian space. The copula function and theory have been introduced in detail in Chap. 2.

8.2.3.1 Prior Density

The prior CDF of H_n given $H_0 = h_0$ can be expressed as

$$G_n(h_n|h_0) = P(H_n \leq h_n|H_0 = h_0) \tag{8.16}$$

where $G_n(h_n/h_0)$ is the conditional CDF, and P is the non-exceedance probability.

The prior density function $g_n(h_n/h_0)$ is the corresponding probability density function (PDF) of $g_n(h_n/h_0)$ and can be defined as

$$g_n(h_n|h_0) = \frac{dG_n(h_n|h_0)}{dh_n} \tag{8.17}$$

Let H_0 and H_n be random variables with marginal CDFs, $U_1 = F_H(H_0)$ and $U_2 = F_H(H_n)$. Then, U_1 and U_2 are uniformly distributed random variables; and u_1 denotes a specific value of U_1, and u_2 denotes a specific value of U_2.

Using the copula function, the joint CDF is expressed by $G_n(h_n, h_0) = C(F_H(h_0), F_H(h_n)) = C(u_1, u_2)$

The conditional CDF $G_n(h_n/h_0)$ and PDF $g_n(h_n/h_0)$ can be rewritten as follows (Zhang and Singh 2006)

$$G_n(h_n|h_0) = P(U_2 \leq u_2|U_1 = u_1) = \frac{\partial C(u_1, u_2)}{\partial u_1} \tag{8.18}$$

$$g_n(h_n|h_0) = \frac{\partial^2 C(u_1, u_2)}{\partial u_1 \partial u_2} \cdot \frac{du_2}{\partial h_n} = c(u_1, u_2) \cdot f_H(h_n) \tag{8.19}$$

where $c(u_1, u_2)$ is the density function of $C(u_1, u_2)$, and $c(u_1, u_2) = \partial^2 C(u_1, u_2)/\partial u_1 \partial u_2$; $f_H(h_n)$ is the PDF of H_n. Equation 8.19 is the expression of the prior PDF.

8.2.3.2 Likelihood Function

It is considered that S_n is a random variable with marginal CDF $u_3 = F_{S_n}(s_n)$ and PDF $f_{S_n}(s_n)$. The conditional CDF of S_n given $H_0 = h_0$ and $H_n = h_n$ can be expressed as

$$F_n(s_n|h_0, h_n) = P(S_n \leq s_n|H_0 = h_0, H_n = h_n) \tag{8.20}$$

where $F_n(s_n|h_0, h_n)$ is the conditional CDF.

The corresponding PDF of $F_n(s_n|h_0, h_n)$ is defined as

$$f_n(s_n|h_0, h_n) = \frac{dF_n(s_n|h_0, h_n)}{ds_n} \tag{8.21}$$

Using the copula function, the joint CDFs of H_0, H_n and S_n, denoted as $F_n(h_0, h_n, s_n)$ can be expressed as $F_k(h_0, h_n, s_n) = C(F_{H_0}(h_0), F_{H_n}(h_n), F_{S_n}(s_n)) =$

$C(u_1, u_2, u_3)$. Thus, the conditional CDF $F_n(s_n|h_0, h_n)$ and PDF $f_n(s_n|h_0, h_n)$ are rewritten as follows, (Zhang and Singh 2007c)

$$F_n(s_n|h_0, h_n) = P(U_3 \le u_3|U_1 = u_1, U_2 = u_2) = \frac{\partial^2 C(u_1, u_2, u_3)/\partial u_1 \partial u_2}{c(u_1, u_2)} \tag{8.22}$$

$$f_n(s_n|h_0, h_n) = \frac{1}{c(u_1, u_2)} \cdot \frac{\partial^3 C(u_1, u_2, u_3)}{\partial u_1 \partial u_2 \partial u_3} \cdot \frac{du_3}{ds_n} = \frac{c(u_1, u_2, u_3)}{c(u_1, u_2)} \cdot f_{S_n}(s_n) \tag{8.23}$$

where $c(u_1, u_2, u_3) = \partial^3 C(u_1, u_2, u_3)/\partial u_1 \partial u_2 \partial u_3$ is the density function of $C(u_1, u_2, u_3)$. From another point of view, given $H_0 = h_0$ and $S_0 = s_0$, the likelihood function of H_n can be calculated by Eq. 8.23.

8.2.3.3 Posterior Density

Substitute Eqs. 8.19 and 8.23 to Eqs. 8.1 and 8.2, then the posterior density function of H_n can be rewritten as follows

$$\phi_n(h_n|h_0, s_n) = \frac{c(u_1, u_2, u_3)}{\int_0^1 c(u_1, u_2, u_3) du_2} \cdot f_H(h_n) \tag{8.24}$$

For fixed realizations $H_0 = h_0$ and $S_n = s_n$, u_1 and u_3 are constants, while u_2 varies from 0 to 1. Since the denominator $\int_0^1 c(u_1, u_2, u_3) du_2$ cannot be obtained directly by an analytic method, the Monte Carlo sampling technique (Yu et al. 2014; Xiong et al. 2014) is applied by following steps: (1) Generate M random numbers u_2 from uniform distribution **U**(0, 1); (2) Compute the value of $C(u_1, u_2, u_3)$; (3) The mean value of the M calculated $C(u_1, u_2, u_3)$ equals to the definite integral $\int_0^1 c(u_1, u_2, u_3) du_2$ approximately (Robert and Casella 2013; Kroese et al. 2013). Subsequently, the posterior density function $\phi_n(h_n|h_0, s_n)$ can also be estimated.

8.2.3.4 Candidate Marginal Distributions and Trivariate Copulas

The main purpose of this study aims to extrapolate the extreme events far beyond the observations. The probability distribution of daily flows refers to the flow duration curve, which gives a summary of flow variability at a site and is interpreted as a relationship between any discharge value and the percentage of time that this discharge is equaled or exceeded during a given period (Vogel and Fennessey 1994; Castellarin et al. 2004; Shao et al. 2009). Flow-duration curve has been widely used by engineers and hydrologists around the world in numerous applications, such as hydropower generation, inflow forecasting, and designing of irrigation systems (Vogel and Fennessey 1995; Yokoo and Sivapalan 2011; Gottschalk et al. 2013).

Even though flow-duration curve can be defined and constructed for different time scales, such as daily, weekly or monthly stream flows, our study will focus on a daily flow-duration curve. If the daily streamflow is assumed to be a random variable, the flow-duration curve may also be viewed as the complement of the cumulative distribution function used in hydrologic frequency analysis when identifying the percentage of time with probability (Castellarin et al. 2004). As a consequence, the flow-duration curve is also a very practical tool used to describe hydrological regimes and represents the relationship between magnitude and frequency of flow (Vogel and Fennessey 1995; Liucci et al. 2014; Xiong et al. 2015).

Six commonly used distributions in hydrology, namely Normal, GMA, Gumbel, P-III, Log-Normal and Log-Weibull, are selected as candidate models for H_0 and S_n ($n = 1, \ldots, N$). These univariate probability distributions are summarized in Table 1.1 of Chap. 1. L-moment method is used to estimate the distribution parameters for given data series (Hosking 1990). The Kolmogorov-Smirnov statistic D is used to measure the goodness of fit between the hypothesized distribution and the empirical distribution (Tsai et al. 2001; Arya et al. 2010). In this study, the 95% confidence level is selected to reject or accept a fitted distribution. The probability distribution which provides the minimum D value is chosen as the best fitting distribution.

To estimate of the posterior density functions expressed in Eq. 8.24, three-dimension joint distributions of H_0, H_n and S_n are needed to be constructed. The symmetric copulas are not considered because the dependence among the three variables pairs (H_0, H_n), (H_0, S_n) and (H_n, S_n) are not the same, which will be tested against data for the case study. Hence, we use three widely used asymmetric trivariate Archimedean copulas, namely Gumbel-Hougaard, Frank and Clayton as candidates. These three trivariate Archimedean copulas are described in Table 2.2 of Chap. 2. Dependence parameters of the trivariate copula functions are estimated using the maximum pseudo-likelihood method (Zhang and Singh 2007b, c; Chen et al. 2010). The RMSE is used to measure the goodness of fit of the copula distribution (Zhang and Singh 2007a). The copula which has the smallest *RMSE* value is preferred.

8.2.4 Evaluation Criteria

8.2.4.1 Performances of Deterministic Forecasts

Two widely applied criteria, namely Nash-Sutcliffe efficiency (*NSE*) and Relative Error (*RE*) are adopted to evaluate the performance of the deterministic forecast model (Xiong and Guo 1999; Liu et al. 2016).

(1) Nash-Sutcliffe efficiency

The first criterion is the Nash-Sutcliffe efficiency (*NSE*) coefficient (Nash and Sutcliffe 1970) which is defined by

$$NSE = \left[1 - \frac{\sum_{t=1}^{T}(h_t - s_t)^2}{\sum_{t=1}^{T}(h_t - \bar{h})^2}\right] \times 100\% \tag{8.25}$$

where t is the time step, T is the total number of time steps; h_t and s_t are the simulated and observed discharges at time t, and $\bar{h}$ is the mean value of the observed discharge. Nash-Sutcliffe efficiency can range from $-\infty$ to 1. An efficiency of 1 ($NSE = 1$) corresponds to a perfect match of simulated discharge to the observed data. An efficiency of 0 ($NSE = 0$) indicates that the model predictions are as accurate as the mean of the observed data, whereas an efficiency less than zero ($\mathrm{NSE} < 0$) occurs when the observed mean is a better predictor than the model. Essentially, the closer the model efficiency is to 1, the more accurate the model is.

(2) Relative error

The second criterion used is the relative error (*RE*) of the total runoff amount fit between the observed and simulated discharge series, defined as (Xiong and Guo 1999)

$$RE = \left[1 - \frac{\sum_{t=1}^{T}(h_t - s_t)}{\sum_{t=1}^{T} h_t}\right] \times 100\% \tag{8.26}$$

RE represents a systematic error of water balance simulation. A value of *RE* closes to zero indicates a good agreement between observed and simulated runoff volume. In this study, we rank *NSE* as the primary criterion, while *RE* is an auxiliary criterion. Only when simulated discharge series yield the same (higher) *NSE* value, the one with the smaller *RE* value is preferred. Otherwise, the simulation with smaller *RE* value does not reveal any superiority (Liu et al. 2016). For instance, the model with all simulated discharges equal to the mean of observed values can easily provide $RE = 0$. Unfortunately, in this case, the $NSE = 0$, which clearly means an undesired simulation.

8.2.4.2 Performances of Probabilistic Forecasts

The probabilistic forecast technique is expected to provide (a) accurate forecast probabilities, further on named reliability; and (b) narrow forecast intervals, further on the named resolution. Several methods, e.g., predictive quantile-quantile (QQ) plot, α-index and π-index have been proposed in the literatures to evaluate probabilistic forecasts (see e.g. Gneiting et al. 2007; Laio and Tamea 2007; Thyer et al. 2009; Engeland et al. 2010; Renard et al. 2010; Madadgar et al. 2014; Smith et al. 2015) and are used in this study.

(1) Predictive QQ plot

The predictive quantile-quantile (QQ) plot provides an overall assessment of whether the total predictive uncertainty is consistent with the observations. This requires a diagnostic approach that compares a time-varying distribution (the predictive distribution at all times t) to a time series of observations (Thyer et al. 2009; Evin et al. 2014). The predictive QQ plot provides a simple, intuitive and informative summary of the performance of probabilistic prediction frameworks (Gneiting et al. 2007; Laio and Tamea 2007).

The predictive QQ plot is constructed as follows: Let F_t be the CDF of the predictive distribution of runoff at time t, and h_t the corresponding observed runoff. If the hypotheses in the calibration framework are consistent with the data, the observed value h_t should be consistent with the distribution F_t. Hence, under the assumption that the observation h_t is a realization of the predictive distribution, the p-value $F_t(h_t)$ is a realization of a uniform distribution on [0,1]. The predictive QQ plot compares the empirical CDF of the sample of p values $F_t(h_t)$ ($t = 1, \ldots, T$) with the CDF of a uniform distribution to assess whether the hypotheses are consistent with the observations.

As illustrated in Fig. 8.1, the predictive QQ plot can be interpreted as follows (Thyer et al. 2009): (1) if all points fall on the 1:1 line, the predicted distribution agrees perfectly with the observations; (2) If the observed p values cluster around the mid-range (i.e., a low slope around theoretical quantile 0.4–0.6), the predictive uncertainty is overestimated; (3) If the observed p values cluster around the tails (i.e., a high slope around theoretical quantile 0.4–0.6), the predictive uncertainty is underestimated; (4) If the observed p values at the theoretical median are higher/

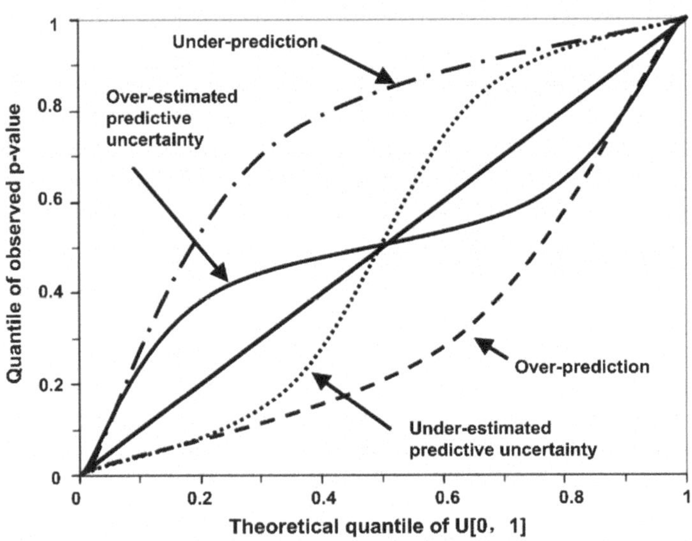

Fig. 8.1 Interpretation of the predictive QQ plot

lower than the theoretical quantiles, the modeled predictions systematically under/over predict the observed data.

Other metrics are the supportive quantitative scores derived from the predictive QQ plot (Laio and Tamea 2007; Thyer et al. 2009; Madadgar et al. 2014). The metrics α-index assesses the reliability of forecasts, and π-index indicates the resolution (precision, sharpness) of the predictive distribution (PD).

(2) Reliability

Reliability means that the forecast should be well calibrated. This can be checked graphically: deviations from the bisector (the 1:1 line) denote interpreted deficiencies (see Fig. 8.1). To simplify the comparison of QQ plots, it is summarized using an index that quantifies the reliability of the PD (Renard et al. 2010; Madadgar et al. 2014):

$$\alpha - \text{ index} = 1 - \frac{2}{T}\sum_{t=1}^{T}\left[\left|q^{em}(p_t) - q^{th}(p_t)\right|\right] \tag{8.27}$$

where p_t is the observed p-value at time t; $q^{em}(h_t)$ is the empirical quantile of $p_t, q_t^{th}(h_t)$ is the theoretical quantile of p_t obtained from the uniform distribution U [0, 1]; T is the number of p_t values.

The α-index measures the closeness of quantile plot of the observations to the corresponding uniform quantiles and reflects the overall reliability of the PD. According to Thyer et al. (2009), as the area between the empirical CDF of the observed p-values and the CDF of the uniform distribution in the predictive QQ plot becomes larger, the value of α-index decreases towards zero. It varies between 0 (worst reliability) and 1 (perfect reliability).

(3) Resolution

"Resolution" denotes the sharpness (effectively, the "average precision") of the PD. Note that two inferences can both yield reliable PDs, but with different resolutions. Sharpness refers to the spread of the forecast PDFs and is a property of the predictions only. The more concentrated the forecast PDF, the sharper the forecast, and the sharper the better, subject to calibration (Gneiting et al. 2005). In this paper, the resolution is quantified by π-index defined as the average relative precision of the predictions (Renard et al. 2010; Madadgar et al. 2014):

$$\pi - \text{ index} = \frac{1}{T}\sum_{t=1}^{T}\frac{E[H_t]}{Sdev[H_t]} \tag{8.28}$$

where $E[H_t]$ and $Sdev[H_t]$ are the expected value and standard deviation of H_t obtained from the predictive distribution at time t.

Greater value of π-index indicates greater resolution (lower uncertainty) of forecasts. However, comparison of sharpness may not be a meaningful approach when the employed methods do not primarily perform equally in the α-index metric. Assuming that precision has lower priority than reliability, given similar forecast reliability, the method with greater resolution (lower uncertainty) is preferred; otherwise, the method with higher resolution does not reveal any superiority. Most of literature rank reliability as the primary criterion, while sharpness is secondary to reliability (Madadgar et al. 2014).

(4) Continuous rank probability score

The goal of probabilistic forecasting is to maximize the sharpness of the forecast PDFs subject to calibration. However, the trade-off between reliability and sharpness have been discussed in previous researches (Xiong et al. 2009; Li et al. 2010a; Kasiviswanathan et al. 2013), which show that these two desirable objectives could not be achieved simultaneously. It is not adequate to judge the performances of probabilistic forecasts only by reliability or sharpness. The continuous rank probability score (*CRPS*) is a standard measure that combines reliability and sharpness (Hersbach 2000; Gneiting et al. 2005) and is used for selecting the preferred model.

The *CRPS* measures the average distance between the predicted and the observed CDFs over the entire period. It is the integral of the Brier scores at all possible threshold values r for the continuous predictand (Hersbach 2000). Specifically, if F is the predictive CDF and h_t is the verifying observation, the *CRPS* is defined as (Hersbach 2000; Gneiting et al. 2007; Pappenberger et al. 2015)

$$CRPS = \frac{1}{T}\sum_{t=1}^{T} \int_{-\infty}^{+\infty} [F_t(r) - H_s(r - h_t)]^2 dr \tag{8.29}$$

where $H_s(r - h_t)$ denotes the Heaviside step function and takes the value 0 when $r < h_t$ and the value one otherwise.

For a deterministic forecast system, the *CRPS* reduces to the mean absolute error (*MAE*). Thus, the *CRPS* is sometimes interpreted as a generalized version of the *MAE* (Zhao et al. 2015). This is an advantage of *CRPS* and consequently allows the comparison of deterministic and probabilistic forecasts (Gneiting et al. 2007; Pappenberger et al. 2015). The smaller the *CRPS* value is, the better the prediction performance. Its minimal value of zero is only achieved in the case of a perfect deterministic forecast.

8.2.5 Case Studies

8.2.5.1 Study Area and Data

Three Gorges Reservoir (TGR) is a vitally important and back-bone project in the development and harnessing of the Yangtze River in China. The annual average discharge and runoff volume at the dam site are 14,300 m^3/s and 4510×10^8 m^3, respectively. The total storage capacity of the TGR is 393×10^8 m^3, of which

221.5×10^8 m^3 is flood control storage. The reservoir has a surface area of about 1080 km^2, an average width of about 1100 m, a mean depth of about 70 m and a maximum depth near the dam of about 170 m. With all the profiles being narrow and deep, the TGR retains the long narrow belt shape of the original river section and is a typical river channel-type reservoir.

As shown in Fig. 8.2, the intervening basin of TGR has a catchment area of 55,907 km^2, about 5.6% of the upstream Yangtze River basin. There are 40 rainfall gauged stations in the intervening basin and two hydrological stations (Cuntan and Wulong), which control the upstream inflow and tributary inflow, respectively. The data set for TGR inflow forecasting includes the daily runoff data of the Cuntan, Wulong and Yichang hydrological stations, arithmetic mean of observed rainfall data in the intervening basin during the flood period (June 1–September 30) from 2003 to 2009. The period 2003–2007 is used for deterministic forecast model calibration and 2008–2009 is used for validation (Li et al. 2010b; Chen et al. 2015).

8.2.5.2 Deterministic Inflow Forecasts of the TGR

The inflow of TGR consists of three components, i.e., the main upstream inflow, the tributary inflow from the Wu River, and the lateral flow from the TGR intervening basin as shown in Fig. 8.2. A multiple-input single-output linear systematic model is chosen for the inflow forecasting of the TGR (Liang et al. 1992). The total inflow to the TGR can be expressed by the following equation

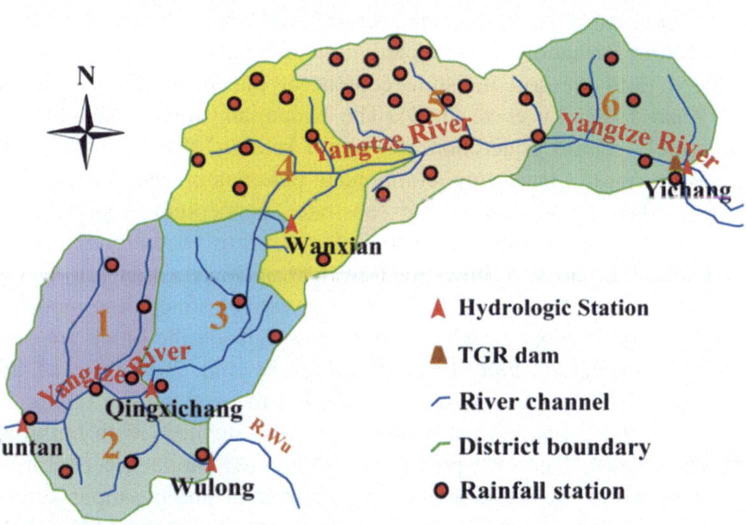

Fig. 8.2 Sketch map of the TGR's intervening basin

$$\widehat{Q}_t = A\sum_{j=1}^{m_1} R_{t-j+1}^{(1)} h_j^{(1)} + \sum_{j=1}^{m_2} R_{t-j+1}^{(2)} h_j^{(2)} \tag{8.30}$$

where $R_j^{(1)}$ is the lateral flow from the TGR intervening basin which is calculated via the Xinanjiang model (Zhao 1992). $R_j^{(2)}$ is the upstream inflow (inflow at Wulong is added to the inflow at Cuntan). A is the area of the TGR intervening basin, m_1, m_2 are the memory length of the system corresponding to $R_j^{(1)}$ and $R_j^{(2)}$, $h_j^{(1)}$ and $h_j^{(2)}$ are the jth ordinates of the pulse response functions relating inputs $R_j^{(1)}$ and $R_j^{(2)}$, which are calculated by the Nash model as follows

$$h_j^{(1)} = \frac{1}{T}\int\limits_{(j-1)T}^{jT} [S_i(t) - S_i(t-T)]/T\,dt \quad (i = 1,2) \tag{8.31}$$

$$S_i(t) = \int\limits_0^t \frac{1}{NK_i\Gamma(NK_i)} e^{-(\tau/NK_i)}(\tau\backslash NK_i)^{N_i-1} d\tau \quad (i = 1,2) \tag{8.32}$$

where $S_i(t)$ is the step response function of the ith input, N_i and NK_i are the parameters, and T is the time-step. $\Gamma(\bullet)$ is the gamma function.

The Xinanjiang model was developed in the middle 1970s for forecasting flows in the Xinanjiang reservoir, China. The model has been widely applied for flood forecasting in a large number of basins all over the world, especially in China. Until now, this model is the most popular rainfall-runoff hydrologic model in China for streamflow forecasting in humid and semi-humid areas. Its main feature is the concept of runoff formation on the repletion of storage, which denotes that runoff is not produced until the soil moisture content of the aeration zone reaches field capacity (Zhao 1992; Xu et al. 2013). The Xinanjiang model includes two components, namely, runoff generation and runoff routing. It has 17 parameters that include seven runoff generating component parameters and 10 runoff routing parameters. These parameters are abstract conceptual representations of non-measurable watershed characteristics that have to be calibrated by an optimization method. Figure 8.3 shows the flowchart of the Xinanjiang model for three water sources. All symbols inside the blocks are variables including inputs, outputs, state variables and internal variables while those outside the block are parameters (Zhao 1992; Cheng et al. 2006; Li et al. 2011; Lin et al. 2014; Si et al. 2015).

The deterministic forecast model is calibrated respectively by taking *NSE* and *RE* as objective functions via automatic calibration methods with multiple objectives (Madsen 2000). Table 8.1 presents the calibrated parameters obtained for the Xinanjiang model of the TGR intervening basin and multiple-input single-output linear systematic model for TGR. The simulation results for the *NSE* and *RE* in the calibration period are 97.72 and −1.04%, respectively. Meanwhile, in the

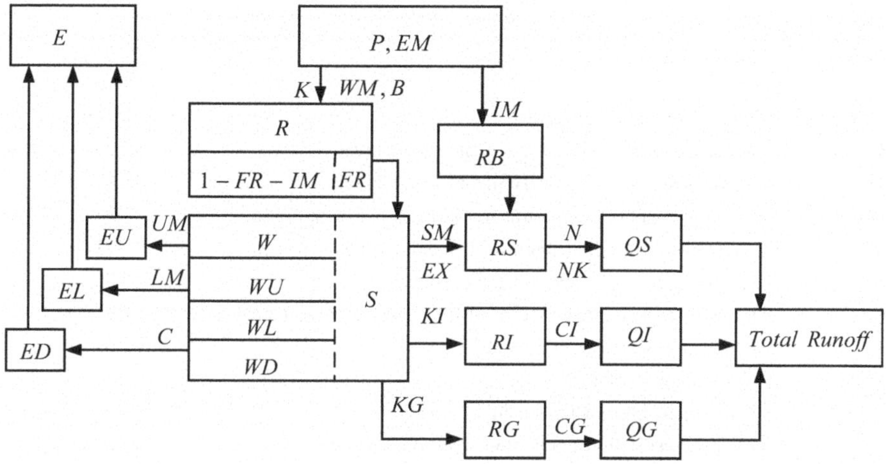

Fig. 8.3 The flow chart of Xinanjiang model for three water sources

verification period, the *NSE* and *RE* are 95.84 and −0.21%, respectively. These results show that the deterministic forecast model is proved to be quite efficient in simulating the inflow series for the TGR. The deterministic forecasts obtained from the well-calibrated deterministic model are subsequently applied to produce the probabilistic forecasts through the meta-Gaussian HUP and copula-based HUP.

8.2.5.3 Determination of Marginal Distributions

In this study, future rainfalls are treated as the case of perfect foreknowledge, rather than using the real forecast rainfalls to obtain simulated flows in the future, when the established deterministic forecast model is operated in the real-time forecasting mode. This is only for the illustration purpose if forecast rainfalls are available in reality and these would be used. The forecast lead times are 24 h ($n = 1$), 48 h ($n = 2$), and 72 h ($n = 3$). Especially, for each forecasting time in the record, the recorded rainfall data of 24-, 48- and 72- later followed by this forecasting time are treated as the "deterministic rainfall forecasts" (i.e., assuming perfectly known rainfalls in the future). Then these perfect forecasts of the rainfalls are input to the well-calibrated deterministic forecast model, which in turn produced model inflows (s_1, s_2, s_3). They are attached to actual inflow (h_0, h_1, h_2, h_3) to obtain one joint realization of the model-actual inflow process. The dataset from 2003 to 2009 are used to calibrate and compare the meta-Gaussian HUP and copula-based HUP.

The sample series of H_0 is taken from June 1 to Sept. 27 every year, S_1 from June 2 to Sept. 28, S_2 from June 3 to Sept. 29, and S_3 from June 4 to Sept. 30, thus all these four variables will have a data length of 833. The parameters of six candidate distributions are estimated by the L-moment method, and the K-S tests are used to verify the null hypothesis. The null hypothesis could not be rejected at

Table 8.1 Estimated parameters of Xinanjiang model in the TGR intervene basin

Parameter	Physical meaning	Estimated value
WM	Mean tension water capacity	149.80
UM	Areal mean water capacity of the upper layer	65.23
LM	Areal mean water capacity of the lower layer	38.64
K	Ratio of potential evapotranspiration to pan evaporation	0.433
B	Parameter in the distribution of tension water capacity	1.471
SM	Areal mean free water storage capacity	25.09
EX	Parameter in the distribution of free water storage capacity	0.984
KI	Coefficient relating *RI*	0.151
KG	Coefficient relating *RG*	0.12
IM	Impervious area of the basin	0.184
C	Evapotranspiration coefficient from deep layer	0.287
CI	Interflow reservoir constant	0.832
CG	Groundwater reservoir constant	0.904
m_1	Memory length of TGR intervening basin	10
N_1	Number of cascade linear reservoirs for TGR intervening basin	2.967
NK_1	Scale parameter of cascade linear reservoirs for TGR intervening basin	6.991
m_2	Memory length of upstream inflow	14
N_2	Number of cascade linear reservoirs for upstream inflow	1.241
NK_2	Scale parameter of cascade linear reservoirs for upstream inflow	1.911

Note The unit of *WM*, *UM*, *LM* and *SM* is mm, the rest of parameters are dimensionless

the 95% confidence level (critical value is 0.0471) for all six candidate distributions except Normal distribution. For the four hydrological variables, Gumbel distribution provides the minimum *D* value and is chosen as the best fitting distribution, respectively. Figure 8.4 shows the empirical CDF values obtained from the Gringorten plotting-position formula (Zhang and Singh 2006) and theoretical CDF values calculated by the Gumbel distributions. It can be seen that the theoretical values fit the empirical values very well. For comparison purposes, the copula-based HUP used the same marginal distributions as the meta-Gaussian HUP.

8.2.5.4 Calibration of Meta-Gaussian HUP

For the given climatic record of actual flows, the joint sample $\{(h_0, h_1)\}$ is formed of realizations on two consecutive days. Each joint realization (h_0, h_1) is processed through the empirical NQT to obtain the transformed joint sample $\{(w_0, w_1)\}$ and this joint sample is used to estimate the Pearson's correlation coefficient c. The advantage of using the empirical distributions in the NQT (instead of the parametric

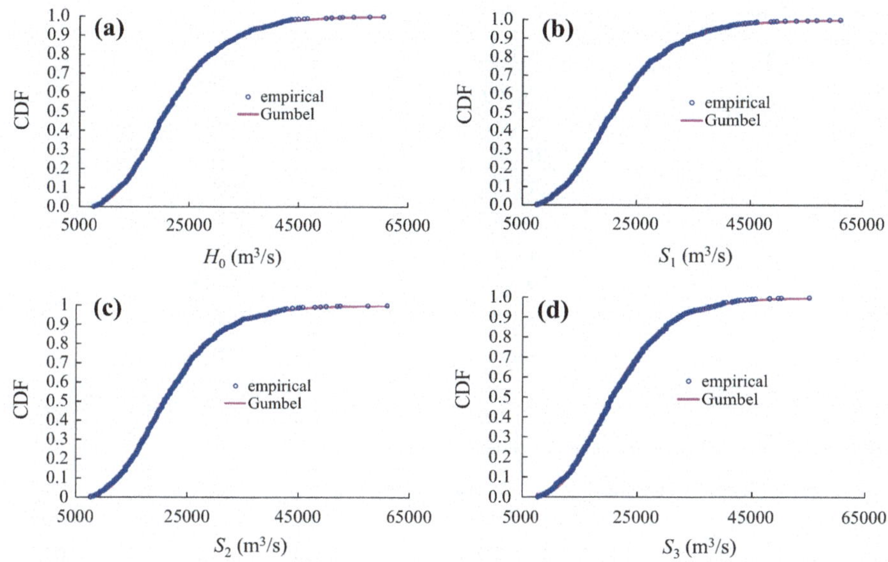

Fig. 8.4 Empirical and theoretical values fitted by Gumbel distributions

distribution) is that the estimate of c remains unaffected by the choice and the goodness of fit of the parametric model. The estimated result of Pearson's correlation coefficient is 0.951.

The procedure for validating the meta-Gaussian dependence structure for the likelihood function parallels the procedure described in the prior density section above. The NQT performs adequately, as the empirical structure of dependence between X_n, W_n and W_0, appears to be linear and homoscedastic. The meta-Gaussian model for the likelihood function captures the nonlinearity and heteroscedasticity of the dependence structure between S_n, H_n and H_0.

8.2.5.5 Calibration of Copula-Based HUP

The rank-based correlation (Kendall's coefficient) matrix of variables, H_0 H_n and S_n are shown in Table 8.2. It is demonstrated that the dependence among the three variables pairs (H_0, H_n), (H_0, S_n) and (H_n, S_n) are not the same. Furthermore, the highest correlation coefficient is exhibited in the variables pair (H_n, S_n). This result indicates that rather than symmetric, the asymmetric trivariate copula functions may be more appropriate to be used to three-dimension joint distributions of H_0, H_n and S_n. When constructing the three-dimension joint distributions using the asymmetric copula functions, the structures (H_n, S_n)H_0 are applied. Specifically, copula was firstly built for (H_n, S_n), and then for H_0 and $C(F_{Hn}(h_n, F_{sn}(s_n)))$.

The three-dimension joint distributions of H_0, H_n and S_n (n = 1, 2, 3) are constructed using the three candidate trivariate copula functions. Dependence

Table 8.2 Ranked based correlation matrix of the variables

Lead times (h)	Variables	τ
24	H_0, H_1	0.823
	H_0, S_1	0.824
	H_1, S_1	0.929
48	H_0, H_2	0.694
	H_0, S_2	0.711
	H_2, S_2	0.883
72	H_0, H_3	0.600
	H_0, S_3	0.658
	H_3, S_3	0.828

Table 8.3 Estimated parameters of the three candidate copulas

Variables	Gumbel-Hougaard		Frank		Clayton	
	$[\theta_1, \theta_2]$	RMSE	$[\theta_1, \theta_2]$	RMSE	$[\theta_1, \theta_2]$	RMSE
H_0, H_1, S_1	[9.08, 10.65]	0.0116	[25.44, 35.23]	0.0103	[15.16, 20.78]	0.0150
H_0, H_2, S_2	[4.25, 7.58]	0.0141	[13.45, 20.38]	0.0112	[6.49, 14.33]	0.0186
H_0, H_3, S_3	[2.98, 6.57]	0.0144	[9.58, 16.21]	0.0117	[4.58, 9.34]	0.0188

parameters of the trivariate copula functions are estimated using the maximum pseudo-likelihood method, and the results are listed in Table 8.3. It is found that Frank copula performs best with the smallest *RMSE* values for the three joint distributions. Empirical CDFs obtained from the Gringorten plotting-position formula and theoretical CDFs calculated from Frank copula for three joint distributions are plotted in Fig. 8.5. An overall satisfactory agreement between the empirical and theoretical CDF is shown. Hence, the asymmetric trivariate Frank copula functions have good performances in modeling the joint distributions of H_0, H_n and S_n.

8.2.5.6 Comparison of the Meta-Gaussian HUP and Copula-Based HUP

(1) Posterior median forecasts

For 24-, 48- and 72 h lead times, the model efficiency *NSE* and relative error *RE* calculated by both the deterministic forecast model and posterior median forecasting associated with the meta-Gaussian HUP and copula-based HUP are listed in Table 8.4. It is shown that both the results of the meta-Gaussian HUP and copula-based HUP are slightly better than those of the deterministic forecast model,

Fig. 8.5 Plots of empirical and theoretical values estimated by Frank copulas for three joint CDFs. *Note* Rank represents number of ordered pair, ranked in the ascending order in terms of theoretical joint CDF, respectively

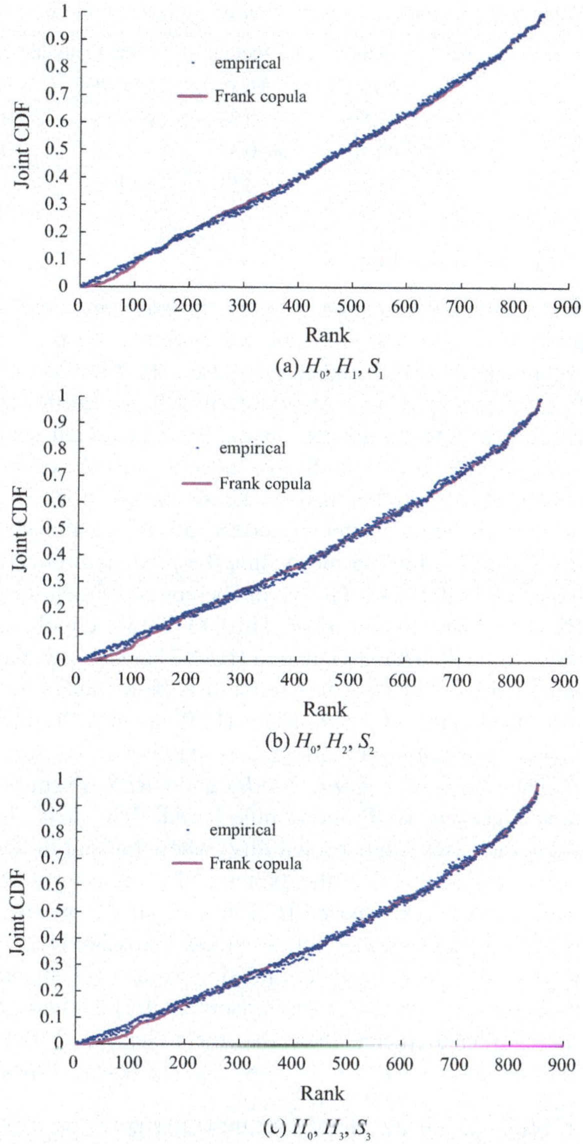

and the copula-based HUP is comparable to the meta-Gaussian HUP. Compared with deterministic forecasts, the NSE and the RE of the copula-based HUP for 24-, 48- and 72 h lead times forecasts are improved by 1.24, 1.26 and 1.26% and reduced by 0.17, 0.57, and 1.72%, respectively. It is also noted that the accuracy of posterior median forecasts of the both HUPs decreases as the lead time increases.

Table 8.4 Comparison of performances evaluation criteria for deterministic forecasts

Lead times (h)	Deterministic model		Meta-Gaussian HUP		Copula-based HUP	
	NSE (%)	*RE* (%)	*NSE* (%)	*RE* (%)	*NSE* (%)	*RE* (%)
24	97.55	−0.34	97.65	−0.31	98.79	−0.17
48	94.10	−0.85	94.54	−0.68	95.36	−0.28
72	88.52	−2.51	89.14	−1.14	89.78	−0.79

(2) Probabilistic forecasts

The predictive QQ plot, α-*index*, π-*index* and *CRPS* are adopted to evaluate the probabilistic forecasts. Figure 8.6 presents the predictive QQ plots regarding the meta-Gaussian HUP and copula-based HUP for 24-, 48- and 72 h lead times. Using Fig. 8.1 as a guide to assess the results, it is clear that the overall performances of all predictive QQ plots are acceptable. Both meta-Gaussian HUP and copula-based HUP systematically under-predict the inflows, since the observed p values at the theoretical median are a bit higher than the theoretical quantiles. In addition, it also shows that the observed p values cluster around the tails (i.e., a high slope around theoretical quantile 0.4–0.6). This finding means that the predictive uncertainty is somewhat underestimated for both HUPs. The overall behaviors of meta-Gaussian HUP and copula-based HUP are found to be similar. The QQ plot for copula-based HUP is slightly closer to the 1:1 line than meta-Gaussian HUP. That is to say, the copula-based HUP performs marginally better regarding reliability. Nonetheless, these underestimations for both meta-Gaussian and copula-based HUPs are in such zones where p values are relatively higher, indicating such differences may not be statistically significant.

The results of α-*index*, π-*index* and *CRPS* are summarized in Table 8.5. For both meta-Gaussian HUP and copula-based HUP, it is clearly shown that the α-*index* value increases (higher reliability) when the lead time increases. However, it should be noted that this is at the expense of decreasing π-*index* value (lower resolution). Besides, the copula-based HUP has slightly larger α-*index* values while smaller π-*index* values compared with the meta-Gaussian HUP. Regarding *CRPS* value, both HUPs outperform the deterministic forecasts which demonstrate the effectiveness of probabilistic forecasts. Comparison results also indicate that the copula-based HUP is marginally better than the meta-Gaussian HUP. The *CRPS* value of the copula-based HUP for 24-, 48- and 72 h lead times is improved (decreased) by 16.6, 21.2, and 23.3%, respectively.

Although that such marginally better performance does not result for each year, for illustration purposes, the observed and median discharges, and 90% inflow prediction intervals estimated by meta-Gaussian HUP and copula-based HUP in 2004 are presented in Figs. 8.7 and 8.8, respectively. It can be seen that most observed inflows are contained within the 90% prediction intervals. This demonstrates that these 90% prediction intervals can effectively capture the forecast uncertainty and provide more information for decision-making in flood control and reservoir operation. As lead time increases, the 90% prediction intervals become wider (i.e., greater uncertainty).

Fig. 8.6 The predictive QQ plots of meta-Gaussian HUP and copula-based HUP

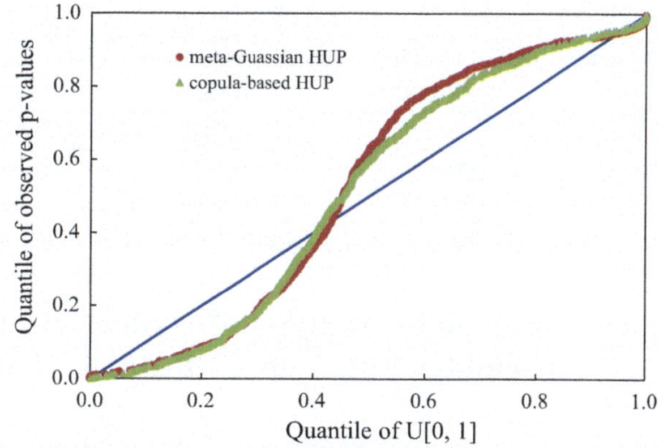

(a) 24h lead time

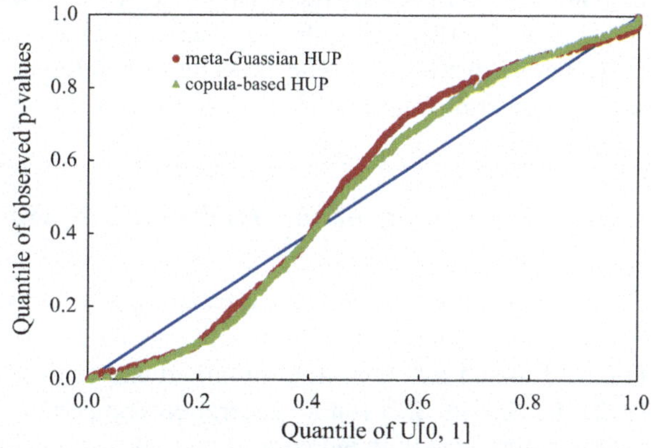

(b) 48h lead time

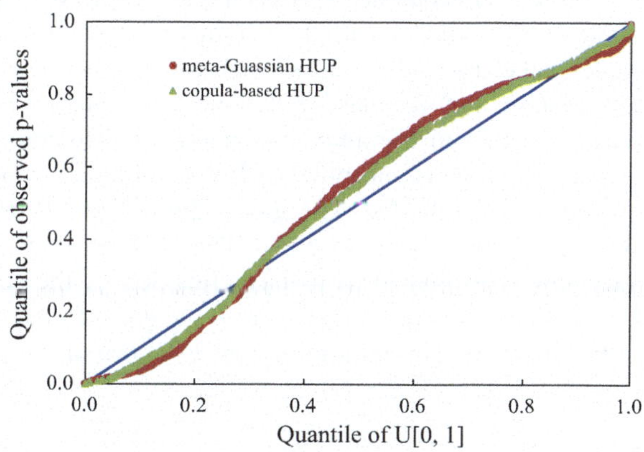

(c) 72h lead time

Table 8.5 Comparison of performances evaluation criteria for probabilistic forecasts

Lead times (h)	Deterministic model	Meta-Gaussian HUP			Copula-based HUP		
	CRPS/MAE	*α-index*	*π-index*	*CRPS*	α-index	π-index	CRPS
24	688	0.8028	18.55	608	0.8507	16.39	574
48	1180	0.8555	12.45	975	0.8916	10.42	930
72	1763	0.8879	9.13	1420	0.9184	7.68	1353

Note α-index and π-index are dimensionless; the unit of *CRPS* is m^3/s

8.3 Uncertainty Analysis of Hydrological Multi-model Ensembles Based on CBP-BMA Method

Inspired by the ideas of Madadgar and Moradkhani (2014), a general framework of the combination of copula Bayesian processor with BMA (CBP-BMA) is proposed by He et al. (2018), where the Bayesian theory is applied in the transformation of the posterior distribution. The flowchart of different probability forecast methods based on deterministic models is described in Fig. 8.9.

8.3.1 Description of the Hydrological Models

Three world-famous conceptual hydrological models are implemented in the Mumahe catchment, including the Xinanjiang (XAJ), HBV and SIMHYD models. The XAJ model has been used in humid and semi-humid region worldwide (Zhao 1992). It consists of a runoff generation component with seven parameters and a routing component with ten parameters. Those model physical parameters represent the abstract conceptual expression of watershed features. The HBV model is a synthetic flow model with 13 parameters needed to be calibrated. Units of HBV model makes up of the routines for snowmelt accumulation, evapotranspiration and soil routine and response function. The core concept assumes runoff volume changes with soil humidity exponentially (Montero et al. 2016). The SIMHYD model is a lumped conceptual hydrological model which contains seven parameters needed to be calibrated. The model divides runoff into three components: surface flow, interflow and base flow. The surface flow is infiltration excess runoff, inter-flow is estimated as a linear function of the soil wetness, and base flow is simulated as a linear recession from the groundwater store (Chiew et al. 2009; Yu and Zhu 2015). The infiltration rate is a core of the model.

8.3.2 Bayesian Model Averaging (BMA)

Raftery et al. (2005) successfully extended BMA to statistical post-processing for forecast ensembles. The BMA method addresses total model uncertainty by

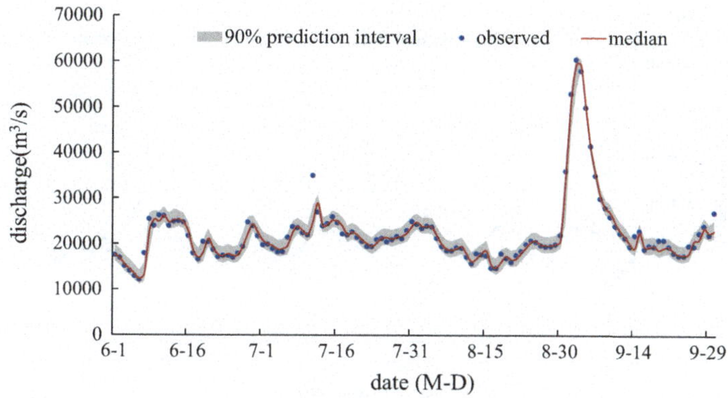

(a) 24h lead time

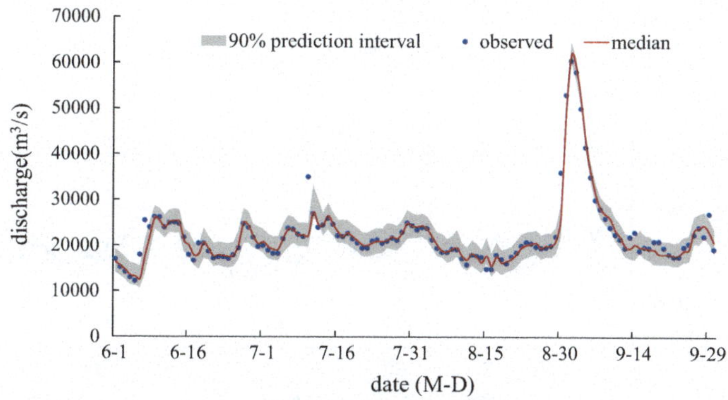

(b) 48h lead time

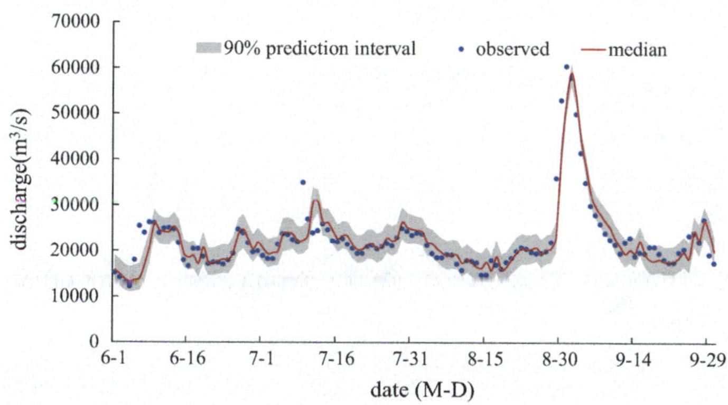

(c) 72h lead time

Fig. 8.7 The 90% prediction intervals, median and observed discharges in 2004 (meta-Gaussian HUP)

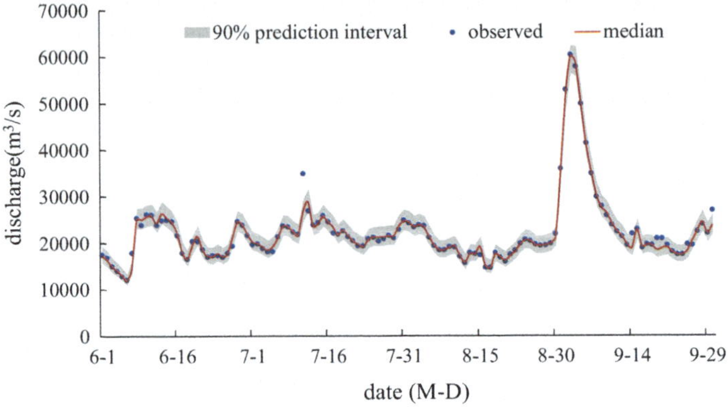

(a) 24h lead time

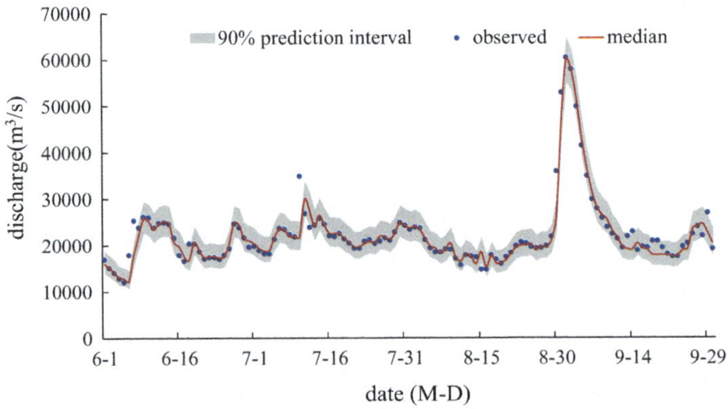

(b) 48h lead time

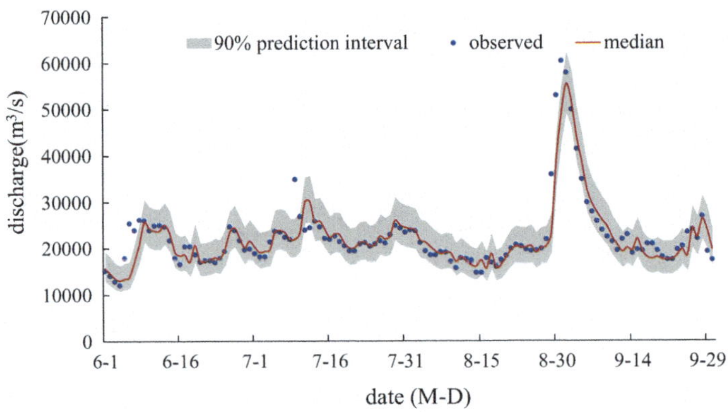

(c) 72h lead times

Fig. 8.8 The 90% prediction intervals, median and observed discharges in 2004 (copula-based HUP)

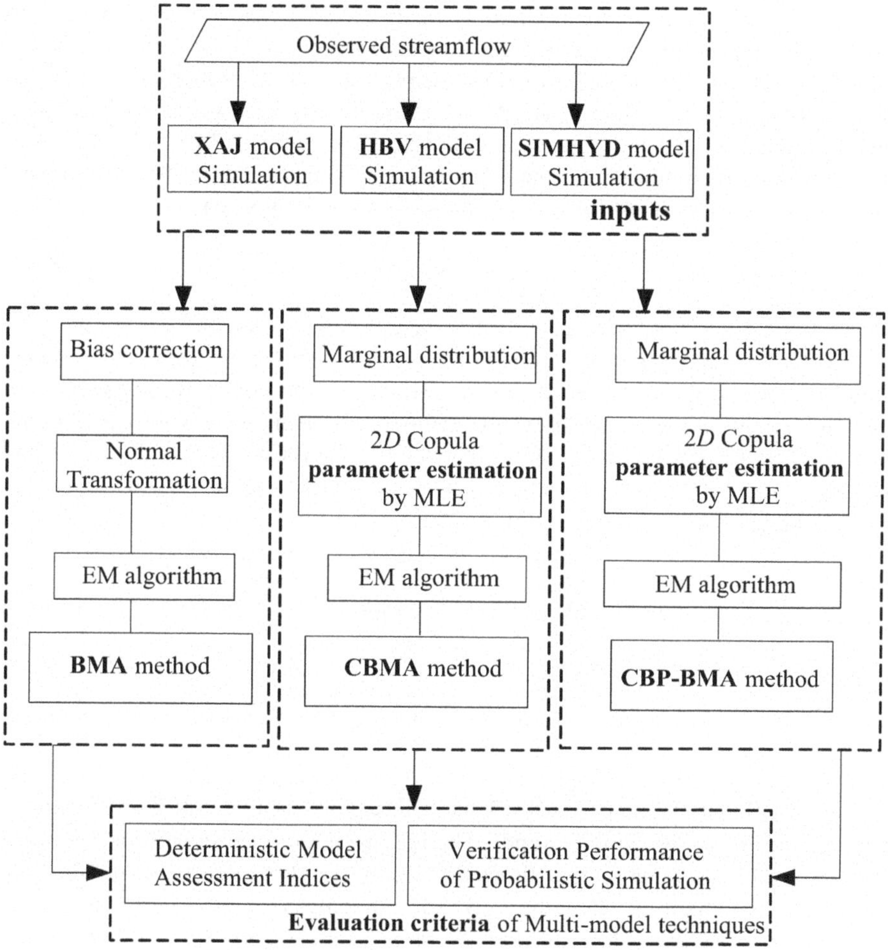

Fig. 8.9 Flowchart of hydrologic multi-model ensembles for uncertainty analysis

conditioning not only on a single outstanding model but on the entire ensemble models. The method was originally proposed as a pathway for method combination of several competing models (Duan et al. 2007; Liang et al. 2011).

According to BMA (Duan et al. 2007), the ensemble predictive density of the actual flow variable q, given the different hydrologic model simulations of K models $[S_1, S_2, \ldots, S_K]$ and the observations during the training period, Q, can be expressed in terms of the law of total probability:

$$p(q|S_1, S_2, \ldots, S_K, Q) = \sum_{i=1}^{K} p(S_i|Q) \cdot p_i(q|S_i, Q) \tag{8.33}$$

where $p(S_i|Q)$ is the posterior probability of ith model prediction. This static term can also be expressed as w_i, reflecting how well the ensemble term fits the observation dataset. It ranges from 0 to 1 since the posterior model probabilities add up to one. Before the implantation of BMA algorithm, the expected value of observation and forecast for each model should be equal zero ($E[q - S_i] = 0$). Any bias-correction method, such as linear regression, should be applied to substitute the bias-corrected forecast (f_i) for the original deterministic forecast:

$$f_i = a_i + b_i \cdot S_i \tag{8.34}$$

where $\{a_i, b_i\}$ are the coefficients of the linear regression model.

The term $p_i(q|f_i, Q)$ is the conditional pdf of h based on the bias-corrected simulation f_i and the observation dataset. Moreover, the power Box-cox transformation is taken for the computational convenience of using a Gaussian distribution. The posterior distribution $p_i(q|f_i, Q)$ is mapped to a Gaussian space with mean f_i and variance s_i^2; i.e., $p_i(q|f_i, Q) \sim g(q|f_i, \sigma_i^2)$. The BMA predictive mean and variance of q are defined as follows (Raftery et al. 2005):

$$E(q|Q) = \sum_{i=1}^{K} p(f_i|Q) \cdot E[p_i(q|f_i, Q)] = \sum_{i=1}^{K} \omega_i f_i \tag{8.35}$$

$$Var(q|Q) = \sum_{i=1}^{K} \omega_i \left(f_i - \sum_{i=1}^{K} \omega_i f_i \right)^2 + \sum_{i=1}^{K} \omega_i \sigma_i^2 \tag{8.36}$$

Successful application of the BMA method requires estimations of the weight ω_i and variance σ_i^2 of the individual pdf. The log maximum likelihood function rather than the likelihood function is optimized for reasons of both numerical stability and algebraic simplicity. If the BMA parameters are estimated by $\theta = \{\omega_i, \sigma_i, i = 1, 2, \ldots, K\}$, the log likelihood function of θ is mathematically denoted as:

$$l(\theta) = \log\left(\sum_{i=1}^{K} \omega_i \cdot p_i(q|f_i, Q) \right) \tag{8.37}$$

After the completion of BMA parameter estimation by the EM algorithm (Duan et al. 2007), another feature of the BMA method is to make use of Monte Carlo method to derive BMA probabilistic ensemble prediction for any time t (Kuczera and Parent 1998). The procedures are described as follows (Zhou et al. 2016).

(1) Select the probabilistic ensemble size, M (M = 100).
(2) Randomly generate a value of k from the numbers [1, 2, …, K] with probabilities $[\omega_1, \omega_2, \ldots, \omega_i]$. The detail processes are shown as follows: (a) Initial the cumulative weight $\omega_0' = 0$ and compute $\omega_i' = \omega_{i-1}' + \omega_i$ for $i = 1, 2, \ldots, K$;

(b) Generate a random number u between 0 and 1; and (c) If $\omega'_{i-1} \leq u \leq \omega'_i$, then the ith member of the ensemble predictions are chosen.

(3) Generate a value of q from the pdf of $p_i(q|f_i, \sigma_i^2)$.

(4) Repeat steps (2) and (3) for M times.

The results are sorted in ascending order, and the 90% confidence interval can be derived within the range of the 5 and 95% quantiles.

8.3.3 The Hybrid Copula-BMA (CBMA)

As illustrated before, the BMA predictive distribution provides a weighted average of simulation pdf which generally complies with a parametric distribution, e.g., Gaussian distribution after the box-cox transformation. Madadgar and Moradkhani (2014) employed copula to estimate the posterior distribution of forecast variables for each model, i.e., $p_i(q|f_i, Q)$, and found that the hydrological forecasts are improved after the integration of copulas and BMA (CBMA). A series of research demonstrates that the procedures of CBMA not only eliminate the prophase bias-correction and the external calculation of variance but also simplify the calculation of the weighted average and the probability model structure by copula (Möller et al. 2013).

Alternatively, in statistical applications, the conditional probability distribution of h given s_i (i = 1, 2, 3) is expressed as (Madadgar and Moradkhani 2014):

$$f(q|s_i) = \frac{f(q, s_i)}{f(s_i)} = \frac{c(u, v_i) \cdot f(q) \cdot f(s_i)}{f(s_i)} = c(u, v_i) \cdot f(q) \qquad (8.38)$$

where $c(u, v_i)$ is computed for each pair of (u, v_i), $f(q)$ represents the marginal distribution of actual flow. Although different copula families have been proposed and described in current studies (Chebana and Ouarda 2007), several families of Archimedean copulas, including Frank, Gumbel, and Clayton, have been popular choices for dependence models in hydrologic analyses due to their simplicity and generation properties.

The predictive distribution of CBMA is modified as follows (Madadgar and Moradkhani 2014):

$$f(q|s_1, s_2, \ldots, s_K) = \sum_{i=1}^{K} \omega_i f(q|s_i) = \sum_{i=1}^{K} \omega_i \cdot c(u, v_i) \cdot f(q) \qquad (8.39)$$

It can be seen from Eq. 8.39 that it relaxes any assumption on the type of posterior distribution $f(q|s_i)$, whose term can be directly inferred with the help of copula functions. Once the term $f(q|s_i)$ is defined, their weights are estimated by the EM algorithm with a few adjustments, which can refer to Madadgar and Moradkhani (2014) for details.

The hybrid CBMA model applies the idea of "pair and ensemble". The pair of observation q and the ith model simulation is established to get the probability task by the well-developed copula theory, while the ensemble is to formulate a consensus probability interval.

8.3.4 *Copula Bayesian Processor Associated with BMA (CBP-BMA) Method*

8.3.4.1 Copula Bayesian Processor (CBP)

Copula Bayesian processor (CBP) is developed as another component of the probabilistic forecasting system in virtue of the integration of Bayesian theory and copula functions. The CBP procedure generates a probabilistic result and quantifies the hydrologic uncertainty under the assumption that input uncertainty is ignored, which refers to hydrologic uncertainty processor (Krzysztofowicz and Kelly 2000). This method also has the advantage of leaving out a data transformation procedure into Gaussian space. The Bayesian procedure based on the law of total probability involves two parts for information revision of uncertainty (Zhang and Singh 2007a, b, c):

(1) The expected conditional density function of deterministic simulation, S_i given $Q = q$ is expressed as:

$$\kappa(s_i|q) = \int f(s_i|q) \cdot g(q)dq \tag{8.40}$$

where $f(q|s_i)$ has the same conception as before, $g(q)$ represents the prior density function.

(2) The posterior density function conditional on a deterministic result $S_i = s_i$ is derived via Bayes' theorem:

$$\phi(q|s_i) = \frac{f(s_i|q) \cdot g(q)}{\kappa(s_i|q)} \tag{8.41}$$

Equations 8.40 and 8.41 could be rewritten by using copula functions, i.e., the CBP form of the right term is mathematically expressed by:

$$\phi(q|s_i) = \frac{f(s_i|q) \cdot g(q)}{\int f(s_i|q) \cdot g(q)dq} = \frac{c(u, v_i)}{\int_0^1 c(u, v_i)du} \cdot g(q) \tag{8.42}$$

The final CBP outputs a posterior distribution of the process, conditional upon the deterministic simulation. Since the analytical solution to the integral term $\int_0^1 c(u, v_i)du$ is very complex, the Monte Carlo technique is used to estimate the posterior density function $\phi(q|s_i)$ (Robert and Casella 2011; Kroese et al. 2013).

8.3.4.2 The CBP-BMA Method

The difference between the CBP-BMA and CBMA methods is the estimation procedure of the posterior density function.

$$\phi(q|s_1, s_2, \ldots, s_K) = \sum_{i=1}^{K} \omega_i \phi(q|s_i) = \sum_{i=1}^{K} \omega_i \frac{c(u, v_i)}{\int_0^1 c(u, v_i)du} g(q) \tag{8.43}$$

It should be rational to assign weights on account of multiple deterministic results. The calculation process of weights is conducted by the EM algorithm (Montanai and Grossi 2008). The three main steps of the presented weights calculating paradigm can be summarized as follows:

$$\begin{aligned} w_i^{Iter} &= \frac{1}{T}\sum_{t=1}^{T} z_{i,t}^{Iter} \\ z_{i,t}^{Iter} &= \frac{w_i^{Iter-1} \cdot \phi(q_t|s_{i,t})}{\sum_{i=1}^{K} w_i^{Iter-1} \cdot \phi(q_t|s_{i,t})} = \frac{w_i^{Iter-1} \cdot c(u_t, v_{i,t}) g(q_t) / \int_0^1 c(u_i, v_{i,t}) du_t}{\sum_{i=1}^{K} w_i^{Iter-1} \cdot c(u_t, v_{i,t}) g(q_t) / \int_0^1 c(u_i, v_{i,t}) du_t} \\ l(\theta_{Iter}) &= \log\left(\sum_{i=1}^{K} w_i^{Iter-1} \cdot \sum_{i=1}^{K} c(u_i, v_{i,t}) g(q_t) / \int_0^1 c(u, v_{i,t}) du_t\right) \end{aligned} \tag{8.44}$$

where T is the length of the training period; and z is a latent variable. Compared with the standard BMA method, the calculation of variance and data transformations are eliminated in Eq. 8.44. The posterior probability of q_t is calculated only once while it need be re-calculated every time in the standard BMA method.

8.3.5 Evaluation Criteria for Multi-model Techniques

8.3.5.1 Deterministic Model Assessment Indices

To evaluate the quality of the deterministic model, three metrics are used.

(1) Nash-Sutcliffe efficiency coefficient (*NSE*), see Eq. 8.25
(2) Daily root mean square error (*DRMS*)

$$DRMS = \sqrt{\frac{\sum_{i=1}^{N} (q_o^i - q_m^i)^2}{T}} \tag{8.45}$$

As the second tool employed is sensitive to the differences between observations and simulations, the values of *DRMS* approaching to stand for better performance.

(3) Kling-Gupta efficiency (*KGE*)

$$
\begin{aligned}
KGE &= 1 - \sqrt{(r-1)^2 + (\beta - 1)^2 + (\gamma - 1)^2} \\
\beta &= \overline{q_m}/\overline{q_o} \\
\gamma &= CV_m/CV_o = \frac{\sigma_m/\overline{q_m}}{\sigma_o/\overline{q_o}}
\end{aligned}
\tag{8.46}
$$

where r is the Pearson correlation between the observation and simulation, b is the bias ratio indicator; g is the variability ratio (Kling et al. 2012). All calculative variables are replaced by the expected values of the estimate predictive distributions.

8.3.5.2 Verification of Probabilistic Simulations

With regard to assessment of assessing the uncertainty analysis of simulation interval, Xiong et al. (2009) and Dong et al. (2013) presented multiple verification indices and applied in hydrologic practice. Three main metrics are selected to evaluate the simulation uncertainty intervals generated by the BMA, CBMA and CBP-BMA methods.

(1) Containing ratio (*CR*)

The containing ratio is utilized as a significant index for assessing the goodness of the uncertainty interval. It is defined as the percentage of observed data points that fall between the prediction bounds, directly reflecting the interval performance.

$$
CR = \frac{C_{i=1}^{N}(q_l^i \le q_o^i \le q_u^i)}{N} \times 100\% \tag{8.47}
$$

where q_l^i is denoted as the lower bound corresponding to 5% quantile at time t, q_u^i is denoted as the upper bound corresponding to 95% of the quantile. $C_{i=1}^{N}$ is the number of the observed data points q_o^i that satisfy the inequality conditions.

(2) Average bandwidth (*BW*)

$$
B = \frac{1}{N}\sum_{i=1}^{N}(q_u^i - q_l^i) \tag{8.48}
$$

where BW is also an index measuring the average width of estimated uncertainty interval just as the definition name indicates. Smaller values of BW show a greater precision. Consider two forecasts with the same containing ratio; the situation with smaller BW is preferred because it has less uncertainty or greater precision.

(3) Average deviation amplitude (DA)

The average deviation amplitude DA is an index to quantify the average deflection of the curve of the middle points of the prediction bounds from the observed streamflow hydrograph. It is defined as

$$D = \frac{1}{N}\sum_{i=1}^{N}\left|\frac{1}{2}(q_u^i - q_l^i) - q_o^i\right| \tag{8.49}$$

where the notations are defined previously.

8.3.6 Case Study

The Mumahe catchment (Fig. 8.10), a sub-basin of Hanjiang River basin in China is selected as a case study. The catchment lies in Shanxi Province with an area of 1224 km^2 and locates in the subtropical monsoon region with a humid climate and fairly plenty of precipitation. The annual mean precipitation and runoff is 1070 and 687 mm, respectively. The available dataset contains daily precipitation, runoff, and evaporation with a length of 11 years (1980–1990). The first year (1980) is used as the spin-up period for each hydrologic model to achieve the best effective model formulation. The remaining years (1981–1990) are divided into two sub-periods, with seven years (1981–1987) for calibration and three years (1988–1990) for validation.

Different multi-model techniques, i.e., BMA, CBMA, and CBP-BMA, are applied to combine the ensemble flow simulation. The structures of three hydrologic models ought to be determined as the deterministic results are crucial to final uncertainty analysis. As mentioned above, the calibration parameters of the first BMA method are ω_k and σ_k^2; In the CBMA method, they are the parameters of marginal distributions, weights ω_k and the parameters of the PDF of the copula. In the CBP-BMA method, the Monte Carlo sampling technique is also used to obtain the integral item.

8.3.6.1 Deterministic Hydrologic Model Simulations

The genetic and simplex algorithms are used for model calibration on account of their flexibility and good convergence. The genetic algorithm can acquire the global optimal value with independent of initial parameter values. The simplex algorithm

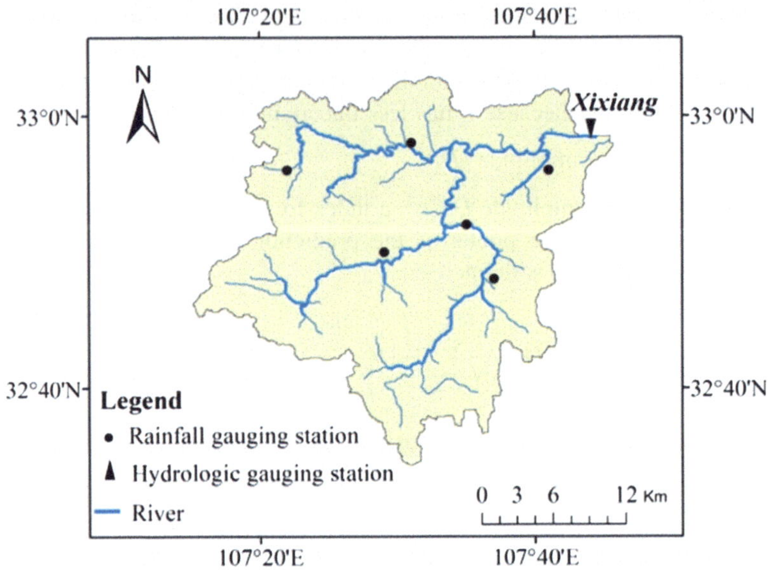

Fig. 8.10 Sketch map of the Mumahe catchment

is of high accuracy with low convergence rate. With the merit of two methods integrated, the approximately optimal values of model parameters are obtained. Three deterministic assessment indices: *NSE*, *DRMS* and *KGE* scores over the calibration period (1981–1987) and the validation period (1988–1990) are calculated for XAJ, HBV, and SIMHYD models. Table 8.6 indicates that the XAJ model has the best results, the HBV model takes the second place, and the SIMHYD model behaves worst among the three. The reason can be attributed to the dissimilar process for the calibration of each model (Nasonova et al. 2009). In practice, it might partially refer to inaccurate estimation of model parameters as one of the error sources of the model structure, abstract formulation of physical processes, and different sources of forcing data set for each model. In general, these simulation results can be used as the input data of multi-model ensemble in terms of the *NSE* and *KGE* values, which are 85% and higher than 0.82 respectively beside the ill value of *KGE* of HBV.

8.3.6.2 Determination of the Marginal Distributions

The marginal distributions of the random variables of H and S_i (i = 1, 2, 3) need to be determined. Five common candidate distributions, namely Normal, Gamma, Gumbel, P-III and Log-Normal, have been fitted to the daily mean streamflow values as well as to the XAJ, HBV and SIMHYD model simulations.

Table 8.6 Deterministic accuracy assessment of different hydrological models

Model	Calibration			Validation		
	NSE (%)	*DRME*	*KGE* (%)	*NSE* (%)	*DRME*	*KGE* (%)
XAJ	88.25	30.06	90.59	84.85	24.06	87.08
HBV	84.81	34.16	52.99	82.24	26.06	42.89
SIMHYD	86.25	32.50	82.51	84.98	23.97	85.01
BMA	88.87	33.79	89.14	85.72	24.81	**90.62**
CBMA	88.93	**26.54**	90.06	86.07	**21.25**	88.45
CBP-BMA	**89.76**	27.63	**90.96**	**86.69**	23.39	89.23

Notes Values in bold represent the optimal result

Regarding random variable H, the parameters of five candidate distributions are estimated by the method of L-moment (Hosking 1990), and the parameter values are listed in Table 8.7. The K-S tests are used to verify the null hypothesis, and the corresponding statistic $D_{K\text{-}S}$ values are also listed in Table 8.7. It is shown that the null hypothesis could not be rejected at the 95% confidence level (threshold value $D_n = 1.36/\sqrt{N}$, N is the number of sampling points) for Log-Normal distribution with providing the minimum $D_{K\text{-}S}$ value. Meanwhile, Fig. 8.11 indicate the Log-Normal is satisfactory on visual inspection that the cumulative distribution function (CDF) plots of the theoretical Log-Normal distributions fitted the empirical CDF values obtained from the Gringorten plotting-position formula (Zhang and Singh 2006) relatively well. The estimation of marginal distributions for S_i had the similar procedures. The Kolmogorov-Smirnov statistics $D_{K\text{-}S}$ indicate that the Log-Normal distribution also gives the best fit in this study.

8.3.6.3 Archimedes Copula Selection and Estimation

In the application of the CBMA and CBP-BMA methods, a copula function to link the CDF of observation and model simulations needs to be defined. The Gumbel, Clayton and Frank copula belonging to Archimedes family are chosen to test for flexibility and universality (Madadgar and Moradkhani 2014; Chen et al. 2015).

For Archimedes copula, the Kendall correlation coefficient τ_i (i = 1, 2, 3) between observed and different simulated flows is firstly derived. The higher τ_i indicator reflects the stronger correlation between observation and model simulation. The corresponding copula parameter θ_i is calculated by the method based on the inversion of τ_i in Table 2.3 of Chap. 2. The parameter estimators and goodness-of-fit test (*RMSE* and *AIC*) are used to determine the best fit copula for integrating the streamflow properties. The results illustrate that copulas have the good performance in exploring the associations of observed and simulated flows. All variables passed the null hypothesis for Gumbel and Frank copulas. Gumbel copula performs with the lowest *RMSE* and *AIC* values.

Table 8.7 Estimated parameters and statistic test $D_{k\text{-}s}$ of five candidate marginal distributions

Marginal parameter distribution and *K-S* test		*H*	S_1	S_2	S_3
Gumbel	σ	40.3	40.6	39.5	37.2
	μ	17.8	14.9	15.1	14.5
	$D_{k\text{-}s}$	0.025	0.024	0.026	0.026
Gamma	α	0.4	0.3	0.3	0.3
	β	103.4	129.7	153.2	106.5
	$D_{k\text{-}s}$	0.148	0.122	0.141	0.153
P-III	α	0.20	0.24	0.20	0.19
	β	0.0056	0.0067	0.0050	0.0063
	γ	6.02	4.43	5.27	4.59
	$D_{k\text{-}s}$	0.034	0.028	0.029	0.031
Log-normal	α	2.83	2.11	2.43	2.57
	γ	1.17	1.46	1.28	1.26
	$D_{k\text{-}s}$	**0.016**	**0.013**	**0.017**	**0.015**
Normal	α	41.08	38.27	38.43	34.69
	γ	49.57	49.85	41.19	43.03
	$D_{k\text{-}s}$	0.074	0.062	0.075	0.064

Notes Values in bold denote that the distribution model passes the goodness-of-fit test at 0.05 significance levels

8.3.6.4 Deterministic Assessment of Three Ensemble Methods

We check the mean simulation of hydrologic multi-model ensembles using three criteria illustrated in Sect. 8.3.5.1. The effectiveness results of BMA, CBMA and CBP-BMA methods are listed in Table 8.6. The performances of different multi-models are better than that of the individual XAJ model regarding *NSE*. The BMA method outperforms the reference model at the cast of *DRMS* and *KGE* indicators, The CBMA and CBP-BMA methods slightly improve in all aspects during the calibration period, which have excellent properties in the validation period. The reason of the CBMA and CBP-BMA methods enhancing the performance can be attributed to that copula functions are efficient tools to remove bias instead of a simple bias correction such as linear regression in the BMA method (Madadgar and Moradkhani 2014). Especially, copula has reliable parameter estimation prior model average procedure. Another reason might be owed to the weight of each individual model, which is directly influenced by the estimation of posterior distributions.

Figure 8.12 illustrates the bar plots of *KGE* score and its components. The *KGE* score might be a little descending through BMA or CBMA application, a little incremental through CBP-BMA application in comparison with the best XAJ model. The correlation coefficients between observation and simulation of individual models are up to 0.93 in the calibration period and 0.92 in the verification

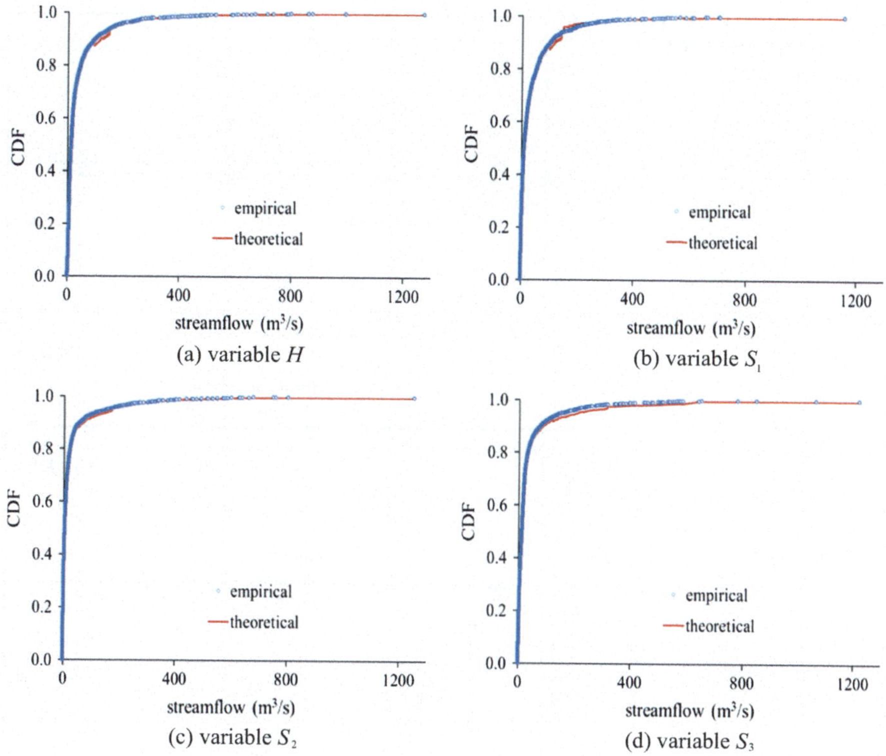

Fig. 8.11 Comparison of the empirical and theoretical cumulative distribution functions

period, which represent stronger correlation for the values are more than 0.9. However, β indictor of deterministic models varies from 0.64 for HBV model to 0.97 for XAJ model. The value less than 1 indicates the total amount of streamflow simulation in any individual model is less than that of observation. It might cause the general underestimation of the mean streamflow (negative bias) in hydrological multi-model ensemble applications. The BMA method is such a promising method for locating simulation to observation for its term β closer to 1. Regarding the variability ratio, all methods except for HBV could perfectly perform, but no particular method is superior to others with all $\gamma \approx 1$.

8.3.6.5 Probabilistic Verification of Three Ensemble Methods

For probabilistic verification of simulation, Figs. 8.13 and 8.14 describe the uncertainty bands of different methods for the representative year during calibration and verification periods with a visual inspection. These two plots indicate that the observed values approximately fall within the 5–95% uncertainty range and fit the

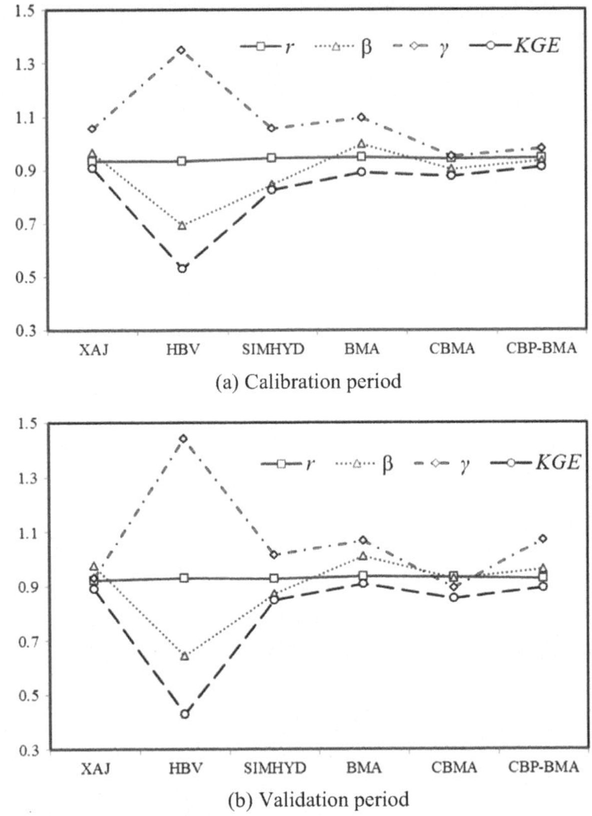

Fig. 8.12 The simulation results of *KGE* score and its components

mean flow hydrograph for all multi-model ensembles. In this case, the 90% confidence interval could capture the flood peaks but miss more low flow values.

Three probabilistic verification measurements (*CR*, *BW*, *DA*) are presented in Table 8.8. It can be seen from these quantitative indices that they have a good performance regarding containing ratio, which is corresponding to the confidence interval. The probability of observed value falling in the range should be in accord with the percentage of confidence interval containing points through many independent statistical experiments. The CBP-BMA method performs better than CBMA method regarding *CR* index because it roughly covers 91% of the sample points, which is more than CBMA does. A combination of *CR* and *BW* possess the power to make a decision on model probabilistic performance. The comparison between the CBMA (and CBP-BMA) and BMA methods exactly illustrates that the CBMA method outperforms the BMA method, either *CR*, *BW* or *DA*, especially, the containing ratios of CBP-BMA method in different periods are up to 91.17 and 91.33%, respectively. Referring to the smaller *BW* result in the CBMA and CBP-BMA methods, the total predictive variance is reduced by relaxing the PDF generated by copula functions rather than the Gaussian posterior distribution via

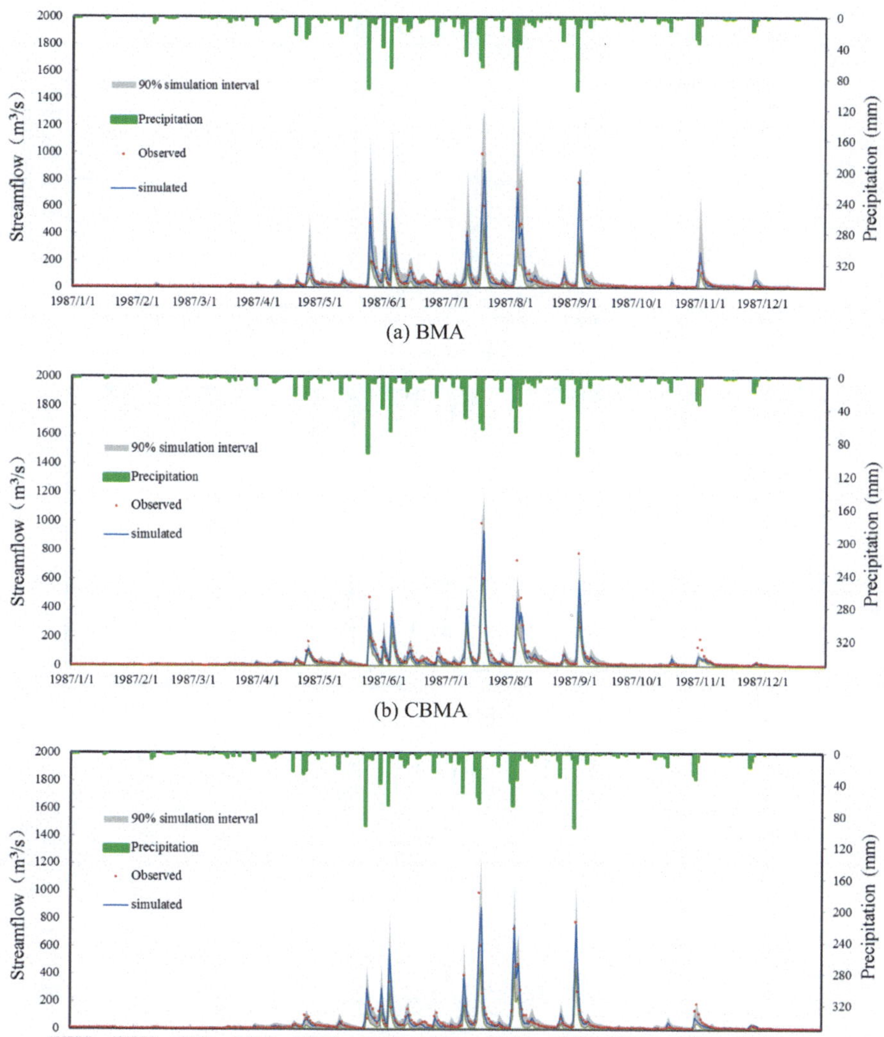

(a) BMA

(b) CBMA

(c) CBP-BMA

Fig. 8.13 The 90% uncertainty interval, observed, mean simulation for the Mumahe catchment in 1987 during the calibration period

box-cox transformation. Since the between-model variance keeps identical after using the same EM algorithm in all three methods, it is inferred that the reduction of within-model variance works.

The CBMA and CBP-BMA methods are two flexible and robust approaches to estimate uncertainty regarding the optimal bandwidth and average deviation amplitude. They have an intuitive and simple structure conditional on several model

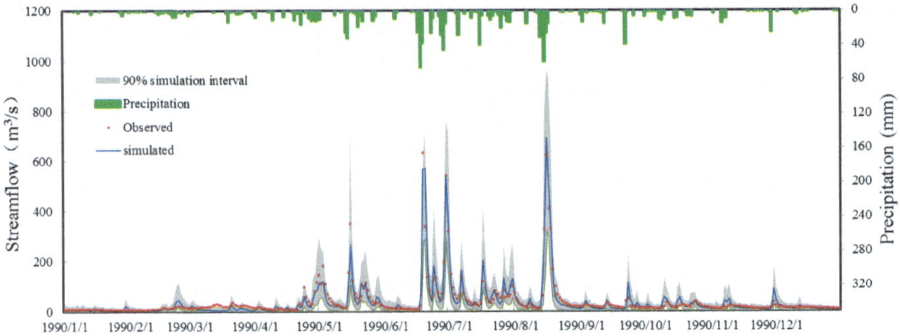

(a) BMA

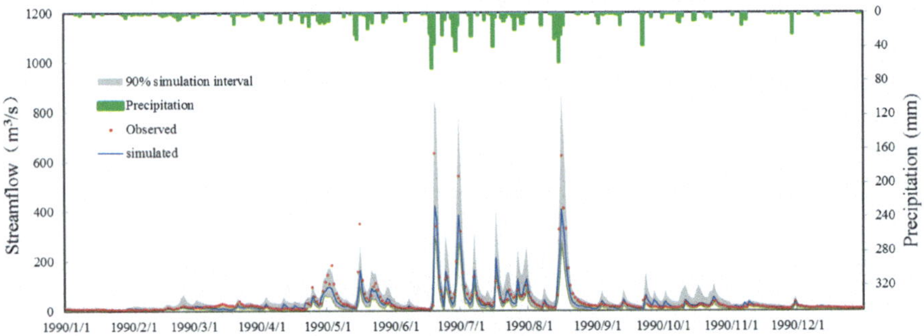

(b) CBMA

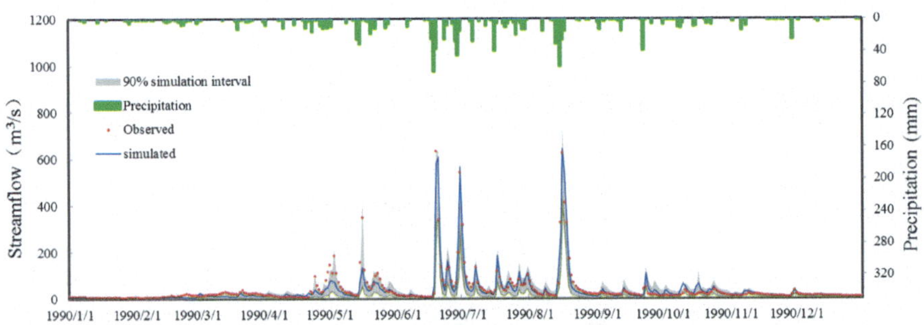

(c) CBP-BMA

Fig. 8.14 The 90% uncertainty interval, observed, mean simulation for the Mumahe catchment in 1990 during the validation period

simulations by the integration of BMA and copula tools, which makes this method promising to derive uncertainty. The difference between them reflected in the procedure of processing posterior distribution. Further improvement might be realized through the weight allocation for each model or the nonparametric posterior distribution.

Table 8.8 Uncertainty assessment of different hydrological multi-model ensembles

Model	Calibration			Validation		
	CR (%)	*BW*	*DA*	*CR* (%)	*BW*	*DA*
BMA	87.34	**56.79**	21.26	88.53	**58.14**	18.79
CBMA	89.23	38.75	16.52	89.76	40.28	12.26
CBP-MA	**91.17**	45.28	**12.35**	**91.33**	42.35	**10.99**

Note Values in bold represent the optimal result

8.4 Conclusion

Hydrological forecasting services are trending toward providing users with probabilistic forecasting, and adequate assessment of uncertainty forecasts is an important issue and task. A copula-based HUP for probabilistic forecasting and CBP-BMA method for evaluating uncertainties of hydrologic multi-model ensembles are proposed. Three Gorges Reservoir (TGR) and Mumahe basins are selected as case studies. The main conclusions are summarized as follows:

(1) The output of the HUP is a posterior distribution of the process, conditional upon the deterministic forecast. This posterior distribution provides the complete and well-calibrated characterization of uncertainty needed by rational decision makers who use formal decision models and by information providers who want to extract various forecast products for their customers (e.g., quantiles with specified exceedance probabilities, prediction intervals with specified inclusion probabilities, probabilities of exceedance for specified thresholds).

(2) Based on copula function, the prior density and likelihood function of the HUP are explicitly expressed, and the corresponding posterior density and distribution can be obtained using the Monte Carlo sampling technique. This copula-based HUP can be implemented in the original space directly without a data transformation procedure into Gaussian space and allows for any form of marginal distribution of predictand and the deterministic forecast variable, and a nonlinear and heteroscedastic dependence structure.

(3) The proposed copula-based HUP is comparable to the meta-Gaussian HUP regarding the posterior median forecasts. It is also shown that probabilistic forecasts produced by the copula-based HUP have slightly higher reliability and lower resolution compared with the meta-Gaussian HUP. According to the CRPS value, it is found that both HUPs are superior to deterministic forecasts which highlight the effectiveness of probabilistic forecasts, and the copula-based HUP is marginally better than the meta-Gaussian HUP.

(4) Deterministic results of different multi-model ensembles outperform those of the individual model. The CBMA and CBP-BMA methods slightly outperform BMA method regarding *NSE*, *DRMS*, and *KGE*. When the CBMA method is used as a reference, the CBP-BMA method can improve the *NSE* and *KGE* and enlarge *DRMS* values. Underestimation of all individual models may cause negative bias of ensemble multi-model.

(5) The combination of containing ratio and bandwidth index demonstrates the probabilistic model performance with the auxiliary index-average deviation amplitude. It is found that containing ratio is approximately equal to the percentage of the confidence interval. The CBMA or CBP-BMA methods outperform BMA method regarding evaluation criteria with a high containing ratio, small uncertainty, and average deviation amplitude.

References

Ajami NK, Duan QY, Sorooshian S (2007) An integrated hydrologic Bayesian multimodel combination framework: confronting input, parameter, and model structural uncertainty in hydrologic prediction. Water Resour Res 43(1):W01403

Arya DS, Goel NK, Dhamy AP (2010) Design flow and stage computations in the Teesta River, Bangladesh, using frequency analysis and MIKE 11 modeling. J Hydrol Eng 16(2):176–186

Ba HH, Guo SL, Wang Y, Hong XJ, Zhong YX (2017) Improving ANN model performance in runoff forecasting by adding soil moisture input and using data preprocessing techniques. Hydrol Res

Bárdossy Götzinger J (2008) A Generic error model for calibration and uncertainty estimation of hydrological models. Water Resour Res 44(12):1393–1442

Bárdossy A, Li J (2008) Geostatistical interpolation using copulas. Water Resour Res 44:W07412

Bergstrand M, As SS, Lindström G (2014) Nationwide hydrological statistics for Sweden with high resolution using the hydrological model S-HYPE. Hydrol Res 45(3):349–356

Biondi D, De Luca DL (2013) Performance assessment of a Bayesian forecasting system (BFS) for real-time flood forecasting. J Hydrol 479(1):51–63

Biondi D, Versace P, Sirangelo B (2010) Uncertainty assessment through a precipitation dependent hydrologic uncertainty processor: an application to a small catchment in southern Italy. J Hydrol 386(1):38–54

Bogner K, Pappenberger F, Cloke HL (2012) Technical note: the normal quantile transformation and its application in a flood forecasting system. Hydrol Earth Syst Sci 16(4):1085–1094

Calvo B, Savi F (2009) Real-time flood forecasting of the Tiber River in Rome. Nat Hazards 50 (3):461–477

Carreau J, Bouvier C (2016) Multivariate density model comparison for multi-site flood-risk rainfall in the French Mediterranean area. Stoch Env Res Risk Assess 30(6):1591–1612

Castellarin A, Vogel RM, Brath A (2004) A stochastic index flow model of flow duration curves. Water Resour Res 40(3). https://doi.org/10.1029/2003wr002524

Chebana F, Ouarda TBMJ (2007) Multivariate L-moment homogeneity test. Water Resour Res 43 (8):199–212

Chen ST, Yu PS (2007) Real-time probabilistic forecasting of flood stages. J Hydrol 340:63–77

Chen L, Guo SL, Yan B, Pan L, Fang B (2010) A new seasonal design flood method based on bivariate joint distribution of flood magnitude and date of occurrence. Hydrol Sci J 55(8):1264–1280

Chen FJ, Jiao MY, Chen J (2013) The meta-Gaussian Bayesian processor of forecasts and associated preliminary experiments. Acta Meteorologica Sinica 27:199–210

Chen L, Singh VP, Guo SL, Zhou J, Ye L (2014a) Copula entropy coupled with artificial neural network for rainfall-runoff simulation. Stoch Env Res Risk Assess 28(7):1755–1767

Chen L, Ye L, Singh VP, Asce F, Zhou J, Guo SL (2014b) Determination of input for artificial neural networks for flood forecasting using the copula entropy method. J Hydrol Eng 19 (11):04014021

Chen L, Zhang Y, Zhou J, Singh VP, Guo SL, Zhang J (2015) Real-time error correction method combined with combination flood forecasting technique for improving the accuracy of flood forecasting. J Hydrol 521:157–169

Cheng C, Zhao M, Chau K, Wu X (2006) Using genetic algorithm and TOPSIS for Xinanjiang model calibration with a single procedure. J Hydrol 316(1):129–140

Chiew FHS, Teng J, Vaze J, Post DA, Perraud JM, Kirono DGC, Viney NR (2009) Estimating climate change impact on runoff across southeast Australia: method, results, and implications of the modeling method. Water Resour Res 45(10):82–90

Coccia G, Todini E (2011) Recent developments in predictive uncertainty assessment based on the model conditional processor approach. Hydrol Earth Syst Sci 15(10):3253–3274

Dong LH, Xiong LH, Yu KX (2013) Uncertainty analysis of multiple hydrologic models using the bayesian model averaging method. J Appl Math 2013:1–11

Duan QY, Ajami NK, Gao X, Sorooshian S (2007) Multi-model ensemble hydrologic prediction using Bayesian model averaging. Adv Water Resour 30(5):1371–1386

Engeland K, Renard B, Steinsland I, Kolberg S (2010) Evaluation of statistical models for forecast errors from the HBV model. J Hydrol 384(1):142–155

Evin G, Thyer M, Kavetski D, Mcinerney D, Kuczera G (2014) Comparison of joint versus postprocessor approaches for hydrological uncertainty estimation accounting for error autocorrelation and heteroscedasticity. Water Resour Res 50(3):2350–2375

Fan YR, Huang GH, Li YP, Wang XQ, Li Z (2016) Probabilistic prediction for monthly streamflow through coupling stepwise cluster analysis and quantile regression methods. Water Resour Manag 30(14):5313–5331

Favre AC, Adlouni SE, Perreault L, Thiémonge N, Bobée B (2004) Multivariate hydrological frequency analysis using copulas. Water Resour Res 40W01101. https://doi.org/10.1029/2003wr002456

Freer J, Beven K, Ambroise B (1996) Bayesian estimation of uncertainty in runoff prediction and the value of data: an application of the GLUE approach. Water Resour Res 32(7):2161–2173

Genest C, Favre A (2007) Everything you always wanted to know about copula modeling but were afraid to ask. J Hydrol Eng 12(4):347–368

Gneiting T, Raftery AE, Westveld AH, Goldman T (2005) Calibrated probabilistic forecasting using ensemble model output statistics and minimum CRPS estimation. Mon Weather Rev 133(5):1098–1118

Gneiting T, Balabdaoui F, Raftery AE (2007) Probabilistic forecasts, calibration and sharpness. J Roy Stat Soc: Ser B (Stat Methodol) 69(2):243–268

Gottschalk L, Yu K, Leblois E, Xiong L (2013) Statistics of low flow: theoretical derivation of the distribution of minimum streamflow series. J Hydrol 481:204–219

Guo SL, Zhang H, Chen H, Peng D, Liu P, Pang B (2004) A reservoir flood forecasting and control system for China. Hydrol Sci J 49(6):959–972

He SK, Guo SL, Liu ZJ, Yin JB, Chen KB, Wu XS. (2018) Uncertainty analysis of hydrological multi-model ensembles based on CBP-BMA method. Hydrol Res (in press)

Hemri S, Lisniak D, Klein B (2015) Multivariate post-processing techniques for probabilistic hydrological forecasting. Water Resour Res 51(9):7436–7451

Hersbach H (2000) Decomposition of the continuous ranked probability score for ensemble prediction systems. Weather Forecast 15(5):559–570

Hosking JRM (1990) L-moments: analysis and estimation of distributions using linear combinations of order statistics. J Roy Stat Soc 52(1):105–124

Kasiviswanathan KS, Cibin R, Sudheer KP, Chaubey I (2013) Constructing prediction interval for artificial neural network rainfall runoff models based on ensemble simulations. J Hydrol 499:275–288

Khajehei S, Moradkhani H (2017) Towards an improved ensemble precipitation forecast: a probabilistic post-processing approach. J Hydrol 546:476–489

Klein B, Meissner D, Kobialka HU, Reggiani P (2016) Predictive uncertainty estimation of hydrological multi-model ensembles using pair-copula construction. Water 8(4):1–22

Kling H, Fuchs M, Paulin M (2012) Runoff conditions in the upper Danube basin under an ensemble of climate change scenarios. J Hydrol 424–425:264–277
Koutsoyiannis D, Montanari A (2015) Negligent killing of scientific concepts: the stationarity case. Hydrol Sci J 60(7–8):1174–1183
Kroese DP, Taimre T, Botev ZI (2013) Handbook of Monte Carlo methods. Wiley, New York
Krzysztofowicz R (1999) Bayesian theory of probabilistic forecasting via deterministic hydrologic model. Water Resour Res 35(9):2739–2750
Krzysztofowicz R, Kelly KS (2000) Hydrologic uncertainty processor for probabilistic river stage forecasting. Water Resour Res 36(11):3265–3277
Kuczera G, Parent E (1998) Monte Carlo assessment of parameter uncertainty in conceptual catchment models: the metropolis algorithm. J Hydrol 211(1–4):69–85
Laio F, Tamea S (2007) Verification tools for probabilistic forecasts of continuous hydrological variables. Hydrol Earth Syst Sci 11(4):1267–1277
Li L, Xia J, Xu CY, Singh VP (2010a) Evaluation of the subjective factors of the GLUE method and comparison with the formal Bayesian method in uncertainty assessment of hydrological models. J Hydrol 390(3):210–221
Li X, Guo SL, Liu P, Chen G (2010b) Dynamic control of flood limited water level for reservoir operation by considering inflow uncertainty. J Hydrol 391:124–132
Li Z, Xin P, Tang J (2011) Study of the Xinanjiang model parameter calibration. J Hydrol Eng 18 (11):1513–1521
Li H, Beldring S, Xu CY (2014) Implementation and testing of routing algorithms in the distributed HBV model for mountainous catchments. Hydrol Res 45(3):322–333
Liang G, Kachroo RK, Kang W, Yu X (1992) River flow forecasting. part 4. applications of linear modelling techniques for flow routing on large catchments. J Hydrol 133(1):99–140
Liang Z, Wang D, Guo Y, Zhang Y, Dai R (2011) Application of Bayesian model averaging approach to multimodel ensemble hydrologic forecasting. J Hydrol Eng 18(11):1426–1436
Lin K, Lv F, Chen L, Singh VP, Zhang Q, Chen X (2014) Xinanjiang model combined with Curve Number to simulate the effect of land use change on environmental flow. J Hydrol 519:3142–3152
Liu ZJ, Guo SL, Zhang HG, Liu DD, Yang G (2016) Comparative study of three updating procedures for real-time flood forecasting. Water Resour Manag 30(7):2111–2126
Liu ZJ, Guo SL, Xiong LH, Xu CY (2017) Hydrological uncertainty processor based on a copula function. Hydrol Sci. J. https://doi.org/10.1080/02626667.2017.1410278
Liucci L, Valigi D, Casadei S (2014) A new application of flow duration curve (FDC) in designing run-of-river power plants. Water Resour Manag 28(3):881–895
Ma Z, Li Z, Zhang M, Fan Z (2013) Bayesian statistic forecasting model for middle-term and long-term runoff of a hydropower station. J Hydrol Eng 18(11):1458–1463
Madadgar S, Moradkhani H, Garen D (2014) Towards improved post-processing of hydrologic forecast ensembles. Hydrol Process 28(1):104–122
Madsen H (2000) Automatic calibration of a conceptual rain-fall model using multiple objectives. J Hydrol 235:276–288
Möller A, Lenkoski A, Thorarinsdottir TL (2013) Multivariate probabilistic forecasting using ensemble Bayesian model averaging and copulas. J Roy Meteorol Soc 139(673):982–991
Montanari A (2007) What do we mean by 'uncertainty'? The need for a consistent wording about uncertainty assessment in hydrology. Hydrol Process 21(6):841–845
Montanari A, Grossi G (2008) Estimating the uncertainty of hydrological forecasts: a statistical approach. Water Resour Res 44:W00B08. https://doi.org/10.1029/2008-wr006897
Montero RA, Schwanenberg D, Krahe P, Lisniak D, Sensoy A, Sorman AA, Akkol B (2016) Moving horizon estimation for assimilating H-SAF remote sensing data into the HBV hydrological model. Adv Water Resour 92:248–257
Nash J, Sutcliffe JV (1970) River flow forecasting through conceptual models part I-A discussion of principles. J Hydrol 10(3):282–290

Nasonova ON, Gusev YM, Kovalev YE (2009) Investigating the ability of a land surface model to simulate streamflow with the accuracy of hydrological models: a case study using MOPEX materials. J Hydrometeorol 10(5):1128–1150

Nelsen RB (2006) An introduction to copulas, 2nd edn. Springer, New York

Pappenberger F, Ramos MH, Cloke HL, Wetterhall F, Alfieri L, Bogner K (2015) How do I know if my forecasts are better? Using benchmarks in hydrological ensemble prediction. J Hydrol 522:697–713

Pokhrel P, Robertson D, Wang QJ (2013) A Bayesian joint probability post-processor for reducing errors and quantifying uncertainty in monthly streamflow predictions. Hydrol Earth Syst Sci 17 (2):795–804

Raftery AE, Gneiting T, Balabdaoui F, Polakowski M (2005) Using Bayesian model averaging to calibrate forecast ensembles. Mon Weather Rev 133(5):1155–1174

Ramos MH, van Andel SJ, Pappenberger F (2013) Do probabilistic forecasts lead to better decisions? Hydrol Earth Syst Sci 17(6):2219–2232

Ravines RR, Schmidt AM, Migon HS, Rennó CD (2008) A joint model for rainfall-runoff: the case of Rio Grande Basin. J Hydrol 353(1):189–200

Reggiani P, Weerts AH (2008) A Bayesian approach to decision-making under uncertainty: an application to real-time forecasting in the river Rhine. J Hydrol 356(1):56–69

Renard B, Kavetski D, Kuczera G, Thyer M, Franks SW (2010) Understanding predictive uncertainty in hydrologic modeling: The challenge of identifying input and structural errors. Water Resour Res 46(5) https://doi.org/10.1029/2009wr008328

Robert C, Casella G (2013) Monte Carlo statistical methods. Springer, New York

Seo DJ, Herr HD, Schaake JC (2006) A statistical post-processor for accounting of hydrologic uncertainty in short-range ensemble streamflow prediction. Hydrol Earth Syst Sci 3(4):1987–2035

Shao QX, Zhang L, Chen YD, Singh VP (2009) A new method for modelling flow duration curves and predicting streamflow regimes under altered land-use conditions. Hydrol Sci J 54(3):606–622

Si W, Bao W, Gupta HV (2015) Updating real-time flood forecasts via the dynamic system response curve method. Water Resour Res 51(7):5128–5144

Sikorska AE, Scheidegger A, Banasik K, Rieckermann J (2012) Bayesian uncertainty assessment of flood predictions in ungauged urban basins for conceptual rainfall-runoff models. Hydrol Earth Syst Sci 16(4):1221–1236

Smith LA, Suckling EB, Thompson EL, Maynard T, Du H (2015) Towards improving the framework for probabilistic forecast evaluation. Clim Change 132(1):31–45

Thyer M, Renard B, Kavetski D, Kuczera G, Franks SW, Srikanthan S (2009) Critical evaluation of parameter consistency and predictive uncertainty in hydrological modeling: a case study using Bayesian total error analysis. Water Resour Res 45(12). https://doi.org/10.1029/2008wr006825

Tsai CN, Adrian DD, Singh VP (2001) Finite Fourier probability distribution and applications. J Hydrol Eng 6(6):460–471

Verkade JS, Werner MGF (2011) Estimating the benefits of single value and probability forecasting for flood warning. Hydrol Earth Syst Sci 15(12):3751–3765

Vogel RM, Fennessey NM (1994) Flow-duration curves. I: new interpretation and confidence intervals. J Water Resour Plann Manag 120(4):485–504

Vogel RM, Fennessey NM (1995) Flow-duration curves. II: a review of applications in water resources planning. J Am Water Resour Assoc 31(6):1029–1039

Vrugt JA, Robinson BA (2007) Treatment of uncertainty using ensemble methods: comparison of sequential data assimilation and Bayesian model averaging. Water Resour Res 43(1):223–228

Weerts AH, Winsemius HC, Verkade JS (2011) Estimation of predictive hydrological uncertainty using quantile regression: examples from the National Flood Forecasting System (England and Wales). Hydrol Earth Syst Sci 15(1):255–265

Wetterhall F, Pappenberger F, Cloke HL, Pozo JT (2013) Forecasters priorities for improving probabilistic flood forecasts. Hydrol Earth Syst Sci 17(11):4389–4399

Wu J, Zhou JZ, Chen L, Ye L (2015) Coupling forecast methods of multiple rainfall–runoff models for improving the precision of hydrological forecasting. Water Resour Manage 29 (14):5091–5108

Wu XS, Wang ZL, Guo SL, Liao W, Zeng Z, Chen X (2017) Scenario-based projections of future urban inundation within a coupled hydrodynamic model framework: a case study in Dongguan City, China. J Hydrol 547:428–442

Xiong LH, Guo SL (1999) A two-parameter monthly water balance model and its application. J Hydrol 216(1):111–123

Xiong LH, Min W, Wei XJ, O'Connor KM (2009) Indices for assessing the prediction bounds of hydrological models and application by generalised likelihood uncertainty estimation. Hydrol Sci J 54(5):852–871

Xiong LH, Yu KX, Gottschalk L (2014) Estimation of the distribution of annual runoff from climatic variables using copulas. Water Resour Res. https://doi.org/10.1029/2008WR006897

Xiong L, Du T, Xu CY, Guo SL, Jiang C, Gippel CJ (2015) Non-Stationary annual maximum flood frequency analysis using the norming constants method to consider non-Stationarity in the annual daily flow series. Water Resour Manag 29(10):3615–3633

Xu H, Xu CY, Chen H, Zhang Z, Li L (2013) Assessing the influence of rain gauge density and distribution on hydrological model performance in a humid region of China. J Hydrol 505:1–12

Yokoo Y, Sivapalan M (2011) Towards reconstruction of the flow duration curve: development of a conceptual framework with a physical basis. Hydrol Earth Syst Sci 15(9):2805–2819

Yu B, Zhu Z (2015) A comparative assessment of AWBM and SimHyd for forested watersheds. Hydrol Sci J 60(7):1–13

Yu KX, Xiong LH, Gottschalk L (2014) Derivation of low flow distribution functions using copulas. J Hydrol 508:273–288

Zhang L, Singh VP (2006) Bivariate flood frequency analysis using the copula method. J Hydrol Eng 11(2):150–164

Zhang L, Singh VP (2007a) Bivariate rainfall frequency distributions using Archimedean copulas. J Hydrol 332(1):93–109

Zhang L, Singh VP (2007b) Gumbel-Hougaard copula for trivariate rainfall frequency analysis. J Hydrol Eng 12(4):409–419

Zhang L, Singh VP (2007c) Trivariate flood frequency analysis using the Gumbel-Hougaard copula. J Hydrol Eng 12(4):431–439

Zhang Q, Chen YD, Chen X, Li J (2011) Copula-based analysis of hydrological extremes and implications of hydrological behaviors in the Pearl River basin, China. J Hydrol Eng 16 (7):598–607

Zhang Q, Li J, Singh VP (2012) Application of Archimedean copulas in the analysis of the precipitation extremes: effects of precipitation changes. Theor Appl Climatol 107(1–2):255–264

Zhang J, Chen L, Singh VP, Cao W, Wang D (2015) Determination of the distribution of flood forecasting error. Nat Hazards 75(2):1389–1402

Zhao RJ (1992) The Xinanjiang model applied in China. J Hydrol 135(1–4):371–381

Zhao T, Wang QJ, Bennett JC, Robertson DE, Shao Q, Zhao J (2015) Quantifying predictive uncertainty of streamflow forecasts based on a Bayesian joint probability model. J Hydrol 528:329–340

Zhou YL, Guo SL, Xu CY, Chen H, Guo J, Lin K (2016) Probabilistic prediction in ungauged basins (PUB) based on regional parameter estimation and Bayesian model averaging. Hydrol Res 47(6):1087–1103

Chapter 9
Copula-Based Uncertainty Evolution Model for Flood Forecasting

9.1 Introduction

Flood is a major natural disaster in many countries and its consequences can be enormous in terms of property loss and fatalities (Pham 2011). From the year 1983–2003, floods caused an average of 98 deaths and $4.5 billion in property damage per year in the United States (Morss et al. 2005). According to the European Commission (2011), in Europe, there were more than 100 major damaging floods within a period of three years between 1998 and 2002, which represents, on average, more than 30 floods per year. Flood forecasting provides much information for future floods and water resources management (Alemu et al. 2011; Boucher et al. 2012). However, the uncertainty in flood forecasting has been identified as the major factor influencing the accuracy of forecasting. Uncertainty in flood forecasting may lead to under-preparation which can cause damages or losses, or over-preparation which can cause unnecessary expense and anxiety (Smith and Ward 1998). Therefore, one important issue is to deal with the uncertainty in flood forecasting (Chen et al. 2015).

Until now, there is a large amount of work on forecasting uncertainty. The uncertainties typically consist of a combination of input, model structural, output, and parameter uncertainty (Schoups and Vrugt 2010), and are usually measured by forecast error series. A multitude of investigations have been carried out to quantify forecast uncertainties (Pokhrel et al. 2013; Li et al. 2010; Montanari and Grossi 2008; Krzysztofowicz 2002). Since flood forecasts are dynamically updated, forecast uncertainty evolves in real-time. Usually, the forecast uncertainty for a certain time period decreases over time as more hydrologic information becomes available (Zhao et al. 2011). However, limited attention has been paid to investigate the relationship of the forecast uncertainty between the adjacent time steps, and study the evolution of uncertainty over time in real-time flood forecasting.

Heath and Jackson (1994) proposed a martingale model to describe the evolution of uncertainty of demand forecasts in the supply chain management. Zhao et al. (2011)

L. Chen and S. Guo, *Copulas and Its Application in Hydrology and Water Resources*, Springer Water, https://doi.org/10.1007/978-981-13-0574-0_9

introduced this method in hydrology and used it to describe the evolution of streamflow forecast uncertainty. However, this method makes assumptions regarding un-biasedness, Gaussianity, temporal independence and stationary, which limits their application in hydrology. Zhang et al. (2015) demonstrated that the assumption of normal distribution was not justified for the flood forecast error series. In order to overcome these limitations, Zhao et al. (2013) modified this method and proposed the generalized martingale model for the evolution of uncertainty in streamflow forecasting to address cases wherein the assumptions are violated. However, the method they proposed is based on a normal quantile transform (NQT), which needs to convert the original variables into a standard Gaussian random variable and an inverse transformation is also needed in the last step. These transformations may bring cumbersome calculation work and potential errors. Chen et al. (2016) proposed a copula-based uncertainty evolution (CUE) model to quantify the evolution of uncertainty in flood forecasting.

Forecasting products predict the inflow of reservoirs and provide hydrologic information to guide reservoir operation. However, flood forecasting is inherently uncertain. That is why some of the reservoir operation risk is unavoidable and the reservoir may sometimes become even an operational failure. Tracing the uncertainties associated with the predicted inflow volumes of reservoirs and propagating them through the reservoir system can help gain information for reservoir decision making, improve reservoir operation efficiency, and reduce flood risk under extreme conditions.

Limited research has been carried out to study the effect of forecast uncertainty on real-time reservoir operation (Zhao et al. 2011). Li et al. (2010) discussed the dynamic control of flood limited water level by considering the inflow uncertainty in the reservoir operation. Zhao et al. (2011) modeled the dynamic evolution of uncertainties involved in the various forecast products and explored their effect on real-time reservoir operation decisions. However, all these studies are based on certain assumptions, e.g., unbiasedness, and Gaussian distribution of flood forecast uncertainty. Zhao et al. (2013) concluded that (1) real-world forecasts can be biased and tend to underestimate actual flood, and (2) real-world forecast uncertainty is non-Gaussian and heavy-tailed. In addition, the stochastic simulation technique has been used to assess the intrinsic risk of reservoir operation (Xu et al. 1997; Li et al. 2010; Zhao et al. 2011, 2013), in which large synthetic flood series are generated. The statistical expectation of an impact is estimated using a process that involves running many hundreds or thousands of simulations (Harvey et al. 2012). These studies show that simulation techniques can provide useful insights into flood processes.

The content of this chapter is therefore to introduce a copula-based uncertainty evolution (CUE) model to describe the evolution of uncertainties in flood forecasting. The flood forecast series with different lead times, that contain forecast uncertainty, are simulated based on the CUE model. Then, the generated flood forecasting sequences are used to evaluate the effect of forecast uncertainty on real-time reservoir operation (Chen et al. 2016).

9.2 Copula-Based Uncertainty Evolution (CUE) Model

First, a copula-based uncertainty evolution (CUE) model is introduced to describe the evolution of uncertainty in flood forecasting. Second, the uncertainty in forecasting is simulated using the proposed CUE model. Third, the predicted flood is generated by adding the simulated uncertainty to the generated flood data. Finally, the effect of flood forecast uncertainty on reservoir operation is evaluated using a flood risk analysis method.

9.2.1 Evolution of Forecast Uncertainty

Heath and Jackson (1994) proposed a martingale model to formulate the evolution of uncertainty of demand forecasts in a supply chain management. Zhao et al. (2011) introduced this method in hydrology. In the following, we describe this uncertainty evolution model for application to flood forecasting.

The rainfall-runoff model is usually used to predict the flood with different lead time. Let Q define the predicted flow, and h the forecasting horizon with forecast lead time from 0 to h. Figure 9.1 shows a schematic of forecasting made at time s for time t. This means that the prediction is made at time s for forecasting future flood at time t ($t = s + 0, s + 1, \ldots, s + h$) simultaneously. For example, assume that today is June 1st and the forecast lead time h is equal to 4. Therefore, s equals June 1st. Since $h = 4$, the future flood occurring on June 2nd, 3rd, 4th and 5th are predicted simultaneously. Those predicted results are organized as a vector $Q_{1,-}$, $\mathbf{Q}_{1,-} = [Q_{1,1}, Q_{1,2}, Q_{1,3}, Q_{1,4}, Q_{1,5}]$, in which for $Q_{1,1}$ we directly substitute the observed value. The general form of vector $Q_{s,-}$ is defined as

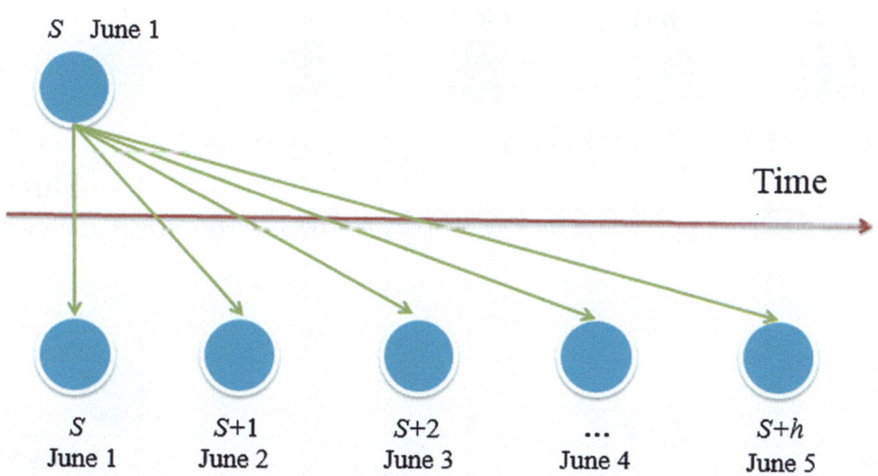

Fig. 9.1 Schematic of flood forecasting at time s

$$\mathbf{Q}_{s,-} = [Q_{s,s}, Q_{s,s+1}, \ldots, Q_{s,s+h}] \quad (9.1a)$$

where $Q_{s,s}$ equals the observed flood at time s, since the flood at time s is known.

Similarly, flood at the coming time t corresponds to multiple predictions made at preceding times, shown in Fig. 9.2. For example, assume that today is June 1st (variable s) and the forecast lead time h is equal to 4. The flood occurring on June 5th (variable t) is predicted today. Actually, the flood occurring on June 5th will be predicted for four times (i.e. it will be predicted on June 1st, 2nd, 3rd, and 4th). We define those predicted flood values for June 5th as $Q_{-,5}$, $\mathbf{Q}_{-,5} = [Q_{1,5}, Q_{2,5}, Q_{3,5}, Q_{4,5}, Q_{5,5}]$. The general form of vector $Q_{-,t}$ is defined as

$$\mathbf{Q}_{-,t} = [Q_{t-h,t}, Q_{t-h+1,t}, \ldots, Q_{t,t}] \quad (9.1b)$$

where $Q_{t,t}$ equals the observed flood at time t.

Let q_t denote the observed flow at time t, and let $Q_{s,t}$ denote the flow predicted at time s for time t ($s \leq t$). At last, a meaningful definition: the absolute forecast uncertainty error is defined by Eq. 9.2.

$$e_{s,t} = Q_{s,t} - q_t \quad (9.2)$$

According to Eqs. 9.1 and 9.2, vectors $\mathbf{e}_{s,-}$ and $\mathbf{e}_{-,t}$ are defined as

$$\mathbf{e}_{s,-} = [Q_{s,s} - q_s, Q_{s,s+1} - q_{s+1}, \ldots, Q_{s,s+h} - q_{s+h}] \quad (9.3a)$$

$$\mathbf{e}_{-,t} = [Q_{t-h,t} - q_t, Q_{t-h+1,t} - q_t, \ldots, Q_{t,t} - q_t] \quad (9.3b)$$

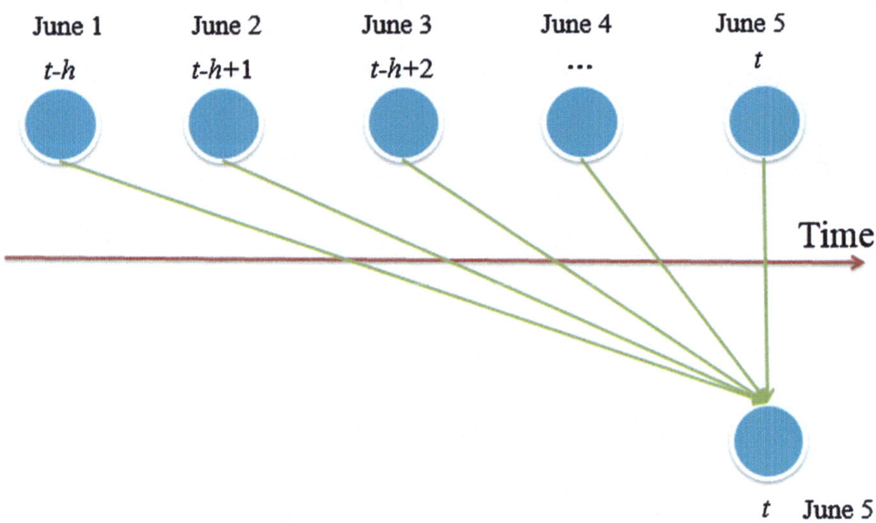

Fig. 9.2 Schematic of flood forecasting at time t

For vector $\mathbf{e}_{s,-}$, the prediction is made at time s, and the future flood at time $s + 1$, $s + 2$,..., $s + h$ are predicted. Thus, their corresponding observed flood are $q_{s+1}, q_{s+2},\ldots, q_{s+h}$, respectively. For vector $\mathbf{e}_{-,t}$, since the future flood at time t is predicted, the observed flow is q_t. The special cases are $e_{s,s}$ and $e_{t,t}$, in which the time that the prediction made at is equal to the time that the prediction is made for. In this case, the error equals 0, because we directly substitute the observed value and there is no prediction.

As shown in Fig. 9.2, the flow at the coming time t corresponds to multiple prediction made at preceding time, the uncertainty decreases with decrease in the lead period in the vector, and uncertainties of predictions generally can sequenced like this $e_{t-h,\ t} > e_{t-h+1,t} > \ldots > e_{t,\ t}$. Therefore, uncertainty reduction of a single period (namely from time $s - 1$ to time s) can be defined as the reduction from $e_{s-1,t}$ to $e_{s,t}$

$$w_{s,t} = e_{s,t} - e_{s-1,t} \tag{9.4}$$

where $w_{s,t}$ is uncertainty reduction of a single period (from time $s - 1$ to time s).

Equation 9.4 demonstrates the changes in uncertainty from time $s - 1$ to time s for forecasting the future flood at time t. As the uncertainty is time-dependent, the uncertainty evolution can be described as a gradual development of uncertainty.

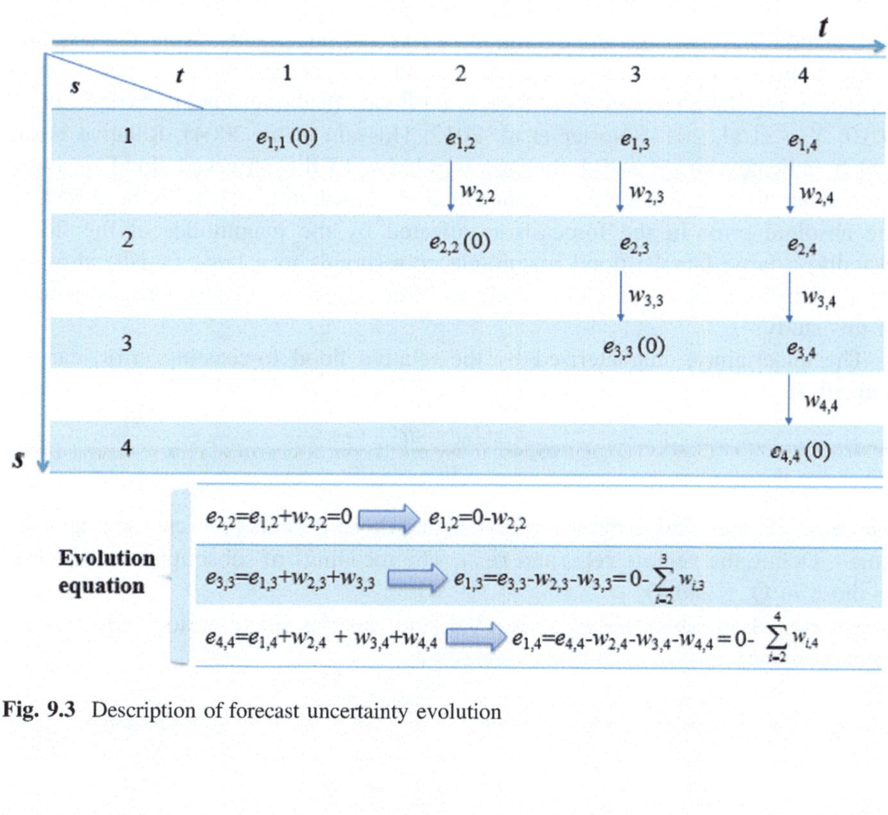

Fig. 9.3 Description of forecast uncertainty evolution

Figure 9.3 illustrates the process of uncertainty evolution in forecasting. The number in the first row represents the time t that the forecasting is made for, and the number in the first column represents the time s that the forecasting is made at. Each row and column represent vector $\mathbf{e}_{s,-}$ and $\mathbf{e}_{-,t}$, respectively.

Take s and t equaling 2 for example. According to Eq. 9.4, uncertainty reduction or the improvement of the predicted results from time $s-1$ to s is equal to $w_{2,2} = e_{2,2} - e_{1,2}$. Thus, the forecast uncertainty $e_{1,2}$ plus the improvement $w_{2,2}$ equals to the uncertainty $e_{2,2}$. Using the same method, we can obtain the equations shown in Fig. 9.3. It can be seen from Fig. 9.3 that the uncertainty $e_{2,3}$ is equal to $e_{1,3} + w_{2,3}$. The single-period uncertainty reduction between $e_{2,3}$ and $e_{3,3}$ is $w_{3,3}$. The uncertainty $e_{3,3}$ can be qualified by $e_{1,3} + w_{2,3} + w_{3,3}$, namely $e_{3,3} = w_{2,3} + w_{3,3} + e_{1,3}$, from which we can derive the $e_{1,3}$, $e_{1,3} = e_{3,3} - w_{2,3} - w_{3,3}$. We have mentioned that $e_{s,s}$ or $e_{t,t}$ equals zero. $e_{1,3}$ can be expressed as $e_{1,3} = 0 - w_{2,3} - w_{3,3}$. Thus, the uncertainty is time-dependent and evolves over time. The earlier the forecast is made, the larger the uncertainty is. The longer the lead time is, the less reliable the forecast is. Based on Eq. 9.4 and Fig. 9.3, a more generalized equation describing the evolution of forecast uncertainty can be given as

$$e_{t-h,t} = -\sum_{i=t-h+1}^{t} w_{i,t} \tag{9.5}$$

Equation 9.5 describes the evolution of forecast uncertainty using the absolute forecast uncertainty. Actually, the relative flood forecast error has been usually used for describing the uncertainties of forecasted floods (Rabuffetti et al. 2008; Li et al. 2010; Yan et al. 2014; Nester et al. 2012; Hossain et al. 2004). Relative error, which indicates how good a forecast is relative to the observed flood, is more appropriate to describe the uncertainty than the absolute one. This is because that the absolute error in the forecasts is affected by the magnitudes of the flood. Usually, a large (small) flood magnitude corresponds to a large (small) absolute error. Therefore, the relative flood forecast error instead of the absolute one is used in this study.

The uncertainty, characterized by the relative flood forecasting error, can be defined as

$$re_{s,t} = \frac{Q_{s,t} - q_t}{q_t} = \frac{e_{s,t}}{q_t} \tag{9.6}$$

where $re_{s,t}$ is the relative flood forecast error, whose flood is predicted at time s for time t. Define the vectors $\mathbf{re}_{s,-}$ and $\mathbf{re}_{-,t}$. The meanings of subscripts are the same as those in $\mathbf{Q}_{s,-}$ and $\mathbf{Q}_{-,t}$.

Dividing both sides by q_t of Eq. 9.4, the relative single-period reduction of forecast uncertainty $rw_{s,t}$ can be expressed as

$$rw_{s,t} = \frac{w_{s,t}}{q_t} = \frac{e_{s,t} - e_{s-1,t}}{q_t} = re_{s,t} - re_{s-1,t} \tag{9.7}$$

where $rw_{s,t}$ is the relative reduction of forecast uncertainty in a single period.

For a special case, $s = t$, $re_{t,t}$ is equal to 0, since the observed value at time t is known and there is no error. According to Eq. 9.7, the relative single-period reduction of forecast uncertainty under this situation can be written as

$$rw_{t,t} = -re_{t-1,t} \tag{9.8}$$

from which it can be seen that when two subscripts of variable rw with the same value, the relative reduction of uncertainty equals the negative relative forecast error $re_{t-1,t}$.

Similarly, dividing both sides of Eq. 9.5 by q_t, the evolution of uncertainty can be expressed as

$$re_{t-h,t} = \frac{-\sum_{i=t-h+1}^{t} w_{i,t}}{q_t} = -\sum_{i=t-h+1}^{t} rw_{i,t} \tag{9.9}$$

where $rw_{i,t} = w_{i,t}/q_t$.

Equation 9.9 is an improvement of the traditional uncertainty evolution model that is developed based on the absolute forecast error. It is also seen from Eq. 9.9 that if the value of rw is known, the forecast uncertainty can be derived. Therefore, in the following, we will give the details for deriving the values of rw.

9.2.2 Derivation of Single-Period Reduction of Forecast Uncertainty

Real-time flood forecast model can dynamically predict flood for different lead times. The rolling process of real-time flood forecasts is shown in Table 9.1, in which the numbers in the first column mean the days that the prediction is made. The numbers in other columns mean the coming day. Take the first line for example. As $h = 4$, on day 1, the future flood on days 2, 3, 4 and 5 are predicted. On day 2, the future flood on days 3, 4, 5, and 6 are predicted. Their corresponding

Table 9.1 The rolling process of real-time flood forecasts

Days	h = 1	h = 2	h = 3	h = 4
1	2	3	4	5
2	3	4	5	6
3	4	5	6	7
4	5	6	7	8

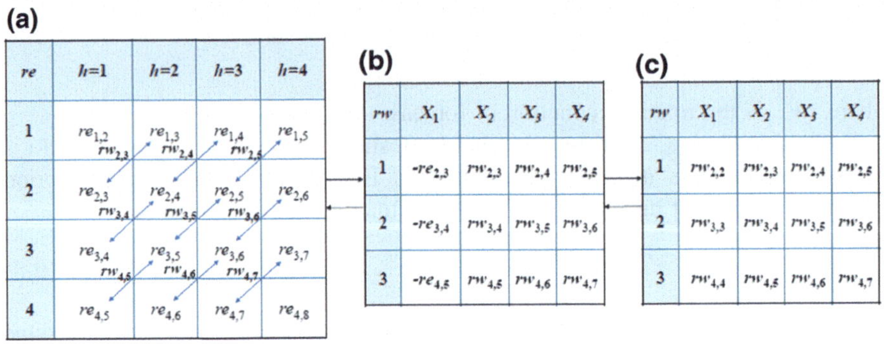

Fig. 9.4 Relationship between uncertainty and single-period uncertainty reduction

errors are given in Fig. 9.4a. Since the error reduction is equal to forecast improvements from $re_{s-1,t}$ to $re_{s,t}$, the rw are calculated in Fig. 9.4a and recorded in Fig. 9.4b. For the lead time equaling to 1 day, based on Eq. 9.8, the value of rw equals to negative re. In Fig. 9.4c, each column represents a variable, and this is a four-dimensional data set. We define the four variables as X_1, X_2, X_3 and X_4. Thus, the relative single-period reduction of forecast uncertainty is finally derived.

9.2.3 *Construction of the Joint Distribution of Uncertainty Reduction*

In this section, the copula function is used to establish the joint distribution of relative single-period reduction of forecast uncertainty by linking the marginal distributions. First, the marginal distribution is established. In order to properly capture the behavior of the relative uncertainty reduction series, several univariate distributions are selected as candidates. Second, the joint distribution is constructed using student copulas.

9.2.3.1 Selection of Marginal Distribution

The exponential (Ex), generalized extreme-value (GEV), generalized logistic (GL), generalized Pareto (GP), generalized normal (GN), Gumbel, Kappa, lognormal (LN), Normal, and Wakeby distributions have been widely used in hydrology. The cumulative probability distributions (CDF, $F(x)$) and their parameters are given in Table 1.1 of Chap. 1. They are selected as alternatives for representing the probability distribution of relative reduction of forecast uncertainty. The L-moments method is employed to estimate their parameters.

9.2.3.2 Construction of the Joint Distribution Using Copulas

The metaelliptical copula and vine-copula are two kinds of copulas, which are employed to establish a higher dimensional probability distribution recently. Compared with the vine-copula, the meta-elliptical copulas can provide a wide range of positive and negative degrees of joint behavior, and model high-dimensional dependence structure with a very simple structure. Especially, for high-dimensional variables, the regular vine-copula embraces a large amount of possible pair-copula decompositions (Aas et al. 2009), which means that we need additional structure to select reasonable and suitable vine specification. Therefore, considering the limitations of vine copulas, the metaelliptical one is employed in this study.

In this study, the variables X_1, X_2, …, X_n are denoted as $rw_{t,t}$, $rw_{t,t+1}$,…, $rw_{t,t+n-1}$, which represent the relative single-period reduction of forecast uncertainty. The $F(rw_{t,t})$, $F(rw_{t,t+1})$,…, and $F(rw_{t,\ t+n-1})$ are the marginal distributions of the variable set. The student copula are used to establish the joint distribution of variables $rw_{t,t}$, $rw_{t,t+1}$,…, $rw_{t,t+n-1}$. The joint distribution $F(rw_{t,t}, rw_{t,t+1},…, rw_{t,t+n-1})$ can be expressed as

$$\begin{aligned} F(rw_{t,t}, rw_{t,t+1}, \ldots, rw_{t,t+n-1}) &= C(F(rw_{t,t}), F(rw_{t,t+1}), \ldots, F(rw_{t,t+n-1})) \\ &= C(u_1, u_2, \ldots, u_n) \end{aligned} \tag{9.10}$$

where $F(rw_{t,t+i}) \sim U(0,1)$ for $i = 0, 1, 2, \ldots, n-1$.

9.3 Generation of Synthetic Predicted Flood Series

Based on the CUE method, the flood forecasting series is generated. The generating process entails three steps. First, the relative single-period uncertainty reduction series are generated based on the established copulas. Second, the uncertainty characterized by relative flood forecast error is obtained according to the uncertainty evolution model. Third, the predicted flood series is simulated by adding the forecast uncertainty to the generated flood series. These series can be applied to analyze the effect of uncertainty on reservoir operation. Information on the process for simulation of the predicted flood series is provided as follows.

Step 1 Generation of relative single-period forecast reduction series

The Student copula is used to generate the uncertainty reduction series. R copula package (http://cran.r-project.org/web/packages/copula/index.html), which provides a carefully designed and easily extensible platform for multivariate modeling with copulas, is used in this generation (Yan 2007).

Assume that the marginal distributions $F(rw_{t,t})$, $F(rw_{t,t+1})$, $F(rw_{t,t+2})$, $F(rw_{t,t+3})$ and joint distribution $F(rw_{t,t}, rw_{t,t+1}, \ldots, rw_{t,t+3})$ are established. The student copula is used to fit this joint distribution. According to the definition of student copula (Demarta et al. 2005), the copula function $C(u_1, u_2, u_3, u_4; \Sigma, v)$ can turn into a student t probability distribution: $C(u_1, u_2, u_3, u_4; \Sigma, v) = t_{\Sigma,v}(t_v^{-1}(u_1), \ldots t_v^{-1}(u_4)) = t_{\Sigma,v}(X_1, X_2, X_3, X_4)$, where Σ is correlation matrix, and v is degrees of freedom. Given Σ and v which are derived from $t_{\Sigma,v}(X_1, X_2, X_3, X_4)$, the procedures for generating of variables $rw_{t,t}, rw_{t,t+1}, rw_{t,t+2}, rw_{t,t+3}$ were summarized by Hörmann and Sak (2010).

Step 2 Generation of forecast uncertainty

Since the relative single-period reduction of forecast uncertainty is known, the uncertainty characterized by the relative forecast error can be generated using Eq. 9.9. A schematic for calculation of uncertainty is given in Fig. 9.4, in which the right figure shows the simulated relative uncertainty reduction. As illustrated before, if the two subscripts of variables *rw* have the same value, the values of uncertainty equal the negative relative uncertainty reduction. Therefore, values in the second column of the left figure are directly equal to $-rw_{t,t}$. The values in the third column are then calculated based on Eq. 9.7. Then the values in the third column of the left figure are derived. A similar method is used to calculate the values in the fourth and fifth columns of the left figure. Finally, the whole series of forecast uncertainty is simulated.

Step 3 Generation of synthetic flood series

Two kinds of data set are used to represent the observed flood series, one of which is the simulated daily flood and the other is the design flood hydrograph.

The daily flood flow is simulated using the copula method proposed by Lee and Salas (2008, 2011). For this simulation, the bivariate copula is used. Considering the advantages of the Archimedean copulas in the bivariate analysis, the Archimedean copula is used to construct the joint distribution. Lee and Salas (2011) concluded that the Gumbel copula showed a better fit than the Frank or Clayton copula. Furthermore, we paid much attention on the flood data, and the Gumbel copula can capture the upper tail dependences very well. Therefore, the synthetic flood series is obtained using the Gumbel copula.

Step 4 Generation of synthetic predicted flood series

Finally, synthetic predicted flood series $Q_{s,t}$ is generated from flood series q_t using the equation

$$Q_{s,t} = q_t(1 + re_{s,t}) \tag{9.11}$$

9.4 Effect of Uncertainty on Reservoir Operation

Real-world forecast uncertainties are complex and depend on the hydrological model, and have a great significant influence on the decision making of reservoir operation. This influence can be quantified by a risk estimate. The propagation of these uncertainties to the reservoir system can be described using a flood risk chain, namely runoff generation-reservoir flood routing-risk analysis.

We have discussed the method for runoff generation before. For flood routing, the pre-release module is applied, which has also been used by Li et al. (2010). This method, making use of the real-time inflow forecast information, aims to alleviate the flood severity by releasing the water stored in the reservoir before the arrival of a large flood.

According to the available real-time inflow forecasting information with h-day lead time, the pre-release flow can be defined as (Li et al. 2010):

$$\bar{O} = \int_{s}^{s+h} I_t \mathrm{d}t / h \tag{9.12}$$

where $\bar{O}$ is the mean release flow during the effective lead time; I_t is the predicted inflow at time t; and s is the time that the prediction made at. Since reservoir operation utilizes a rolling-horizon process to incorporate the dynamically updated flood forecasting into decision making (Zhao et al. 2013), the release flow is real-time updated for each period.

After reservoir routing, the corresponding risk is estimated. In the reservoir operation, the undesirable event is that the reservoir storage (or water level) exceeds Vc (Zc). This corresponding risk can be defined as (Yan et al. 2014)

$$R = P(max\{V(t), t = 1, 2, \ldots, L\} > V_C) \tag{9.13a}$$

or

$$R = P(max\{Z(t), t = 1, 2, \ldots, L\} > Z_C) \tag{9.13b}$$

where L is the length of the inflow series. $V(t)$ and $Z(t)$ are the reservoir storage and water level at time t, respectively. V_c or Z_c is a water storage (water level) corresponding to a flood control criterion.

Stochastic simulation is taken as an effective approach for estimating risks and uncertainty, which has been widely used to assess the likelihood of specified outcomes for studying different kinds of uncertainty. Using the proposed method, the simulated daily forecasting inflow series are generated to simulate uncertainties in flood forecasting and flood hydrograph shape. Equation 9.13 is used to assess the corresponding flood risk.

9.5 Case Study

9.5.1 Data

Three Gorges Reservoir (TGR), located upstream of the Yangtze River, is selected as a case study. The inflow of TGR stems from the mainstream, the tributary of the Wu River, and the rainfall-runoff in the TGR intervening basin. The catchment area of the intervening basin is 55,907 km^2, about 5.6% of the upstream Yangtze River basin. Two hydrological stations (Cuntan and Wulong) control the flood from the mainstream and tributary (Wu River), respectively. In the intervening basin, the data from 40 precipitation stations are used to calculate the areal average rainfall. The current inflow forecasting is updated daily with a forecast horizon of 4-day (Zhao et al. 2013). The predicted and observed inflow data of TGR from 2003 to 2009 in the flood season is used in this study. According to the current operation policies, the flood season of TGR is from June 1st to September 30th.

The parametric characteristic of TGR is given in Table 9.2, in which the flood control water level (FCWL) is the operation water level in the flood season in order to offer adequate storage for flood prevention. The determination of FCWL is usually according to the annual maximum flood frequency analysis. From June to September, the water level of TGR cannot always be higher than FCWL, because of the possible incidences of large floods. In order to illustrate the influence of uncertainty on predicted flow, the observed, predicted flow and their corresponding absolute forecast errors during the flood season of the year 2003 are given in Fig. 9.5, which shows that the absolute forecast errors are usually from −8000 to 8000 m^3/s, and these uncertainties cannot be neglected.

Table 9.2 Characteristic parameters of TGR

FCWL (m)	Normal water level (m)	Height of the dam (m)	Flood protection storage (10^8 m^3)	Total reservoir storage (10^8 m^3)
145	175	185	221.5	393

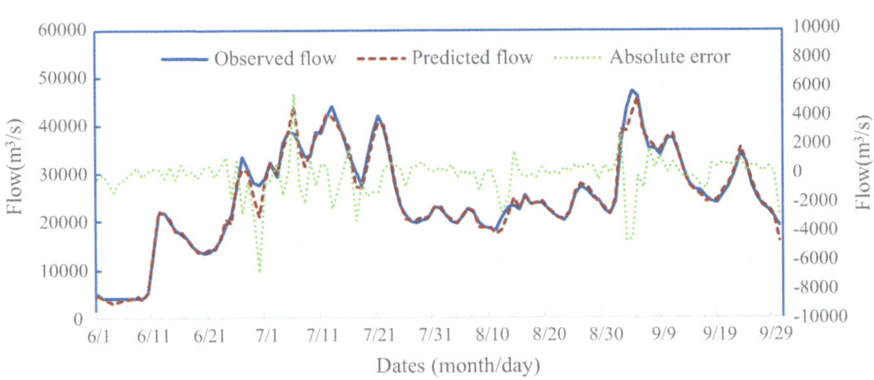

Fig. 9.5 Observed, predicted flow and forecast errors during the flood season of the year 2003

9.5.2 Generation of Forecast Uncertainty

First, the uncertainty of the predicted flood from 2003 to 2009 characterized by the flood forecast error is calculated. Then the relative single-period reduction of forecast uncertainty is derived using Eq. 9.7. Zhao et al. (2013) tested the stationary of the forecast uncertainty reduction series of TGR and concludes that the whole series is non-stationary. They divided the whole flood season into two sub-seasons, namely pre-flood season (June) and main flood season (from July to September). For each sub-season, different marginal and joint distributions were constructed.

The exponential, generalized extreme-value, generalized logistic, generalized pareto, generalized normal, Gumbel, Kappa, lognormal, normal, and Wakeby distributions are selected as candidates for the probability distribution of forecast uncertainty reduction. The L-moments method is employed to estimate distribution parameters. Two criteria, namely bias and *RMSE*, are used to measure the performance of these distributions. The bias and *RMSE* between the observed and theoretical values for the pre-flood and main flood seasons are calculated and listed in Tables 9.3 and 9.4, respectively. The Kolmogorov-Smirnov (K-S) test is employed for the goodness-of-fit test. Their p-values are calculated and also given in Tables 9.3 and 9.4. If the p-values are small, the K-S test is rejected. In other words, the null hypothesis (H_0) is accepted for all the p-values greater than α. The fixed values of α (0.01, 0.05, and 0.1) are usually used to assess the null hypothesis (H_0). In this chapter, α is equal to 0.05, except for variable X_2 in the main flood season. For X_2, a lower α ($\alpha = 0.005$) is applied, since the p-values for all of the distributions are very small. According to the results of bias, *RMSE,* and K-S test, it indicates that the generalized logistic (GL) distribution give a better fit than the other distributions. It is seen from Table 9.4 that the normal distribution is rejected for variable X_1, X_2, and X_3 in the main flood season. In order to show the performance of normal distribution for those variables, the marginal distribution and histogram fitted by normal distribution are given in Fig. 9.6, in which the left section shows the marginal distributions, and the right shows the histogram. Figure 9.6 demonstrates that the normal distribution is found to be inappropriate in these cases and the GL distribution fits better than the normal distribution. Therefore, the assumption of the Gaussianity is not justified all the time in the practical application. The marginal distributions of single-period forecast reduction fitted by GL distribution are shown in Fig. 9.7 for the pre-flood and main flood seasons, respectively, in which the line represents the theoretical distribution and the circles the empirical frequencies of observations. Figure 9.7 indicates that the GL distribution fitted the empirical distribution well. Thus, the GL distribution is used as the marginal distribution hereafter.

After determining the marginal distributions, the joint distribution is constructed. Since copula function requires that each variable is independent, we must make sure that there is no autocorrelation in the variables for building copulas. The autocorrelation functions for each variable in copula functions for both pre-flood and main flood seasons are calculated and shown in Fig. 9.8. The lags are from 1 to 23.

Table 9.3 Results of bias, RMSE and K-S test of the marginal distributions in pre-flood season

Order	Distributions	1			2			3			4		
		Bias	RMSE	K-S	Bias	RMSE	K-S	Bias	RMSE	K-S	Bias	RMSE	K-S
1	Exponential	0.052	0.085	0.005	0.050	0.087	0.002	0.051	0.085	0.007	0.052	0.092	0.002
2	Generalized extreme-value	0.006	0.045	0.263	0.002	0.033	0.734	0.002	0.027	0.924	0.000	0.013	0.996
3	Generalized logistic	0.004	0.034	0.581	0.001	0.023	0.924	0.001	0.015	0.985	0.000	0.015	0.996
4	Generalized pareto	0.006	0.045	0.263	0.002	0.032	0.734	0.001	0.025	0.924	0.000	0.011	0.999
5	Generalized normal	0.011	0.072	0.028	0.004	0.059	0.218	0.004	0.055	0.263	−0.002	0.041	0.373
6	Gumbel	0.017	0.046	0.218	0.020	0.042	0.218	0.023	0.039	0.373	0.029	0.050	0.146
7	Kappa	−0.500	0.577	0.000	−0.500	0.577	0.000	−0.500	0.577	0.000	−0.500	0.577	0.000
8	Three-parameter lognormal	−0.500	0.577	0.000	−0.500	0.577	0.000	−0.500	0.577	0.000	−0.500	0.577	0.000
9	Normal	−0.018	0.056	0.075	−0.011	0.037	0.373	−0.007	0.028	0.658	0.003	0.012	0.999
10	Pearson type III	0.006	0.047	0.218	0.002	0.033	0.734	0.001	0.025	0.924	0.000	0.011	0.999
11	Wakeby	0.007	0.053	0.146	−0.004	0.057	0.095	0.048	0.108	0.000	0.055	0.154	0.000

Table 9.4 Results of bias, RMSE and K-S test of the marginal distributions in the main flood season

Order	Distributions	1			2			3			4		
		Bias	RMSE	K-S	Bias	RMSE	K-S	Bias	RMSE	K-S	Bias	RMSE	K-S
1	Exponential	0.081	0.126	0.000	0.062	0.087	0.000	0.051	0.087	0.000	0.046	0.079	0.000
2	Generalized extreme-value	0.001	0.063	0.001	0.004	0.033	0.008	0.004	0.040	0.075	0.002	0.021	0.623
3	Generalized logistic	−0.001	0.048	0.008	0.002	0.023	0.075	0.002	0.028	0.366	0.000	0.011	0.978
4	Generalized pareto	0.000	0.058	0.002	0.003	0.032	0.011	0.004	0.039	0.086	0.001	0.020	0.670
5	Generalized normal	0.002	0.090	0.000	0.006	0.059	0.000	0.008	0.066	0.000	0.005	0.048	0.022
6	Gumbel	0.046	0.080	0.000	0.028	0.042	0.000	0.019	0.043	0.013	0.020	0.031	0.147
7	Kappa	−0.500	0.577	0.000	−0.500	0.577	0.000	−0.500	0.577	0.000	−0.500	0.577	0.000
8	Three-parameter lognormal	−0.500	0.577	0.000	−0.500	0.577	0.000	−0.500	0.577	0.000	−0.500	0.577	0.000
9	Normal	0.009	0.058	0.002	−0.008	0.037	0.011	−0.014	0.047	0.009	−0.009	0.028	0.238
10	Pearson type III	0.000	0.059	0.002	0.003	0.033	0.011	0.004	0.040	0.075	0.002	0.020	0.670
11	Wakeby	0.000	0.061	0.001	−0.004	0.057	0.167	0.025	0.052	0.000	0.047	0.110	0.000

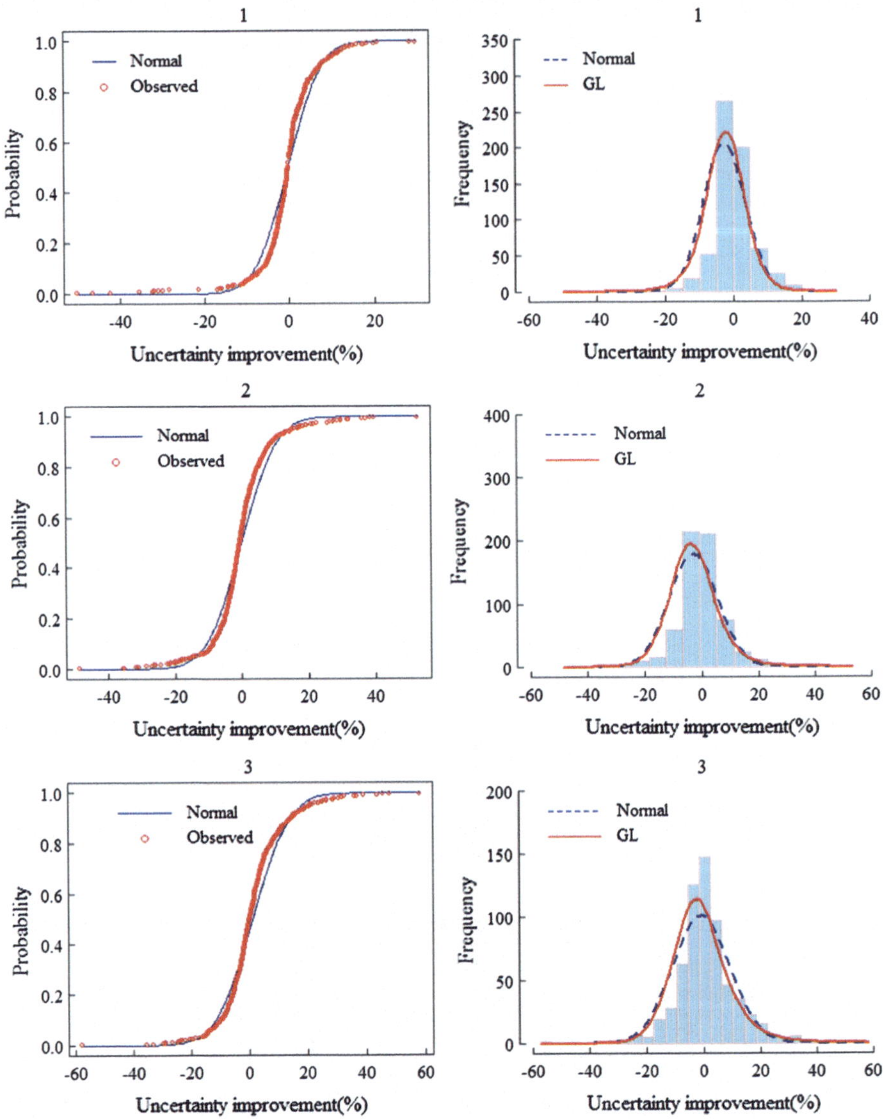

Fig. 9.6 Marginal distributions fitted by normal distribution and histograms fitted by normal and GL distributions for variables X_1, X_2, and X_3 in the main flood season

Results show that the autocorrelations are not large and usually within the range of −0.2 to 0.2. Therefore, each variable is slightly time-dependent, and we can take it as independent. A number of recent papers, such as Breymann and Dias (2003) and Demarta and McNeil (2005), have shown that the empirical fit of the Student copula is generally superior to that of the so-called Gaussian copula. In addition, Student

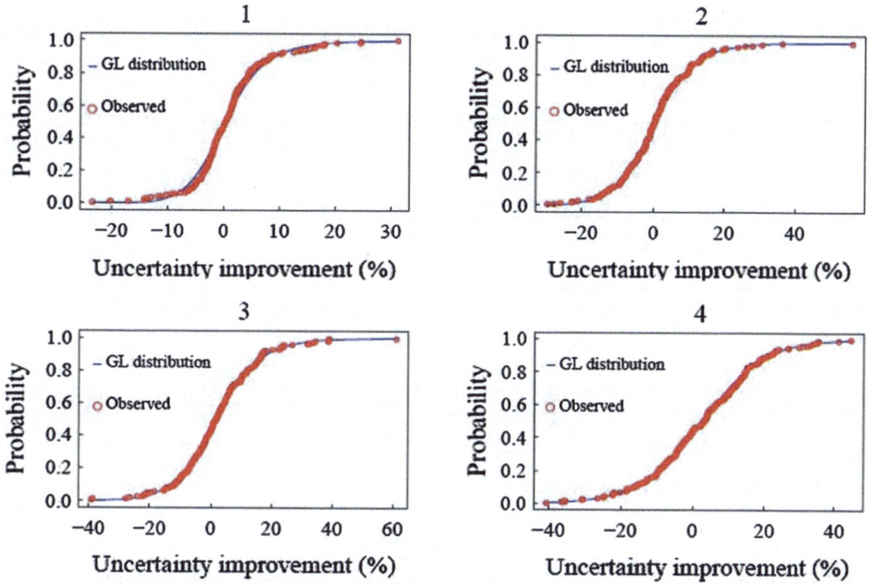

(a) Pre-flood season

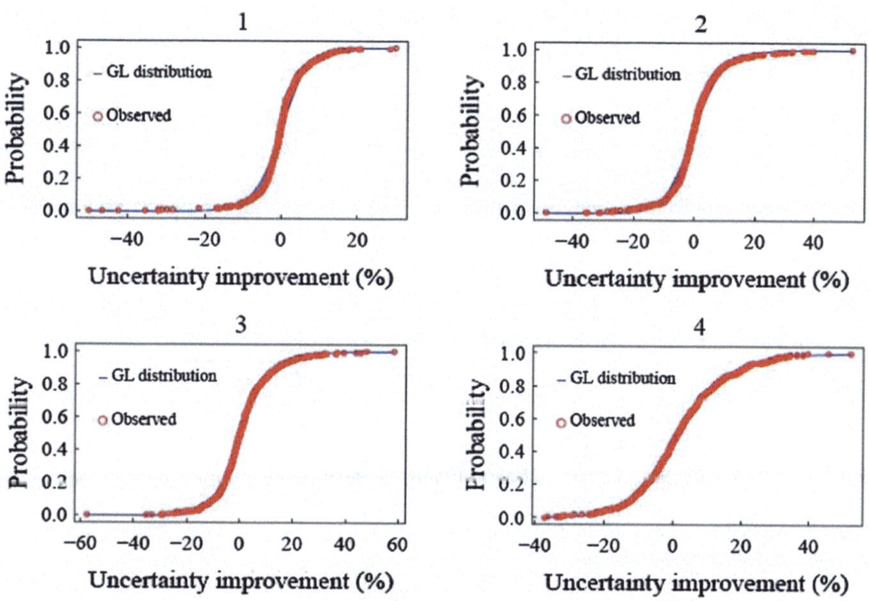

(b) Main flood season

Fig. 9.7 Marginal distributions of single-period reduction of forecast uncertainty fitted by the GL distributions for the main season

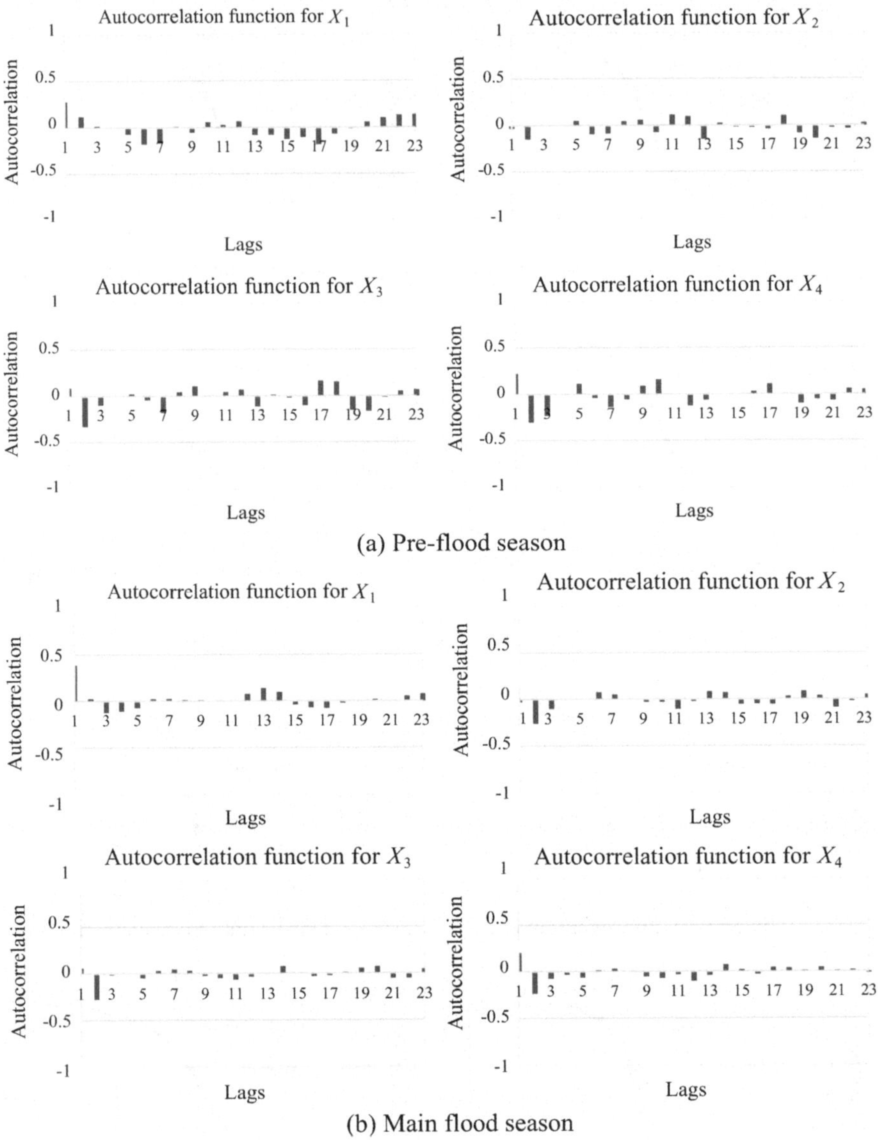

Fig. 9.8 Autocorrelation functions for the variables X_1, X_2, X_3, and X_4 in copula function for pre-flood and main flood seasons

copula can capture better the phenomenon of dependent extreme values (Demarta and McNeil 2005), which is often observed in hydrological data. Therefore, the four-dimensional Student copula is used for modeling the dependence among forecast uncertainty reduction for the pre-flood and main flood seasons, respectively. The maximum likelihood method is used to estimate the parameters of those two copulas. The estimated parameters of Student copulas for the pre-flood and main flood seasons are given in Table 9.5. The p-values are also calculated. Results demonstrate that the selected bivariate copulas can be accepted. The empirical and theoretical joint probabilities are calculated and plotted as shown in Fig. 9.9, in which the x-axis represents the empirical copula value, and the y-axis represents the theoretical copula value. If the copula model gives a better fit, the values calculated by empirical copulas should be equal to those calculated by theoretical copulas. Figure 9.9 indicates that there are no significant differences between empirical and theoretical probabilities in both pre-flood and main flood seasons.

Based on the established Student copulas, the forecast uncertainty series are generated step by step. First, the relative reduction of forecast uncertainty in a single period rw is generated for both the pre-flood and main flood seasons using the CUE model. Since the forecast horizon is 4, four variables are simulated at each time. 1000 samples of rw are generated finally. Statistics of the observed and simulated single-period reduction of forecast uncertainty, including the mean, standard deviation and skewness, are compared. In order to illustrate the distribution

Table 9.5 Parameters of Student t copulas in June and main flood season

Periods	θ_1	θ_2	θ_3	θ_4	θ_5	θ_6	df	p-value
June	0.337	0.034	0.032	0.490	0.078	0.500	3	0.21
Main flood season	0.418	0.087	−0.030	0.600	0.200	0.546	3	0.10

Note df means the degree of freedom

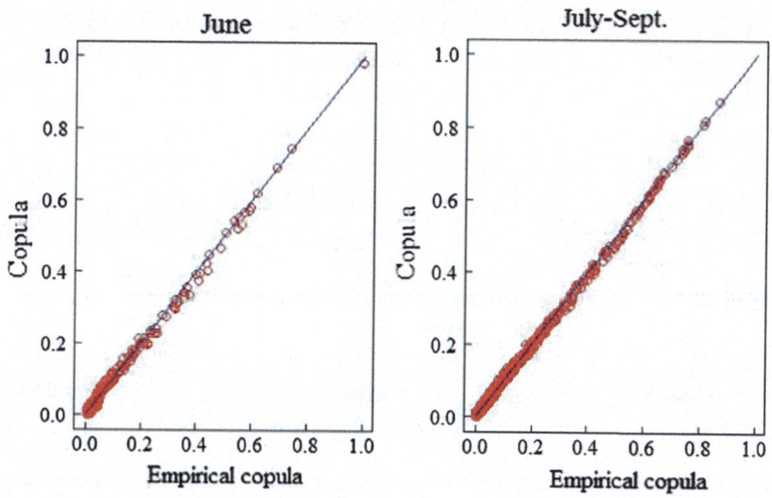

Fig. 9.9 Joint distributions and empirical probabilities of observed combinations for the pre-flood season (June) and main flood seasons (July to Sept.)

characteristic of mean value, standard deviation and skewness, the box figures are drawn and shown in Fig. 9.10. The box figure can give the information of minimum, lower quartile (25% of data less than this value), median, upper quartile (25% of data greater this value) and maximum values. If the stochastic simulated model performs well, the observed values should be around the median value in the box. It can be seen from Fig. 9.10 that the generated statistics fit the observed statistics well. Therefore, the generated single-period reduction of forecast uncertainty can be used for calculation hereafter.

Second, knowing the single-period uncertainty reduction, the forecast uncertainty is calculated based on the uncertainty evolution method. The uncertainties with a lead-time from 1-day to 4-day are computed. The performance of the proposed method is evaluated by comparing mean, standard deviation and skewness of the generated uncertainty values with those of the observed uncertainty. These statistics are presented by boxplots in Fig. 9.11. It can be seen that the generated forecast uncertainty fitted the observed one well. The simulated forecast uncertainty captures all the statistics of the observed uncertainty, which can be used for risk analysis. The generated forecast uncertainty hydrograph in flood season is shown in Fig. 9.12.

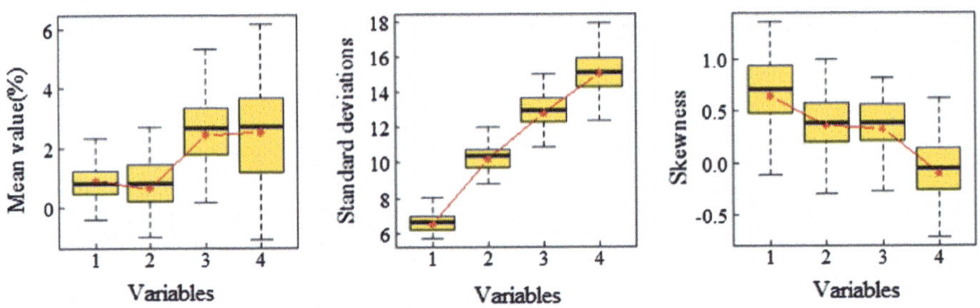

(a) Pre-flood season

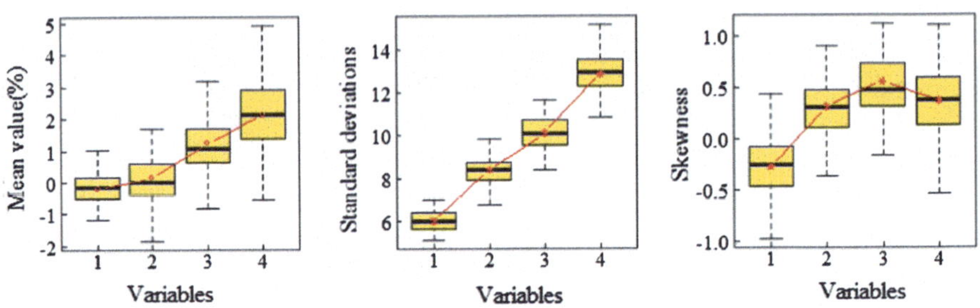

(b) Main flood season

Fig. 9.10 Observed and simulated mean value, standard deviation and skewness of single-period forecasting reduction in the pre-flood and main flood seasons

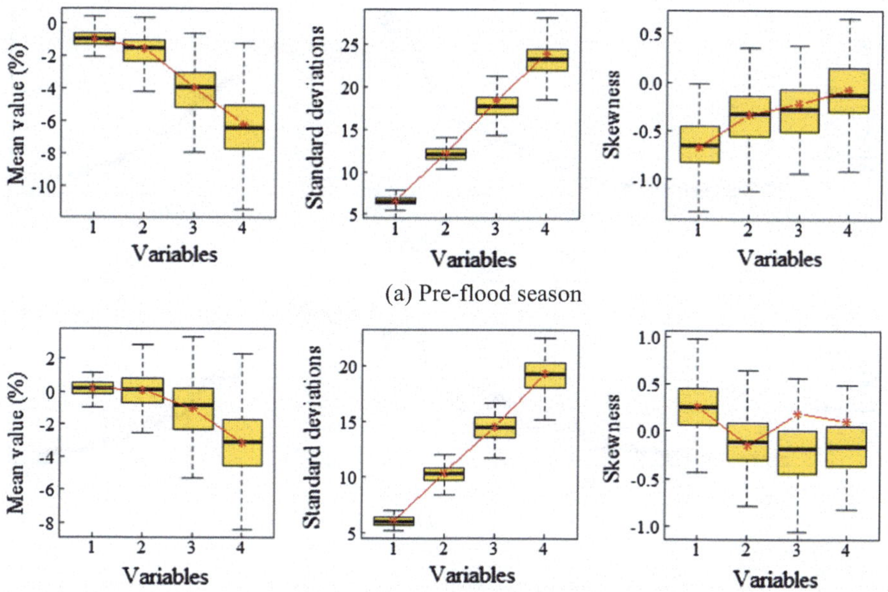

(a) Pre-flood season

(b) Main flood season

Fig. 9.11 Means, standard deviations, and coefficients of skewness of the observed and simulated forecast uncertainty in the pre-flood and main flood seasons

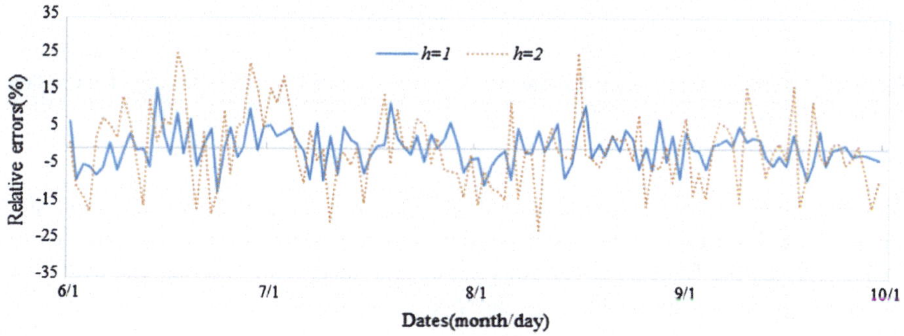

Fig. 9.12 The simulated relative flood forecast errors during the flood season

9.5.3 Generation of Predicted Inflow Series

The observed daily inflow of the Three Gorges Reservoir is simulated using the copula method described by Lee and Salas (2011) who concluded that the Gumbel copula showed a better fit than the Frank or Clayton copula. Therefore, the Gumbel copula is used to generate the inflow series. In order to demonstrate the performance of synthetic daily flood simulation, basic statistics, including mean value, standard

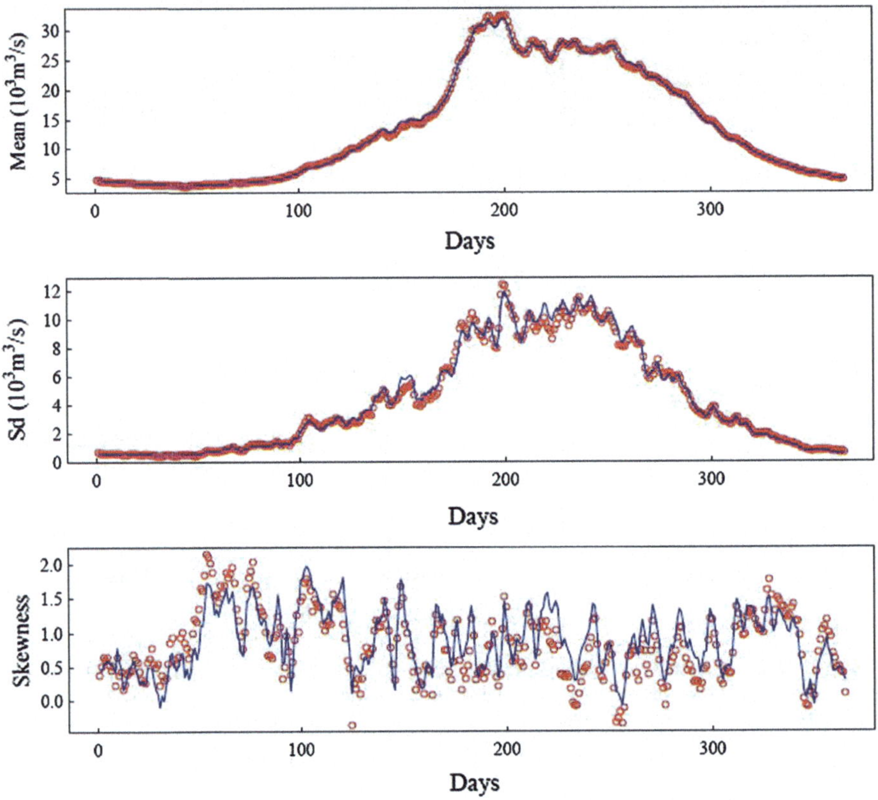

Fig. 9.13 Observed and simulated mean flood, standard deviation and skewness of the inflow data of TGR

deviation, and skewness, are calculated and shown in Fig. 9.13. Results show that the generated series can be used as the observed data in forecast uncertainty analysis. Since the forecast uncertainty series and observed flood data are available, the predicted daily inflow series is obtained using Eq. 9.11.

9.5.4 Flood Risk Analysis

In order to analyze the flood risk of TGR, we first need to know its operation policy that is summarized as follows. From May 20 to the beginning of June, the water level of TGR needs to be dropped to 145 m. During the flood season (from June to September), the water level should not be higher than 145 m, if there is no large flood occurring. In October, the water level needs to increase to the normal water level, 175 m. From November to the end of April the following year, the water level of TGR should be kept as high as possible in order to have more water to generate electricity. For flood control, we only focus on the reservoir operation in the flood season, when the water level should be kept at FCWL 145 m.

The most important area for flood control is the Jinjiang River, which is part of the Yangtze River from the Zhicheng to Chenglingji stations. The water level at Shashi, which is the control gauging station of the Jingjiang River, should not exceed 44.5 m during the flood season. Usually the TGR can control 95% flood of the Jingjiang River reach. If the release of TGR is equal to or less than 53,900 m^3/s, flood discharge in the Jingjiang River should not exceed 56,700 m^3/s, and its corresponding water level will not surpass 44.5 m. The reservoir release is calculated by Eq. 9.12 for each period. If it is greater than 53,900 m^3/s, it will be equal to 53,900 m^3/s in order to keep the safety of the Jingjiang River reach. When the water level is higher than 175 m and below 180.4 m, the calculated reservoir release should be not greater than 79,000 m^3/s. If it is, it will be equal to 79,000 m^3/s. If the water level is higher than 180.4 m, the reservoir release will be determined by the reservoir's release capacity.

The predicted daily flow data with the lead time from 1-day to 4-day is simulated based on the proposed method. The system state of interest is the 10,000-year flood, since the highest water level, 180.4 m, is determined by 10,000-year design flood. This extreme flood event is not contained in the available observations. Therefore, a stochastic simulation method is used to estimate the flood risk. Then, the risk of 10,000-year daily flow data with forecast uncertainty is generated.

After calculation using the observed data coupled with regular reservoir flood routing method, when the 10,000-year design flood $q_{10,000}$ occurs, the water level will reach 180.4 m. Therefore, the water level of 180.4 m corresponds to the 10,000-year design flood. In other words, when Z_c is equal to 180 m, the occurrence probability $P(\max\{Z(t), t = 1, 2, \ldots, T\} > 180.4)$ in the flood season is 0.01%.

The risk of $P(\max\{Z(t), t = 1, 2, \ldots, T\} > 180.4)$ is estimated using the generated 10,000-year predicted daily inflow data with the lead time from 1-day to 4-day and the pre-release model. Results show that the risk is 0, which is less than 0.01%. This means that the risk calculated by the predicted daily flood coupled with the pre-release model is lower than that by the observed data coupled with the regular reservoir routing method.

9.6 Conclusion

The uncertainty in flood forecast has been recognized as a major factor that impacts the accuracy of forecasting. One important issue is to deal with the uncertainty in forecasting when predicting flood. This chapter introduces an approach to estimate the forecast uncertainty evolution and assesses its effect on real-time reservoir operation. A copula-based uncertainty evolution (CUE) model is used to generate the uncertainty in flood forecasting. Then, the effects of forecast uncertainty on real-time reservoir operation are assessed using a flood risk analysis method. The Three Gorges reservoir is selected as a case study. The main conclusions are summarized as follows.

(1) The forecast uncertainty evolution model decomposes the total forecast uncertainty into uncertainty reductions of individual single periods. The proposed CUE method overcomes the disadvantages of the traditional uncertainty evolution model, including the assumptions of unbiasedness, Gaussianity, and stationarity of forecast uncertainty or errors. The performance of the proposed method for simulating the forecast uncertainty is evaluated by comparing mean, standard deviation and skewness of the generated uncertainty series with those of the observed series. It can be seen that the generated forecast uncertainty series captures all the statistics of the observed uncertainty series.
(2) The synthetic predicted daily flow is simulated to represent the predicted inflow of TGR. Results show that using the forecasted inflow coupled with the pre-release module for real-time reservoir operation of TGR in the flood season will not increase flood risks.

References

Aas K, Czado C, Frigessi A, Bakken H (2009) Pair-copula constructions of multiple dependence. Insur Math Econ 44(2):182–198

Alemu ET, Palmer RN, Polebitski A, Meaker B (2011) Decision support system for optimizing reservoir operations using ensemble streamflow predictions. J Water Res PL-ASCE 137(1): 72–82

Boucher MA, Tremblay D, Delorme L, Perreault L, Anctil, F (2012) Hydro-economic assessment of hydrological forecasting systems. J Hydrol 416:133–144

Breymann W, Dias A, Embrechts P (2003) Dependence structures for multivariate high-frequency data in finance. Quant. Finance 3:1–14

Chen L, Singh VP, Guo S, Zhou J (2015) Copula-based method for multisite monthly and daily streamflow simulation. J Hydrol 528:369–384

Chen L, Singh VP, Lu W, Zhang J, Zhou J, Guo S (2016) Streamflow forecast uncertainty evolution and its effect on real-time reservoir operation. J Hydrol 540(2016):712–726

Demarta S, McNeil AJ (2005) The t copula and related copulas. Int Stat Rev 73:111–129

European Commission (2011) Flood and their impacts. http://ec.europa.eu/environ-ment/water/flood_risk/impacts.htm. Retrieved date: 12 April 2011

Harvey H, Hall J, Peppé R (2012) Computational decision analysis for flood risk management in an uncertain future. J Hydroinform 14(3):537–561

Heath DC, Jackson PL (1994) Modeling the evolution of demand forecasts with application to safety stock analysis in production distribution systems. IIE Trans 26:17–30

Hörmann W, Sak H (2010) t-copula generation for control variates. Math Comput. Simulat 81 (4):782–790

Hossain F, Anagnostou EN, Dinku T (2004) Sensitivity analyses of satellite rainfall retrieval and sampling error on flood prediction uncertainty. IEEE Trans Geosci Remote Sens 42(1): 130–139

Krzysztofowicz R (2002) Bayesian system for probabilistic river stage forecasting. J Hydrol 268:16–40

Lee T, Salas J (2011) Copula-based stochastic simulation of hydrological data applied to Nile River flows. Hydrol Res 42(4):318–330

Li L, Xia J, Xu CY, Singh VP (2010) Evaluation of the subjective factors of the GLUE method and comparison with the formal Bayesian method in uncertainty assessment of hydrological models. J Hydrol 390:210–221

Montanari A, Grossi G (2008) Estimating the uncertainty of hydrologic forecasts: a statistical approach. Water Resour Res 44:W00B08. https://doi.org/10.1029/2008wr00687

Morss RE, Olga V, Wilhelmi MW, Downton Eve Gruntfest (2005) Flood risk, uncertainty, and scientific information for decision making: lessons from an Interdisciplinary Project. Bull Am Meteor Soc 86:1593–1601

Nester T, Komma J, Viglione A, Blöschl G (2012) Flood forecast errors and ensemble spread—a case study. Water Resour Res 48:WR011649

Pham TV (2011) Tracking the uncertainty in streamflow prediction through a hydrological forecasting system. http://essay.utwente.nl/61064/

Pokhrel P, Robertson DE, Wang QJ (2013) A Bayesian joint probability post-processor for reducing errors and quantifying uncertainty in monthly streamflow predictions. Hydrol Earth Syst Sci 17:795–804

Rabuffetti D, Ravazzani G, Corbari C, Mancini M (2008) Verification of operational quantitative discharge forecast (QDF) for a regional warning system—the AMPHORE case studies in the upper Po River. Nat Hazards Earth Syst Sci 8:161–173

Schoups G, Vrugt JA (2010) A formal likelihood function for parameter and predictive inference of hydrologic models with correlated, heteroscedastic, and non-Gaussian errors. Water Resour Res 46:W10531. https://doi.org/10.1029/2009WR008933

Smith K, Ward R (1998) Floods: physical processes and human impacts. Wiley 24(13)

Xu ZX, Ito K, Liao S, Wang L (1997) Incorporating inflow uncertainty into risk assessment for reservoir operation. Stochastic Hydrol Hydraul 11(5):433–448

Yan J (2007) Enjoy the joy of copulas: With a package copula. J Stat Softw 21(4):1–21

Yan B, Guo S, Chen L (2014) Estimation of reservoir flood control operation risks with considering inflow forecasting errors. Stoch Env Res Risk A 28(2):359–368

Zhang J, Chen L, Singh VP, Cao W, Wang D (2015) Determination of the distribution of flood forecasting error. Nat Hazards 1:1389–1402

Zhao T, Zhao J, Yang D, Wang H (2013) Generalized martingale model of the uncertainty evolution of streamflow forecasts. Adv Water Resour 57:41–51

Zhao T, Cai X, Yang D (2011) Effect of streamflow forecast uncertainty on real-time reservoir operation. Adv Water Resour 34(4):495–504

Chapter 10
Flood Forecasting Using Copula Entropy Method

10.1 Introduction

There are many models used for flood forecasting. A data-driven technique that has gained significant attention for its effectiveness in function approximation characteristics is artificial neural network (ANN) modeling (de Vos and Rientjes 2005; Kasiviswanathan and Sudheer 2013). Artificial neurons (AN), first introduced in 1943 (McCulloch and Pitts 1943), which mimic the functioning of a human brain by acquiring knowledge through a learning process that involves finding an optimal set of weights for the connections and threshold values for the nodes. Many studies focusing on flood forecasting have shown that ANN is superior to traditional regression techniques and time-series models, including autoregressive (AR) and autoregressive moving average (ARMA) (Raman and Sunilkumar 1995; Jain et al. 1999; Thirumalaiah and Deo 2000; Abrahart and See 2002; Castellano-Méndeza et al. 2004). Hsu et al. (1995) showed that the ANN model provided a better representation of the rainfall–runoff relationship than the ARMAX time series model or the conceptual SAC-SMA (Sacramento soil moisture accounting) models. Shamseldin (1997) examined the effectiveness of rainfall-runoff modeling with ANNs by comparing their results with the Simple Linear Model (SLM), the seasonally based Linear Perturbation Model (LPM) and the Nearest Neighbor Linear Perturbation Model (NNLPM), and concluded that ANNs provided more accurate discharge forecasts than some of the traditional models. Birikundavyi et al. (2002) investigated the ANN models for daily streamflow prediction and also showed that ANNs outperformed the classic autoregressive model coupled with a Kalman filter (ARMAX-KF) and a conceptual model (PREVIS). Therefore, ANNs have proved to be an excellent tool for flood forecasting.

One of the most important steps in the ANN development is the determination of significant input variables (Bowden et al. 2005a; Fernando et al. 2009). In most water resources applications of ANNs, little attention has been given to the task of selecting appropriate model inputs (Maier and Dandy 2000). In general, not all of

L. Chen and S. Guo, *Copulas and Its Application in Hydrology and Water Resources*, Springer Water, https://doi.org/10.1007/978-981-13-0574-0_10

the potential input variables will be equally informative, since some may be correlated, noisy or may have no significant relationship with the output variable being modeled (Bowden et al. 2005a). Including a large number of inputs in ANN models and relying on the network to determine the critical model inputs usually increase the network size (Maier and Dandy 2000). This also brings a number of disadvantages, such as decreasing processing speed and increasing the amount of data required to efficiently estimate the connection weights (Lachtermacher and Fuller 1994).

Fernando et al. (2009) indicated that the task of an input selection algorithm is to determine the strength of the relationship between potential model inputs and outputs. However, the real systems are generally complex and mostly associated with nonlinear processes. Therefore, the dependencies between output and input variables are difficult to measure. Bowden et al. (2005a) reviewed the methods for input determination in water resources ANN applications. Three most commonly used approaches are methods that rely on the use of a priori knowledge of the system being modeled, methods based on linear correlation, and methods that utilize a heuristic approach. The prior knowledge method which depends on an expert's knowledge is very subjective and case dependent. The drawbacks of the linear correlation method are summarized as (a) it only applies to linear correlation, and (b) it tends to focus on the degree of dependence, and ignore the structure of dependence (Zhao and Lin 2011). For a heuristic approach, various ANN models are trained using different subsets of inputs. The main disadvantage of these approaches is that they are based on trial-and-error, and as such there is no guarantee that they will find the globally best subsets. Another disadvantage of stepwise approaches is that they are computationally intensive (Bowden et al. 2005a).

Maier and Dandy (2000) indicated that there were distinct advantages in using analytical techniques to help determine the inputs for multivariate ANN models. Mutual information is an analytical and non-linear method to measure the dependences, which has been successfully employed by many researchers (e.g., Mishra and Singh 2009; Angulo et al. 2011; Jeong et al. 2012; Tongal et al. 2013; Mishra et al. 2013). However, there is a disadvantage when using MI to select inputs of ANN. Although a candidate model input might have a strong relationship with the model output, this information might be redundant if the same information is already provided by another input (Fernando et al. 2009). Sharma (2000) proposed an input determination method based on the partial mutual information (PMI) criterion. In a review of approaches used to select the inputs to ANN models, Bowden et al. (2005a) concluded that the partial mutual information (PMI) algorithm of Sharma is superior to methods commonly used to determine the inputs to ANN models, as it is model-free, uses a non-linear measure of dependence (mutual information), and able to cater to input redundancy and has a well-defined stopping criterion (Fernando et al. 2009). May et al. (2008a) used this model for forecasting water quality in water distribution systems. Furthermore, Fernando et al. (2009) modified PMI input selection algorithm to increase its computational efficiency, while maintaining its accuracy. They introduced the average shifted histograms (ASHs) as an alternative to kernel-based methods for the estimation of mutual information (MI).

However, there are several disadvantages of the methods of PMI algorithm mentioned above. First, hydrological events, such as rainfall and runoff, are continuous. Some methods mentioned above used the discrete version to calculate PMI. Therefore, a method for the continuous variable should be used. Second, these methods need to estimate both the marginal and joint probability density distributions, which involves a product of two terms and lead to a complex calculation work. In order to overcome this problem, the copula and entropy-based method, named copula entropy (CE) method was proposed by Chen et al. (2014a, b) to calculate the MI and PMI values.

Furthermore, when using an ANN model, the appropriate ANN model needs to be selected. Lekkas et al. (2004) indicated that it is preferable for every new application to test different types of ANNs rather than using a pre-selected one. A large number of ANN architectures and algorithms have been developed so far, such as multilayer feedforward networks (Rumelhart et al. 1986), self-organizing feature maps (Kohonen 1982), Hopfield networks (Hopfield 1987), counter propagation networks (Hecht-Nielsen 1987), radial basis function networks (Powell 1987), general regression neural network (GRNN) (Specht 1991), and recurrent ANNs (Elman 1988). Of these networks, the most commonly used are feedforward networks and radial basis function networks (Karunanithi et al. 1994). Multi-layer feed forward networks have been found to perform best when used in hydrological applications (Hsu et al. 1995), and as such, they are by far the most commonly used (Maier and Dandy 2000; Lekkas et al. 2004). Jayawardena et al. (1997) compared multilayer Perceptions (MLP), radial basis function (RBF) and artificial neural network (ANN) approaches in flood forecasting. Results showed that the RBF network-based models gives predictions comparable in accuracy to those from the MLP based models. Park and Sandberg (1991) proved theoretically that the RBF type ANNs are capable of universal approximations and learning without local minima, thereby guaranteeing convergence to globally optimum parameters. In addition, Bowden et al. (2005a, b) recommended the general regression neural network (GRNN), a class of ANN that was introduced firstly by Specht (1991), for hydrological prediction, because of its advantages of nonlinear modeling between inputs and output, fixed network architecture and quicker training than other ANNs. Therefore, the multi layer feed (MLF) forward networks, RBF networks and GRNN are considered in this chapter.

This chapter aims to build and improve the accuracy of the hydrological forecast model established by an ANN method. The input technique, copula entropy method, is first used to select optimal model inputs. Three representative models, namely MLF forward networks, RBF networks and GRNN, are then applied to conduct streamflow prediction. The upper Yangtze River and Jinsha River are selected as case studies. The performance of these models is analyzed and compared. Finally, the model with best-predicted results is determined.

10.2 Flood Forecasting Based on Artificial Neural Networks (ANN)

10.2.1 ANN Models

Therefore, in this study, a representative ANN model for flood forecasting model can be defined as

$$\hat{Q}_t = f(Q_{t-l_1}, R_{t-l_2}, X_{t-l_3}) \tag{10.1}$$

where $\hat{Q}_t$ stands for the predicted flow at time instance t; Q_{t-l_1} is the antecedent flow (up to $t - l_1$ time steps); R_{t-l_2} is the antecedent rainfall, and X_{t-l_3} represents the observed runoff at a neighboring sub-basin in this study.

In this chapter, three ANN models are used. They are multi-layer feedforward networks, radial basis function (RBF) networks, and general regression neural network (GRNN).

10.2.1.1 GRNN Model

The general regression neural network (GRNN), developed by Specht (1991), is a simple yet very effective local approximation based neural network in the sense of estimating a probability distribution function (Islam et al. 2001). The GRNN network is a three-layer network with one hidden layer. The GRNN paradigm is briefly outlined below, and details can be found in Specht (1991). Assume that $f(x, y)$ represents a known joint continuous probability density function of a vector random variable x, and a scalar random variable y. Let x be a specific measured value of the random variable X. The conditional mean of y given x, also called the regression of y on x, is given by Specht (1991):

$$E[y|\mathbf{x}] = \frac{\int_{-\infty}^{\infty} yf(\mathbf{x}, y)\mathrm{d}y}{\int_{-\infty}^{\infty} f(\mathbf{x}, y)\mathrm{d}y} \tag{10.2}$$

where $f(\mathbf{x}, y)$ is not known, then a sample of observations of x and y is used to obtain an estimate $\hat{f}(\mathbf{x}, y)$. The GRNN is an estimate of $E[y|\mathbf{x}]$, which is the conditional expectation of y given $\mathbf{x}$.

10.2.1.2 Multi-layer Feed-Forward Networks (MLF)

The feed-forward neural network is the first and is arguably the simplest type of artificial neural network devised. In this network, the information moves in only one direction, forward, from the input nodes, through the hidden nodes (if any), to the output nodes. There are no cycles or loops in the network. MLF neural networks, trained with a back-propagation learning algorithm, are the most popular neural networks (Zupan and Gasteiger 1993).

An MLF neural network consists of neurons, which are ordered into layers. The first layer is called the input layer, the last layer is called the output layer, and the layers between are hidden layers (Svozil et al. 1997). Each neuron in a particular layer is connected with all neurons in the next layer. The connection between the i-th and j-th neuron is characterized by the weight coefficient w_{ij} and the i-th neuron by the threshold coefficient ϑ_i. The weight coefficient reflects the degree of importance of the given connection in the neural network. The output value (activity) of the i-th neuron x_i is determined by

$$x_i = f(\xi_i) \tag{10.3}$$

$$\xi_i = \vartheta_i + \sum \omega_{ij} x_j \tag{10.4}$$

where ξ_i is the potential of the i-th neuron, and $f(\bullet)$ is so-called transfer function. In multi-layer feed forward neural networks, the most popular non-linear transfer function used in neural network studies is the logistic function, defined by

$$f(\xi) = \frac{1}{1 + e^{-\xi}} \tag{10.5}$$

The supervised adaptation process varies the threshold coefficients ϑ_i and weight coefficients w_{ij} to minimize the sum of the squared differences between the computed and required output values (Svozil et al. 1997).

10.2.1.3 Radial Basis Function (RBF)

An RBF network is a three-layer feed-forward type network. The three layers include the input layer, the hidden layer and the output layer. The input of RBF is transformed by the basic functions at the hidden layer. At the output layer, linear combinations of the hidden layer node responses are added to form the output (Jayawardena et al. 1997).

The name RBF comes from the fact that the basic functions in the hidden layer nodes are radially symmetric. The most common choice, however, is the Gaussian function which can be defined by a mean U and a standard deviation σ. For an input X, the j-th hidden node produces a response given as (Jayawardena et al. 1997):

$$h_j = \exp\{\frac{-\|X_i - U_j\|}{2\sigma_j^2}\} \tag{10.6}$$

where $X_i - U_j$ is the distance between the point representing the input X and the center of the hidden node as measured by some norm. In RBF networks, the connections between input and hidden layers are not weighted. The inputs therefore reach the hidden layer nodes unchanged.

The output y_i of the network at the output node is given as:

$$y_i = \sum_{i=1}^{m} h_i w_{ij} \tag{10.7}$$

Parameters of an RBF type neural network are the mean U and standard deviation σ of the basic functions at the hidden layer nodes, and the synaptic weights w_{ij} of the output layer nodes.

10.2.2 Performance Indexes

The performance indexes are used to evaluate the established ANN model, and the one with the best performance is finally selected for the streamflow forecasting.

The performance of the hydrological forecasting models is assessed by the criteria specified by the Ministry of Water Resources of China (MWR 2006). These are the coefficient of efficiency (i.e., Nash–Sutcliffe efficiency), which is a measure of the goodness-of-fit between recorded and predicted discharge time series data, and the 'qualified rate' (α) of predicted individual flood event peak discharges and volumes (Li et al. 2010). A forecast peak discharge or flood volume is termed 'qualified,' when the difference between the predicted and the recorded values is within ±20% of the recorded value. The root-mean-square error (*RMSE*) between observed and predicted flood values is also used as a performance criterion in this study. The formula for Nash–Sutcliffe efficiency (*NSE*) is given in Eq. 8.25 of Chap. 8.

10.3 Determination of Inputs of ANN Using the CE Theory

In this section, a new method for input identification is introduced. First, the relation between PMI and CE is discussed, and the theory of CE is applied for input identification. Second, according to the calculated value of CE, a reliable and efficient criterion to decide when to stop the addition of new inputs to the list of selected inputs is developed. Third, a detailed procedure of the proposed method is given.

10.3.1 Application of CE to Input Identification

In the following, first, the definition of PMI (partial mutual information) is introduced. Second, the relation between CE and PMI is discussed.

MI, which is equal to the negative CE, can be used to identify the non-linear dependence between candidate input and output variables (Fernando et al. 2009). However, the MI method is not directly able to deal with the issue of redundant inputs (Bowden et al. 2005a). To overcome this problem, Sharma (2000) introduced the concept of partial mutual information (PMI), which represents the information between two observations that is not contained in a third one and provides a measure of partial or additional dependence the new input can add to the existing prediction model (Bowden et al. 2005a).

The PMI between the output (dependent variable) y and the input (independent variable) x, for a set of pre-existing inputs z, can be given by (Bowden et al. 2005a):

$$PMI = \iint f_{X',Y'}(x',y') \ln[\frac{f_{X',Y'}(x',y')}{f_{X'}(x')f_{Y'}(y')}]\mathrm{d}x'\mathrm{d}y' \tag{10.8}$$

where $x' = x - \mathrm{E}[x|z]$; and $y' = y - \mathrm{E}[y|z]$, where E denotes the expectation operation. Variables x' and y' only contain the residual information in variables x and y after considering the effect of already selected input z (Fernando et al. 2009), and can be calculated based on the GRNN model. MATLAB is used to realize for GRNN modelling.

From the discussions in Chap. 2, PMI also can be described using the CE:

$$PMI = \iint f_{X',Y'}(x',y') \ln[\frac{f_{X',Y'}(x',y')}{f_{X'}(x')f_{Y'}(y')}]\mathrm{d}x\mathrm{d}y' = -H_c(X',Y') \tag{10.9}$$

Equation 10.9 shows that PMI is equal to the negative CE of variables of X' and Y'. Therefore, the CE can be used to determine the inputs instead of MI and PMI method.

10.3.2 Termination Criterion

The CE algorithm requires a reliable and efficient criterion to decide when to stop the addition of new inputs to the list of selected inputs (Fernando et al. 2009). There are several methods that can be used to achieve this. Sharma (2000) and Bowden et al. (2005a, b) used the 95th percentile confidence limit for the sample PMI to decide whether the PMI of a candidate input is significantly different from zero and should therefore be added to the set of already selected inputs. May et al. (2008b) and Fernando et al. (2009) indicated that this method, which used the bootstrap with 100 bootstraps to estimate the 95th percentile confidence limit, places a significant computational burden on the algorithm. And a bootstrap so small might not provide a reliable estimation of the confidence bound, which could lead to unreliable and/or sub-optimal input variable selection (May et al. 2008b; Fernando et al. 2009). Fernando et al. (2009) introduced the Hampel identifier which is proposed by

Davies and Gather (1993) as a termination criterion. Using this method, May et al. (2008b) and Fernando et al. (2009) recommended a termination algorithm. The details of the Hampel test criterion are described in the following.

The Hampel identifier is an outlier detection method for determining whether a given value x is significantly different from others within a set of values X. Assume that a set of candidates will initially contain some proportion of redundant variables, and the significant variable will be detected. The Hampel distance begins by calculating the absolute deviation from the median negative CE for all candidates and defined as (Fernando et al. 2005; May et al. 2008b):

$$d_j = \left| NH - NH^{(50)} \right| \tag{10.10}$$

where d_j is the absolute deviation; NH represents the negative CE values; and $NH^{(50)}$ denotes the median NH for the candidate set.

Then the Hampel distance is calculated by Fernando et al. (2005) and May et al. (2008b):

$$Z_j = \frac{d_j}{1.4826 d_j^{(50)}} \tag{10.11}$$

where d_j denotes the Hampel distance; and $d_j^{(50)}$ denotes the median absolute deviation dj. If the Hampel distance, Z_j is greater than 3, namely $Z_j > 3$, then the candidates are added to the selected input set.

10.3.3 Procedures of Input Variable Selection

A stepwise input selection algorithm is now formulated for determining the inputs of an ANN using the CE method described above. First, determine the set of variables that can be taken as potential input of the ANN. This variable set is defined as the vector $\mathbf{I_{in}}$. Denote the vector that will store the final identified input as **I**. The algorithm is as follows.

(1) Based on Eq. 10.10, use the copula-entropy method to calculate the PMI between the output and each of the potential new inputs in $\mathbf{I_{in}}$, conditional on the preexisting input set I. The conditional expectations are computed using the GRNN method (Bowden et al. 2005a).
(2) Calculate the Hampel distance Z_j corresponding to the PMI obtained in step (1).
(3) If the Hampel distance for the highest PMI is greater than 3, then move the candidate to the selected input set I.
(4) Repeat steps (1) to (3) until all significant inputs have been selected.

10.4 Evaluation of the Proposed Method

In order to assess the accuracy and performance of the proposed method, two tests are carried out. One test is based on the Gaussian variables whose MI values are known beforehand. The other is based on a range of synthetically generated datasets, whose dependence attributes are known beforehand.

10.4.1 Accuracy Test

The copula-entropy method is used to calculate the PMI values. Two calculation methods, namely multiple integration method and Monte Carlo method introduced in Chap. 2, are employed to calculate PMI values. In order to test the accuracy of those two methods, the estimated MI values are compared with the theoretical ones. The theoretical PMI values for the normal (Gaussian) copula are given as follows (Calsaverini and Vicente 2009).

$$T_{gauss}(X, Y) = -\frac{1}{2}\log(1 - \rho^2) \tag{10.12}$$

where ρ is the Pearson linear correlation coefficient between Gaussian variables X and Y.

Assuming Person correlation coefficient ρ ranges from −0.9 to 0.9 with step 0.1, the MI is calculated according to the Eq. 10.12 and the two proposed methods, respectively. The errors between the exact value of MI and the MI estimates obtained using the copula-entropy method is therefore given by:

$$E = T_{Gauss}(X, Y) - T(X, Y) \tag{10.13a}$$

$$R = (I_{Gauss}(X, Y) - I(X, Y))/I_{Gauss}(X, Y) \tag{10.13b}$$

where E represents the absolute error, and R the relative error.

Assuming that the Pearson correlation coefficient ρ ranges from −0.9 to 0.9 with a step size of 0.1, the exact and estimated MI values are calculated by Eq. 10.12 and the proposed method, respectively. Multiple integration and Monte Carlo methods are used for calculating the CE, respectively. For the first method, the multiple integration method proposed by Berntsen et al. (1991) is applied. For the second method, 10,000 pairs of $\mathbf{u}$ are generated, and average values of $\ln[c(\mathbf{u})]$ are calculated. The absolute and relative errors are calculated and listed in Table 10.1 and results of the calculation are also shown in Fig. 10.1. It is indicated that the MI values calculated by the three methods are very close and the values calculated by the multiple integration method are more accurate than that by Monte Carlo method. Therefore, the proposed method is satisfactory and the multiple integration method can be used for calculations hereafter.

Table 10.1 Absolute and Relative Errors between the Estimates and Theoretical MI value

ρ	E_I	R_I	E_M	R_M
−0.9	0.000	−0.01	−0.009	−1.10
−0.8	0.008	1.59	−0.004	−0.88
−0.7	0.009	2.70	−0.001	−0.18
−0.6	0.000	0.00	0.002	0.67
−0.5	0.000	0.00	0.013	9.18
−0.4	0.000	0.00	0.002	1.83
−0.3	0.000	0.00	0.000	−0.42
−0.2	0.000	0.00	0.002	10.29
−0.1	0.000	0.00	−0.001	−10.00
0	0.000	–	0.000	–
0.1	0.000	0.00	0.000	−6.00
0.2	0.000	0.00	0.002	7.35
0.3	0.000	0.00	0.002	5.08
0.4	0.000	0.00	0.002	2.18
0.5	0.000	0.00	−0.003	−2.43
0.6	0.000	0.00	−0.002	−0.94
0.7	0.009	2.70	0.009	2.70
0.8	0.008	1.59	0.008	1.59
0.9	0.000	−0.01	0.000	−0.01

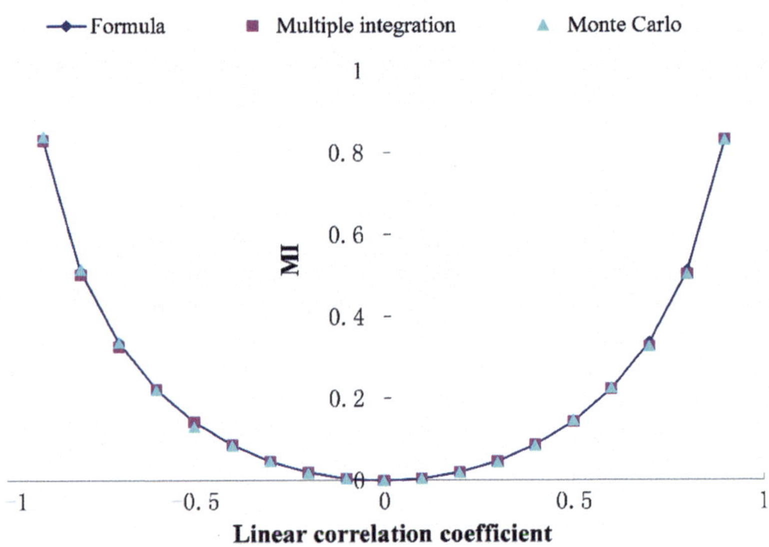

Fig. 10.1 Comparisons of estimated and theoretical MI values, which is calculated by copula entropy method and Eq. 10.13a, b, respectively

10.4.2 Function Text

Before applying the proposed method to a real-world case study, it is necessary to carry out a statistical test based on the generated synthetic data. Bowden et al. (2005a), May et al. (2008b) and Fernando et al. (2009) used several models for testing, four of which are applied in this chapter. These included three time-series models and one a nonlinear system. The four models are given as follows.

(1) AR(1)

$$x_t = 0.9x_{t-1} + 0.866e_t \tag{10.14}$$

where e_t is Gaussian random noise with a zero mean and unit standard deviation for both models. x_t is the time series, and 1 denotes the number of lags.

(2) AR(9)

$$x_t = 0.3x_{t-1} - 0.6x_{t-4} - 0.5x_{t-9} + e_t \tag{10.15}$$

where 1, 4, and 9 represent the number of lags.

(3) TAR(2) Threshold Autoregressive order 2

$$x_t = \begin{cases} -0.5x_{t-6} + 0.5x_{t-10} + 0.1e_t & \text{if } x_{t-6} \leq 0 \\ 0.8x_{t-10} + 0.1e_t & \text{if } x_{t-6} > 0 \end{cases} \tag{10.16}$$

(4) ADD(15)

$$f(x_1, \ldots, x_{15}) = 10\sin\left(\prod x_1 x_2\right) + 20(x_3 - 0.5)^2 + 10x_4 + 5x_5 + \varepsilon \tag{10.17}$$

where ε is Gaussian noise with zero mean and unit variance; and x_1, x_2, x_3, x_4, and x_5 can be generated from a uniform distribution

1020 data points from each of the above synthetic models are generated with the first 20 points being discarded to reduce the effect of an arbitrary initialization (Bowden et al. 2005a). For these models, the first 15 lags are chosen as potential model inputs. The GRNN model with two hidden layers is used to calculate E[x|z] and E[y|z]. A trial and error method is employed to determine a suitable number of hidden layer nodes for each time. The final input subset is obtained based on the improved PMI method. The tested results for each of these models are shown in Tables 10.2, 10.3, 10.4 and 10.5. Take AR(9) model as an example. For the first iteration, the MI value is calculated. The highest MI value occurs in lag 4, which is 0.239 with a Z_j of 4.43. Since the Z_j is greater than 3, the lag 4 is selected. The highest PMI value for the second, third and fourth iterations show up in lags 9, 1 and 8 with the Z_j of 7.00, 5.86 and 1.99, respectively. Therefore, lags 9 and 1 are selected and lag 8 is discarded. The final input sets for these test functions are given

Table 10.2 Test results based on generated data for AR(1) model

First iteration			Second iteration		
Lags	MI	Zj	Lags	PMI	Zj
1	0.935	4.47	2	0.001	0.64
2	0.594	2.43	3	0.003	0.49
3	0.430	1.45	4	0.004	0.71
4	0.329	0.84	5	0.006	1.76
5	0.277	0.53	6	0.004	1.00
6	0.248	0.36	7	0.002	0.42
7	0.213	0.15	8	0.003	0.44
8	0.188	0.00	9	0.000	1.00
9	0.155	0.20	10	0.000	1.18
10	0.128	0.36	11	0.000	1.21
11	0.110	0.47	12	0.002	0.22
12	0.075	0.67	13	0.003	0.20
13	0.064	0.74	14	0.002	0.20
14	0.048	0.83	15	0.006	1.74
15	0.031	0.94			

Table 10.3 Test results based on generated data for AR(9) model

First iteration			Second iteration			Third iteration			Fourth iteration		
Lags	MI	Z_j	Lags	PMI	Z_j	Lags	PMI	Z_j	Lags	PMI	Z_j
1	0.081	1.03	1	0.051	1.37	1	0.142	5.86	2	0.000	0.72
2	0.002	0.67	2	0.006	0.50	2	0.019	0.15	3	0.000	0.66
3	0.037	0.09	3	0.000	0.73	3	0.003	0.60	5	0.001	0.15
4	0.239	4.43	5	0.013	0.20	5	0.070	2.49	6	0.001	0.02
5	0.011	0.47	6	0.006	0.50	6	0.002	0.67	7	0.002	0.82
6	0.002	0.67	7	0.003	0.62	7	0.004	0.55	8	0.004	1.99
7	0.007	0.55	8	0.058	1.67	8	0.016	0.00	10	0.000	0.69
8	0.030	0.06	9	0.188	7.00	10	0.039	1.05	11	0.001	0.02
9	0.090	1.22	10	0.093	3.10	11	0.010	0.27	12	0.004	1.74
10	0.068	0.75	11	0.014	0.15	12	0.001	0.72	13	0.002	0.46
11	0.018	0.32	12	0.010	0.32	13	0.002	0.68	14	0.002	0.45
12	0.001	0.69	13	0.117	4.07	14	0.059	1.97	15	0.000	0.74
13	0.200	3.60	14	0.078	2.49	15	0.031	0.67			
14	0.126	2.00	15	0.021	0.15						
15	0.033	0.00									

in Table 10.6. It is seen from Tables 10.2, 10.3, 10.4, 10.5 and 10.6 that the proposed method is rational and can be applied to both time series and non-linear models. The proposed method is capable of choosing inputs in their correct order of significance.

Table 10.4 Test results based on generated data for TAR(2) model

First iteration			Second iteration			Third iteration		
Lags	PMI	Z_j	Lags	PMI	Z_j	Lags	PMI	Z_j
1	0.006	0.72	1	0.000	0.72	1	0.000	0.53
2	0.014	0.70	2	0.003	0.70	2	0.000	0.69
3	0.006	0.63	3	0.003	0.63	3	0.003	0.87
4	0.021	0.37	4	0.001	0.37	4	0.003	1.34
5	0.004	0.67	5	0.000	0.67	5	0.001	0.00
6	0.041	15.94	6	0.029	15.94	7	0.000	0.67
7	0.004	0.79	7	0.000	0.79	8	0.003	1.17
8	0.012	0.31	8	0.001	0.31	9	0.001	0.16
9	0.009	0.11	9	0.001	0.11	11	0.001	0.00
10	0.359	0.11	11	0.002	0.11	12	0.000	0.60
11	0.009	0.69	12	0.003	0.69	13	0.003	1.03
12	0.018	0.68	13	0.003	0.68	14	0.004	1.90
13	0.011	0.92	14	0.003	0.92	15	0.001	0.23
14	0.012	0.34	15	0.001	0.34			
15	0.005	0.72						

10.5 Flood Forecasting for Three Gorges Reservoir

10.5.1 Study Area

The map and introductions of the study area are given in Chap. 6. A total of six gauging stations are taken into accounts. From upstream to downstream, they are Pingshan, Gaochang, Lijiawan, Beibei, Wulong, and Yichang, each with concurrent mean daily flow data from the year 1998 to 2007. The flow of each gauging station is taken as a variable. The past values of Pingshan, Gaochang, Lijiawan, Beibei, Wulong and Yichang stations are taken as potential input candidate variables and the runoff of Yichang station at time t as the output variable. The data used at Yichang gauging station is naturalized, and the storage effects of Three Gorges Reservoir (TGR) can be removed. We use this data to represent the input flow of Three Gorges Reservoir, which cannot measure directly. Therefore, this flood forecasting model aims to predict the input flow of Three Gorges Reservoir. The CE algorithm with the Hampel distance outlier detection approach as the termination criterion is used to identify the significant inputs.

10.5.2 Selection of Input Variables for ANN Model

Bowden et al. (2005b) proposed a two-stage procedure for input selection. The same method is used in this case study. The first step is called bivariate stage, which aims to determine the significant lag of each variable. The second step is called

Table 10.5 Test results based on generated data for ADD(15) model

First iteration			Second iteration			Third iteration		
Lags	PMI	Z_j	Lags	PMI	Z_j	Lags	PMI	Z_j
1	0.001	0.00	1	0.001	0.01	1	0.002	0.23
2	0.001	0.15	2	0.001	0.26	2	0.007	2.67
3	0.190	172.88	4	0.516	493.78	5	0.255	135.51
4	0.132	119.90	5	0.144	137.10	6	0.004	1.28
5	0.051	45.66	6	0.001	0.28	7	0.002	0.00
6	0.001	0.05	7	0.001	0.20	8	0.002	0.23
7	0.001	0.22	8	0.001	0.01	9	0.000	0.82
8	0.000	0.86	9	0.002	0.95	10	0.000	0.79
9	0.000	0.67	10	0.000	0.94	11	0.000	0.67
10	0.000	0.83	11	0.002	0.72	12	0.003	0.51
11	0.002	0.54	12	0.000	0.81	13	0.000	0.86
12	0.000	0.42	13	0.001	0.40	14	0.001	0.43
13	0.000	0.71	14	0.006	4.48	15	0.001	0.48
14	0.005	3.37	15	0.000	0.63			
15	0.001	0.29						
Fourth iteration			Fifth iteration			Sixth iteration		
Lags	PMI	Z_j	Lags	PMI	Z_j	Lags	PMI	Z_j
1	0.007	8.81	2	0.002	3.56	6	0.000	0.03
2	0.000	0.53	6	0.000	0.00	7	0.000	0.69
6	0.001	0.23	7	0.000	0.67	8	0.000	1.01
7	0.000	0.91	8	0.000	0.21	9	0.000	0.03
8	0.000	0.54	9	0.000	0.56	10	0.001	0.80
9	0.000	0.79	10	0.000	1.11	11	0.001	0.09
10	0.000	0.23	11	0.001	2.50	12	0.000	0.61
11	0.002	2.25	12	0.000	0.57	13	0.000	0.66
12	0.001	0.76	13	0.000	0.44	14	0.002	2.37
13	0.002	1.51	14	0.001	1.76	15	0.001	1.58
14	0.000	0.58	15	0.001	0.74			
15	0.001	0.59						

Table 10.6 Final selected input sets for the four models

Model	Final selected input sets
AR(1)	x_{t-1}
AR(9)	$x_{t-4}, x_{t-9}, x_{t-1}$
TAR(2)	x_{10}, x_6
TR	x_3, x_4, x_5, x_1, x_2

multivariate stages, in which the significant lags selected in the previous step are combined to form a subset of candidates. Then, the final set of significant input can be obtained using the same PMI method as in step 1.

Details of the method are described as follows. If the number of candidate variables is (i.e., x_1, x_2, x_i, …, x_d) and the output variable is y_t, then their own past values ($x_{i,t-1}$, $x_{i,t-2}$, …, $x_{i,t-k}$) and (y_{t-1}, y_{t-2}, …, y_{t-k}) are potential inputs, where k refers to the maximum lag that has been included as a potential input. Bowden et al. (2005b) indicated that if prior knowledge about the relationship between the input and output time series is available, then k can be chosen such that the lags of the input variable that exceed k are not likely to have any significant effect on the output time series. Noting that 3-day and 7-day flood volumes have usually been employed in flood analysis, a flood event lasts less than two weeks. The period of two weeks, which is double the time of 7-day, is taken into account in this study. Except for time t, the first 13 lags of each variable are used as candidate inputs.

First, the MI values between each of ($x_{i,t-1}$, $x_{i,t-2}$, …, $x_{i,t-k}$) and y_t and between (y_{t-1}, y_{t-2}, …, y_{t-k}) and y_t are calculated. The significant lags, which have the highest MI values, are selected. Then the PMI value and its Z_j are calculated in each iteration, the maximum of which with its Z_j greater than 3 is selected. The final selected results of all the stations in step one are summarized in Table 10.7. During this stage, the original 78 inputs are reduced to 12 inputs. The past runoff (y_{t-1}, y_{t-2}, …, y_{t-k}) at Yichang station has a great impact on y_t. Therefore, several values of past runoff at Yichang station are also selected. Only one input is selected for Gaochang, Beibei, Lijiawan and Wulong, and the selected lag times of these stations matched the flood travel times. The travel times from Pingshan, Gaochang, Lijiawan, Beibei, Wulong to Yichang station are about 3, 3, 3, 2 and 2 days. From this point of view, this method is adequate.

Second, the significant lags selected in step one are combined to form a subset of candidates. The PMI values are calculated based on the CE method. During this stage, 12 inputs are reduced to 7. The final input set for ANN consisted of $X_{yc,t-1}$, $X_{yc,t-2}$, $X_{ps,t-1}$, $X_{gc,t-3}$, $X_{bb,t-2}$, $X_{wl,t-2}$, $X_{ljw,t-3}$, in which subscript yc, ps, gc, bb, wl, ljw mean Yichang, Pingshan, Gaochang, Beibei, Wulong, Lijiawan gauging station, respectively.

Table 10.7 Final selected inputs in step one

Stations	Selected inputs
Pingshan	Lag $t-1$, $t-4$, $t-2$
Gaochang	Lag $t-3$
Beibei	Lag $t-2$
Lijiawan	Lag $t-3$
Wulong	Lag $t-2$
Yichang	Lag $t-1$, $t-2$, $t-3$, $t-4$, $t-5$

10.5.3 Flood Forecasting Results Based on the Selected Inputs

The selected variables based on CE method are used as inputs of the ANN model. As mentioned above, we use 10-year (1998–2007) data series, eight years of which are used for training the ANN model, and two of which are used for model validation. A cross-validation is conducted to evaluate the performance of the proposed model, which avoids problems of arbitrarily dividing data into calibration and validation sets.

The GRNN method with two hidden layers is used to establish the ANN model. Since the cross-validation are used in the case study, it is not possible to use the same hidden nodes for each data set. A trial-and-error method is employed to determine a suitable number of hidden layer nodes for each time.

The performance of the hydrological forecasting models is assessed based on the coefficient of efficiency (i.e., Nash–Sutcliffe efficiency) and qualified rate given before. Results of flood forecasting are shown in Table 10.8. Time series plots of observed and predicted flood values obtained with seven inputs selected using the copula-entropy method are shown in Fig. 10.2. It can be seen that from Table 10.8 and Fig. 10.2 that the model performs quite well.

10.5.4 Comparisons with Other Methods

10.5.4.1 Comparison with Inputs Obtained by the Linear Correlation Coefficients

Bowden et al. (2005a) pointed out that the linear correlation method is the most popular analytical technique for selecting appropriate inputs. The proposed method is compared with the linear correlation coefficient in this section.

Table 10.8 The cross-validation results based on the inputs selected by the CE method

Validation period	Periods	Qualified rate	*RMSE*	*NSE*
1998–1999	Training	0.961	2026	0.954
	Validation	0.954	3064	0.948
2000–2001	Training	0.962	2242	0.963
	Validation	0.960	2182	0.932
2002–2003	Training	0.977	2159	0.964
	Validation	0.945	2390	0.938
2004–2005	Training	0.977	1999	0.970
	Validation	0.933	2802	0.906
2006–2007	Training	0.973	2017	0.966
	Validation	0.937	2423	0.960

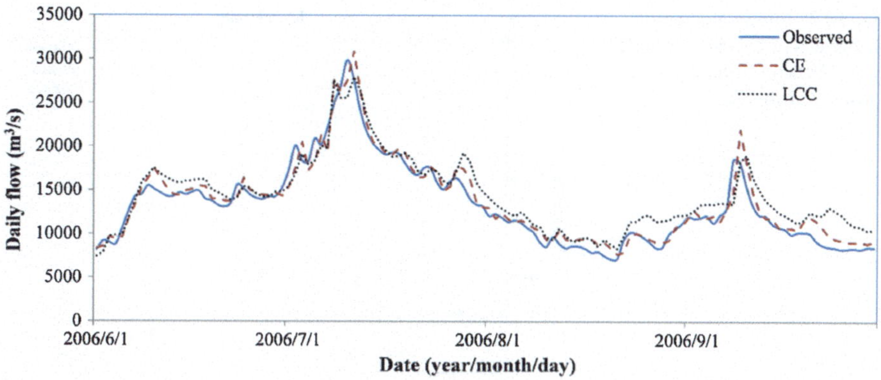

Fig. 10.2 Comparison of observed daily flow series with flood forecasting results based on the proposed input identification method and the traditional linear correlation method

Two assumptions need to be satisfied for the Pearson correlation coefficient. One is that the variable must follow the multivariate normal distribution, and the other is that the pairwise dependency is linear. We presume the normal distribution for the marginal variables. The marginal probability density functions estimated by the normal assumption and the principle of maximum entropy (POME) method for the five rivers are shown in Fig. 10.3. It is seen that the distribution estimated by POME fitted the empirical distribution better than the normal distribution, especially, for the data of Min, Tuo and Wu Rivers which show high kurtosis and skewness. The assumption of normality is found to be inappropriate in this case. However, the notation that this parameter can be measured by the usual linear correlation relies on the additional assumption that marginal functions are also normal (Calsaverini and Vicente 2009).

In order to test the validity of the assumption that the pairwise dependence is linear, the time series of flow data is divided into two segments. Pearson's correlations are calculated for each segment. The calculated results of the Gaochang gauging and Wulong gauging stations are listed in Table 10.9, which indicate that Pearson's correlation is changing over time, and therefore linear correlation coefficients are not valid for these stations.

In the following, we discuss if the linear correlation coefficient is used, which inputs are finally selected. Pearson's correlation coefficients are computed, and the results are given in Table 10.10. The variable at lag *t* with the largest correlation coefficients is definitely selected as an input of ANN. The partial correlation coefficients (PCC) are used to remove the effect of the selected input variable and measure the true correlation between potential inputs and output. The theory of PMI is employed to calculate the PCC between each potential input and output given the selected input lag *t*. The calculated results are shown in Fig. 10.4, where the value for the selected variable is the linear correlation coefficient, and for the other is the PCC value given the selected input lag *t*.

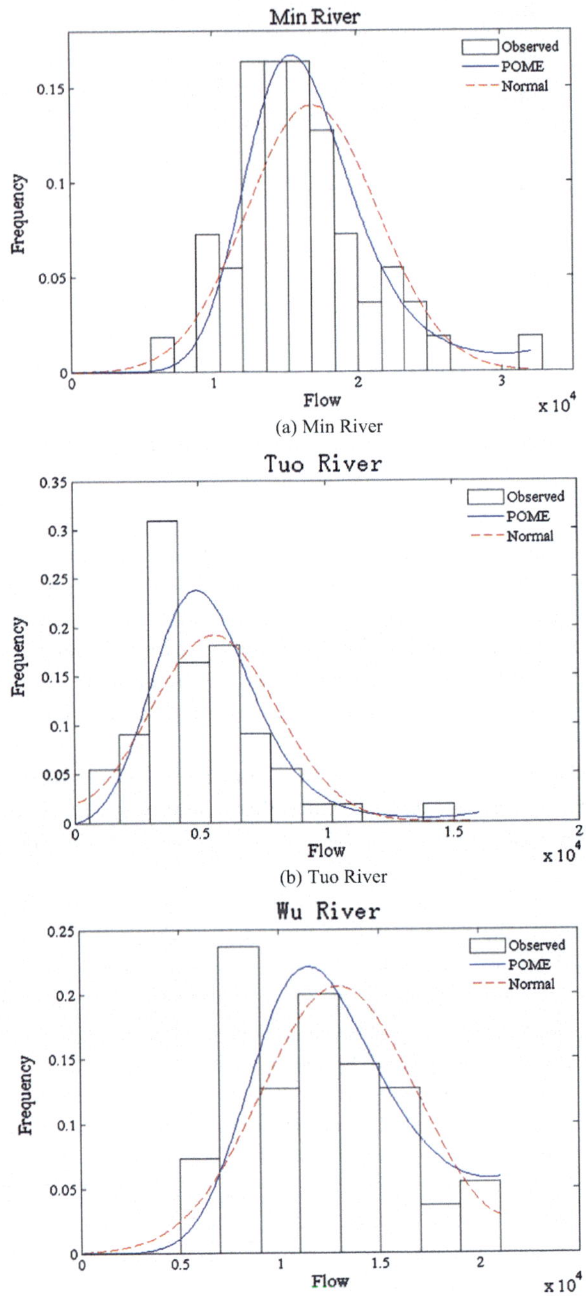

(a) Min River

(b) Tuo River

(c) Wu River

Fig. 10.3 Fitting frequency histograms of flood magnitude by the POME method and normal distribution

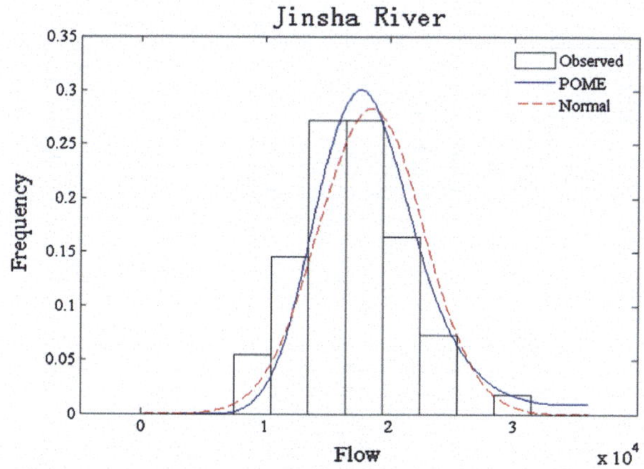

(d) Jinsha River

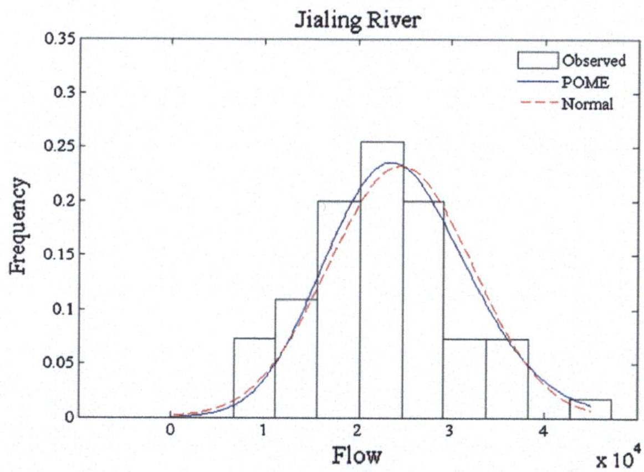

(e) Jialing River

Fig. 10.3 (continued)

It can be seen that compared with the value of the correlation coefficients, which are the largest one in Fig. 10.4, PCC is not large. Take the Beibei gauging station for example. The largest linear correlation coefficient occurred at lag 2, which is the largest one in the figure and equals 0.634. Therefore, the variable at lag 2 is definitely taken as the input of the ANN model. The PCC, which removes the effect of the selected variable at lag t-2, is calculated and shown in Fig. 10.4. The second highest value for Beibei is 0.26, and nearly one-third of the largest one. Therefore, only one variable is selected for each gauging station. The selected inputs, based on the linear correlation method, are listed in Table 10.11, and the selected input set of the proposed method is also shown. Results show that the input set selected by the proposed method and correlation coefficient is somewhat different. For example, the proposed method selected the Gaochang lag 3 as input, and the correlation method

Table 10.9 The correlation coefficients of Gaochang and Wulong gauging station

	Period	Yichang t	Lag $t-1$	Lag $t-2$	Lag $t-3$	Lag $t-4$	Lag $t-5$
Gaochang	1	1.000	0.289	0.325	0.385	0.463	0.488
	2	1.000	0.445	0.505	0.565	0.586	0.564
Wulong	1	1.000	0.363	0.421	0.456	0.444	0.417
	2	1.000	0.514	0.521	0.489	0.442	0.400
			Lag t - 6	**Lag t - 7**	**Lag t - 8**	**Lag t - 9**	**Lag t - 10**
Gaochang	1	1.000	0.457	0.403	0.365	0.341	0.324
	2	1.000	0.531	0.507	0.502	0.496	0.472
Wulong	1	1.000	0.407	0.414	0.428	0.432	0.418
	2	1.000	0.358	0.319	0.288	0.267	0.246
			Lag t - 11	**Lag t - 12**	**Lag t - 13**		
Gaochang	1	1.000	0.313	0.308	0.305		
	2	1.000	0.444	0.413	0.381		
Wulong	1	1.000	0.395	0.368	0.335		
	2	1.000	0.218	0.177	0.132		

Note Consider that the time series of flow data was divided into two segments. “1” represents the correlation coefficients of the first segmentation. And “2” represents the correlation coefficients of the second segmentation

Table 10.10 The correlation coefficients between potential inputs and output of ANN model

Stations	Lag $t-1$	Lag $t-2$	Lag $t-3$	Lag $t-4$	Lag $t-5$	Lag $t-6$	Lag $t-7$
Yichang	0.968	0.901	0.829	0.764	0.714	0.673	0.640
Pingshan	0.698	0.695	0.684	0.661	0.629	0.593	0.557
Lijiawan	0.417	0.487	0.523	0.503	0.459	0.414	0.385
Gaochang	0.435	0.492	0.538	0.540	0.507	0.470	0.443
Beibei	0.603	0.636	0.604	0.526	0.441	0.381	0.350
Wulong	0.498	0.504	0.478	0.446	0.424	0.413	0.407
Stations	**Lag t - 8**	**Lag t - 9**	**Lag t - 10**	**Lag t - 11**	**Lag t - 12**	**Lag t - 13**	
Yichang	0.612	0.585	0.558	0.529	0.499	0.468	
Pingshan	0.523	0.492	0.462	0.434	0.407	0.380	
Lijiawan	0.372	0.367	0.362	0.358	0.355	0.345	
Gaochang	0.424	0.405	0.386	0.369	0.352	0.339	
Beibei	0.333	0.320	0.309	0.302	0.298	0.292	
Wulong	0.401	0.385	0.361	0.331	0.295	0.264	

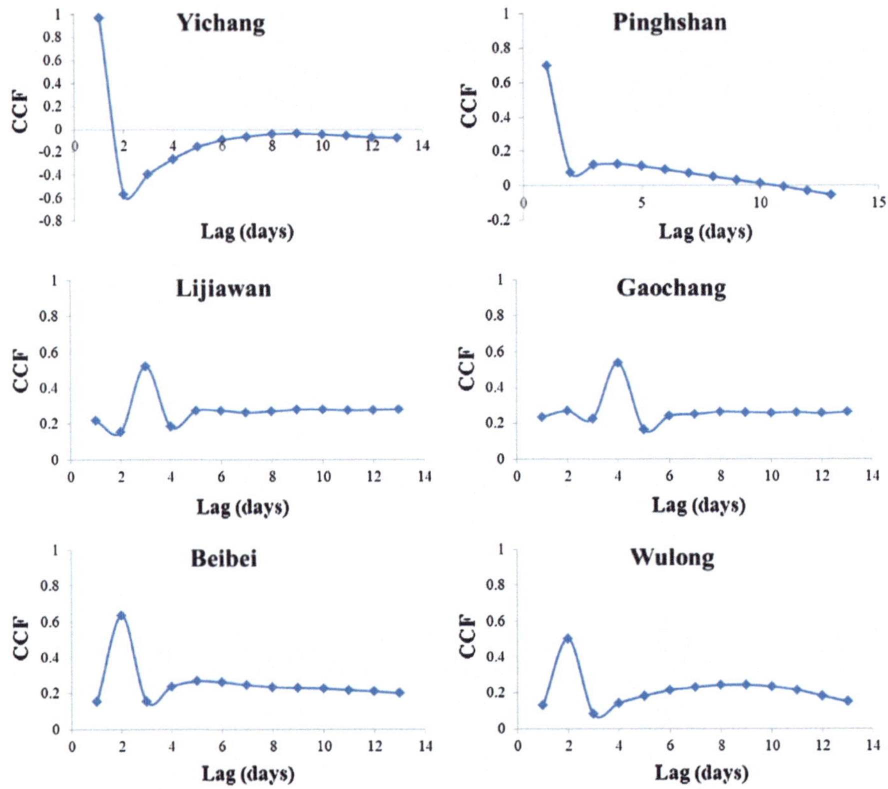

Fig. 10.4 Correlation coefficients between inputs and output of the ANN model (for the selected, the value is the linear correlation coefficient, and for the other potential inputs, it is the PCC given the selected variable)

selected the Gaochang lag 4 as input. Actually, the travel time between Gaochang sand Yichang is three days.

Both of those two input sets, which are selected by the proposed method and Pearson linear correlation coefficients, are employed to forecast the flood at the Yichang station. The same data sets are used to predict the input flow of the Three Gorges Reservoir. The performance criteria are calculated, and the results, given in Table 10.12, indicate that the network trained with the inputs selected by the PMI method has a higher *NSE* and α, and smaller *RMSE* values. In addition, the ANN model results based on the inputs selected by the LCC method are also shown in Fig. 10.2, where indicates that the predicted results based on the inputs selected by the CE method are superior to those based on the inputs selected by the LCC method. Therefore, the flood forecasting model with the selected inputs set based on PMI is better.

Table 10.11 Comparisons between the selected input sets based on the Pearson linear correlation coefficients and proposed PMI method

Rivers	Stations	Pearson correlation coefficient	PMI
Jinsha	Pingshan	Lag $t-1$	Lag $t-1$
Min	Gaochang	Lag $t-4$	Lag $t-3$
Tuo	Lijiawan	Lag $t-3$	Lag $t-3$
Jialing	Beibei	Lag $t-2$	Lag $t-2$
Wu	Wulong	Lag $t-2$	Lag $t-2$
Yangzte	Yichang	Lag $t-1$	Lag $t-1$, $t-2$

Table 10.12 Comparison of results obtained with different input variables

Methods	Nash–Sutcliffe efficiency		RMSE (m^3/s)		Qualified rate	
	Training	Validation	Training	Validation	Training	Validation
Linear correlation	0.9231	0.9036	1476	2932	0.9857	0.8566
CE	0.9402	0.9341	1281	2423	0.9898	0.9590

10.5.4.2 Comparisons with the Current Flood Forecasting Model of TGR

The current flood forecasting method is used to predict the flow of 2006 and 2007 of TGR (Liang et al. 1992). The *RMSE*, *NSE* and α values, calculated using the current flood forecasting model, are 2425 m^3/s, 0.9340 and 0.95; and those of the proposed model are 2423 m^3/s, 0.937 and 0.96, respectively. The current regression method performs well, and the results of the proposed method based on ANN model are comparable to the current method.

10.6 Flood Forecasting for the Jinsha River

In the above section, the flood forecasting model of TGR is built based on the runoff in the upstream mainstream and its tributaries. In this sub-section, the rainfall-runoff relationship is simulated using the ANN models. In order to compare different ANN models, the GRNN, MLF, and RBF ANNs are used hereafter.

10.6.1 Study Area

The Yangtze River rises in the Tanggula Mountains and the Qinghai-Tibet plateau in southwestern China. The reach from Yushu in Qinghai province to Yibin in

Sichuan province is called the Jinsha River, lying on the eastern edge of the Plateau and influenced by a variety of monsoons, e.g., tropical monsoon, subtropical monsoon, and Qinghai-Tibetan plateau monsoon. Jinsha River is the westernmost of the major headwater streams of the Yangtze River. It flows through the Qinghai, Sichuan and Yunnan provinces in western China. The Jinsha River basin is divided into nine sub-basins controlled by the Pingshan gauging station, shown in Fig. 10.5. The rainfall stations in those sub-basins used in this study are listed in Table 10.13. The areal average rainfall of each sub-basin is calculated.

The first and most important steps for building an ANN model is the determination of potential input variables (Bowden et al. 2005a; Fernando et al. 2009). Figure 10.6 demonstrates the areal average rainfall and daily runoff time series in the upper Jinsha basin. It can be seen from Fig. 10.6 that the discharge at Pingshan station has a strong relationship with that at Tongzilin and Shigu stations, and rainfall in some sub-basins has a significant impact on the daily discharge at Pingshan station, such as rainfall in sub-basins 1, 2, 3 and 4. Therefore, the areal average rainfall of each sub-basin and the previous discharge of Tongziling, Shigu and Pingshan stations with different lags are taken as potential inputs of ANN models. The discharge of Pingshan station is predicted based on the established rainfall-runoff model.

10.6.2 Selection of Model Inputs

The two-stage procedure, proposed by Bowden et al. (2005b) for an input selection using PMI, is used in this section as well. First, the copula entropy values between each of $(x_{i,t-1}, x_{i,t-2}, \ldots, x_{i,t-k})$ and y_t, and $(y_{t-1}, y_{t-2}, \ldots, y_{t-k})$ and y_t are calculated.

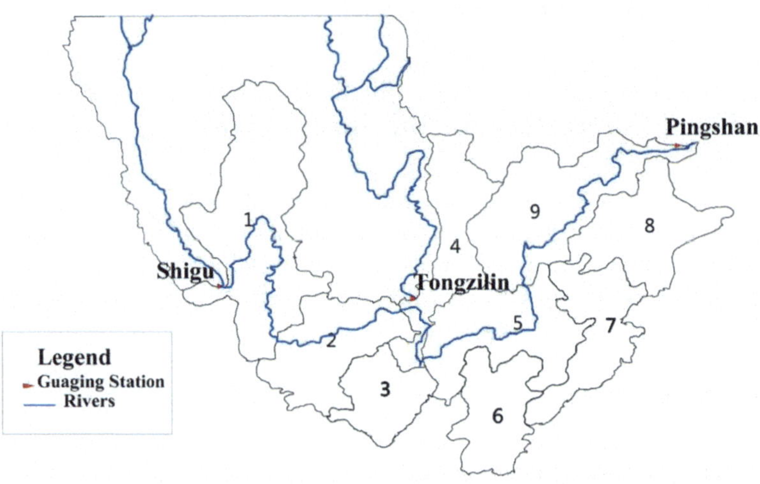

Fig. 10.5 The sub-basins of the Jinsha River basin

Table 10.13 Rainfall stations used in this study for the rainfall-runoff simulation

Sub-basin	Stations	Longitude	Latitude	Length of Record
1	Shigu	99°56′00″E	26°54′00″N	1990–2010
	Daqiaotou	102°54′00″	26°37′00″	
	Luoji	102°54′00″	28°48′00″	2004–2010
	Jinmian	101°35′59″	27°11′0″	2004–2010
	Zongguantian	100°34’33”	26°48’23”	2002–2010
	Huangping	100°23′12″	26°05′25″	2004–2010
2	Jinjiangjie	100°32′32″	26°13′22″	2001–2010
	Dahuizhuang	100°32′11″	25°57′41″	2004–2010
	Renli	101°00′00″	26°29′00″	2000–2010
3	Fengtun	101°22′42″	25°17′51″	2000–2010
	Baihe	101°12′00″	26°40′0″	2004–2010
	Duoke	101°51′44″	25°49′43″	2004–2010
	Doubashi	102°45′40″	25°23′11″	2004–2010
4	Panzhihua	101°40′29″	26°34’50”	1998–2010
	Tongzilin	101°50′15″	26°41′28″	1998–2010
	Xiaohuangua	101°52′00″	25°50′00″	2000–2010
5	Sanduizi	101°50′30″	26°29′0″	2004–2010
	Gaoqiao	102°10′00″	26°38′0″	2004–2010
	Fengguo	102°22′00″	26°9′0″	2004–2010
6	Yunlong	102°23′30″	25°5′16″	2004–2010
	Chemuhe	102°22′00″	25°37′0″	2004–2010
	Caijiacun	102°28′54″	25°18′53″	2004–2010
7	Dacun	103°02′32″	27°04′38″	2004–2010
	Dashuijing	103°34′00″	27°04′02″	2004–2010
	Jinle	103°28′11″	26°29′03″	2004–2010
8	Xiaohe	103°12′00″	27°13′00″	2004–2010
	Huanggeshu	104°34′00″	28°0′0″	2004–2010
	Moshiyi	103°44′06″	27°49′38″	2004–2010
	Malucun	104°02′00″	27°38′36″	1998–2010
	Doushaguan	104°07′42″	28°02′00″	1995–2010
9	Ningnan	102°43′07″	27°02′33″	1992–2010
	Zhaojue	102°51′09″	28°00′41″	1991–2010
	Huapingzi	103°9′7″	27°26′3″	1980–2010
	Xiluodu	103°41′39″	28°13′31″	2004–2010
	Xinhua	103°56′54″	28°34′18″	2000–2010

Then the variable with the maximum negative copula entropy value and its Z_j greater than 3 are listed in Table 10.14.

During this stage, the original 180 inputs are reduced to 17 inputs. The past runoff ($y_{t-1}, y_{t-2}, \ldots, y_{t-k}$) at Pingshan Station has great impacts on y_t. Therefore, lags $t-1$, $t-2$, $t-3$, $t-4$ and $t-5$ are selected for Pingshan station. Only one

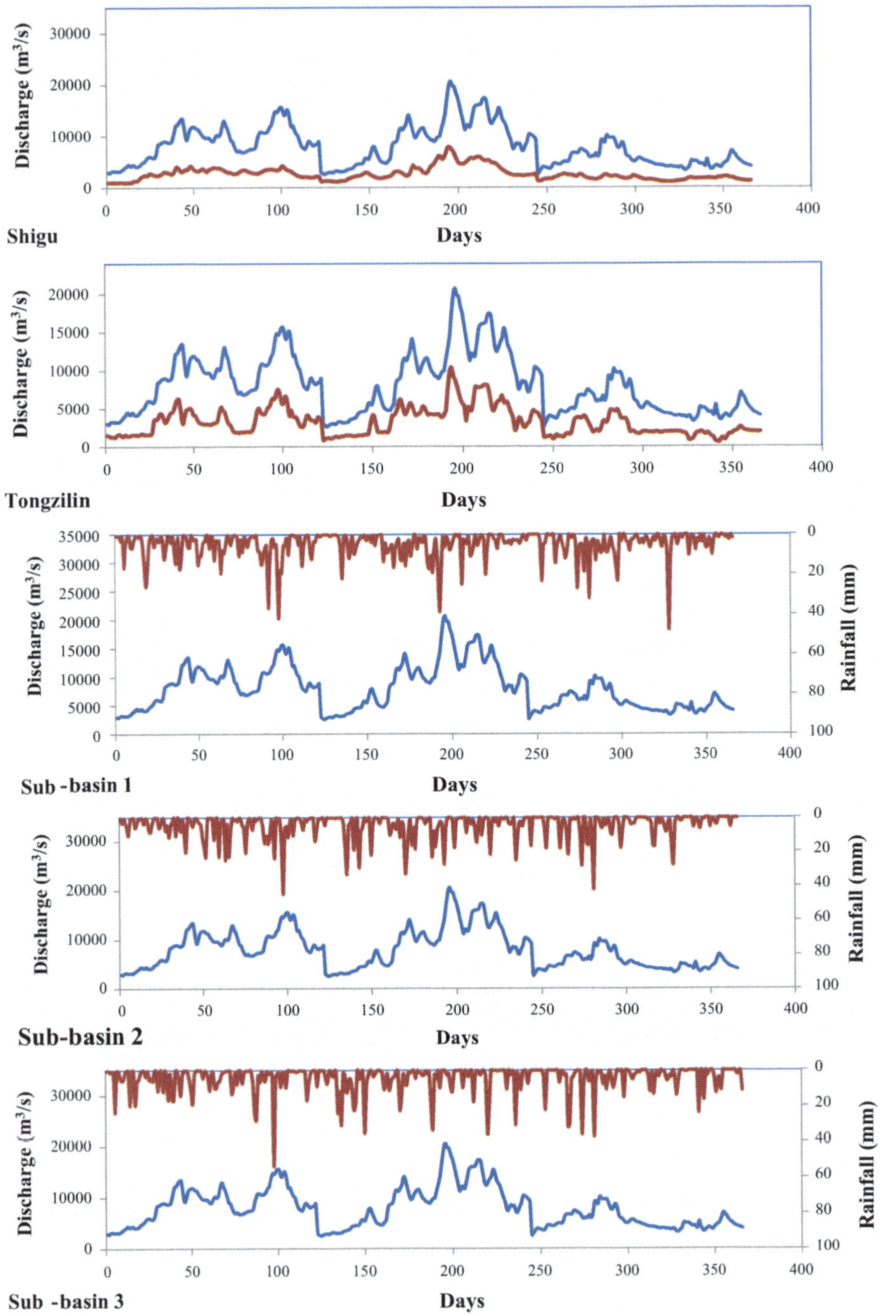

Fig. 10.6 Daily rainfall-runoff (runoff-runoff) time series of the Jinsha River basin

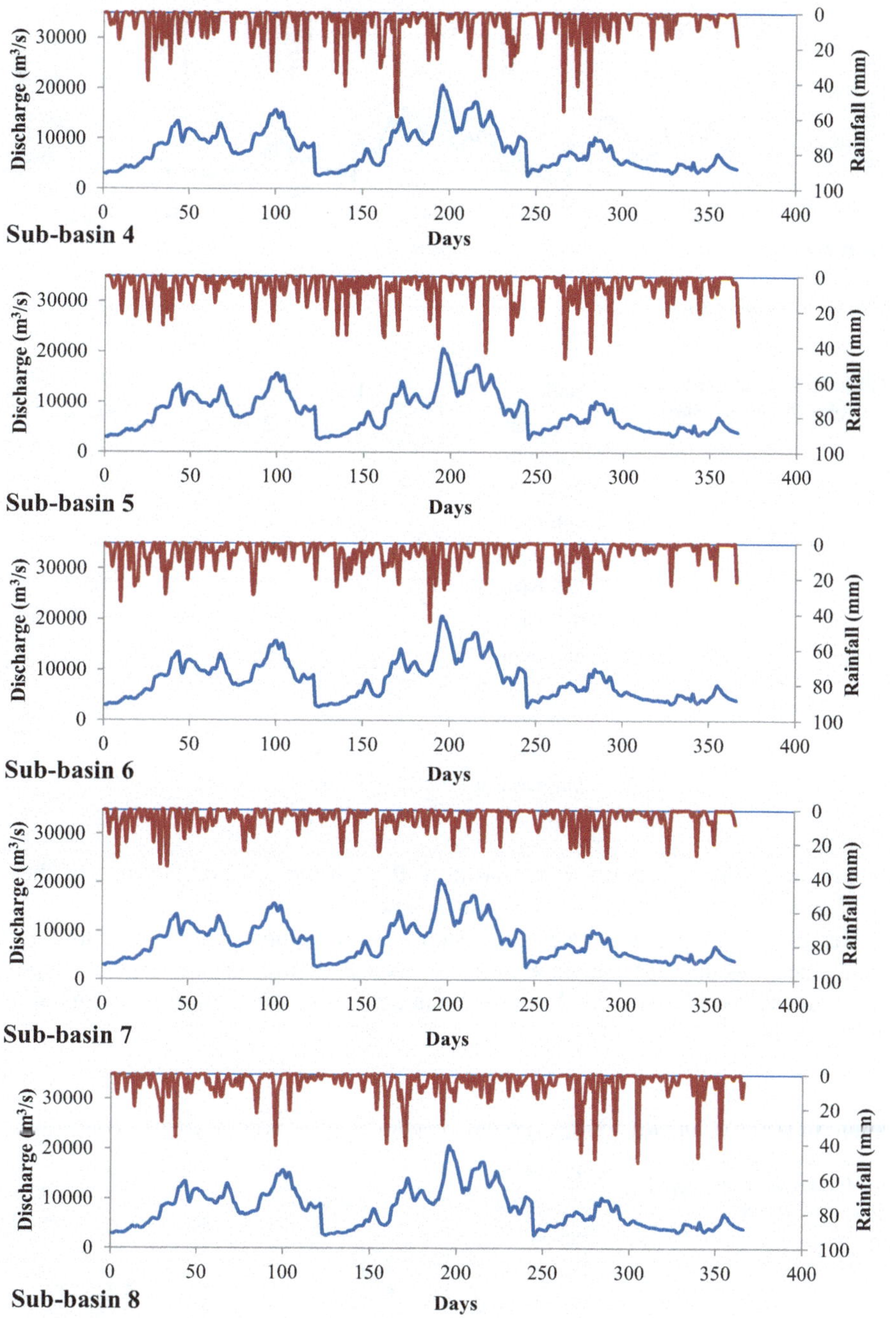

Fig. 10.6 (continued)

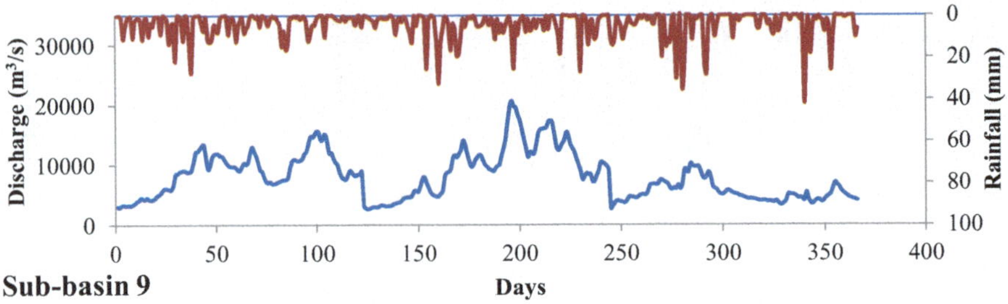

Fig. 10.6 (continued)

Table 10.14 Selected input variables in the first stage

Stations	Lags
Pingshan	t − 1, t − 2, t − 3, t − 4, t − 5
Tongziling	t − 2
Shigu	t − 3
Sub-basin 1	t − 5
Sub-basin 2	t − 4, t − 6
Sub-basin 3	t − 4
Sub-basin 4	t − 4
Sub-basin 5	t − 5
Sub-basin 6	t − 4
Sub-basin 7	t − 5
Sub-basin 8	t − 6
Sub-basin 9	t − 6

or two inputs are selected for sub-basins 1–9. And the selected lag time of these stations basically matched the flood travel time.

Second, significant lags selected in step one are combined to form a subset of candidates and then the copula entropy is calculated. During this stage, 17 inputs are reduced to 11. The final selected variables for ANN model are given in Table 10.15.

Table 10.15 Final selected input variables for the ANN model based on the copula entropy method

Stations	Lags
Pingshan	t − 1, t − 2, t − 3, t − 4, t − 5
Tongzilin	t − 2
Shigu	t − 3
Sub-basin 1	t − 5
Sub-basin 2	t − 4
Sub-basin 3	t − 4
Sub-basin 4	t − 4
Sub-basin 5	t − 5

Bowden et al. (2005a) pointed out that the linear correlation analysis (LCA) method is the most popular analytical technique for selecting appropriate inputs. The Pearson linear correlation coefficients are calculated and shown in Fig. 10.7. It can be seen that the dependencies between rainfall of sub-basin 1, sub-basin 2, Tongzilin and Shigu and the flow of Pingshan station are high. The lags of these stations are selected by the copula entropy method as inputs of the model. There are still some differences between these two methods. For both the Tongzilin and Shigu stations, the highest correlation coefficient values occur at lag $t-2$. However, the input selected by the copula entropy method is lag $t-3$ for the Shigu station. According to Fig. 10.5, the Shigu station is farther than Tongzilin station. From this point of view, the inputs selected by the copula entropy method are more appropriate.

10.6.3 Identification of Models

The inputs obtained by the copula entropy method are used for rainfall-runoff modeling. The identification of a prediction model is to determine the structure by using training data to optimize relevant model parameters, once model inputs are already obtained (Wu and Chau 2011). Three ANN models, namely MLF, RBF, and GRNN, are used in this study. The identification of ANN models is to find the model which performs best when the model inputs have been determined. Three performance criteria are used to assess these models. The initial data set consisted of 7 years, from which data for 2004 to 2008 for model calibration and those for 2009 and 2010 are used for model validation.

Table 10.16 comprises the results obtained using different ANN models. The fitting curves between observed and predicted daily flows at Pingshan station are given in Fig. 10.8. It can be seen from Table 10.16 and Fig. 10.8 that the MLF ANN model performs better with small *NSE*, *RMSE* and Qualified rate than any other ANN model. Therefore, the MLF ANN model with the inputs derived by the copula entropy method gives the best results for predicting the flow at Pingshan station. The values of *NSE*, *RMSE* and Qualified rate for validation period calculated by the MLF ANN model are 0.9524, 751 m^3/s and 0.9786, which indicate that the proposed model can predict daily flow at Pingshan station extremely well.

10.6.4 Comparisons of Predicted Results with Different Input Sets

Two input sets are used for establishing the ANN rainfall-runoff model, one of which is obtained by the linear correlation coefficient (LCC) method and the other by the copula entropy method. The forecasting performances of those two input sets are shown in Table 10.17. It can be seen that the *RMSE* values based on the inputs

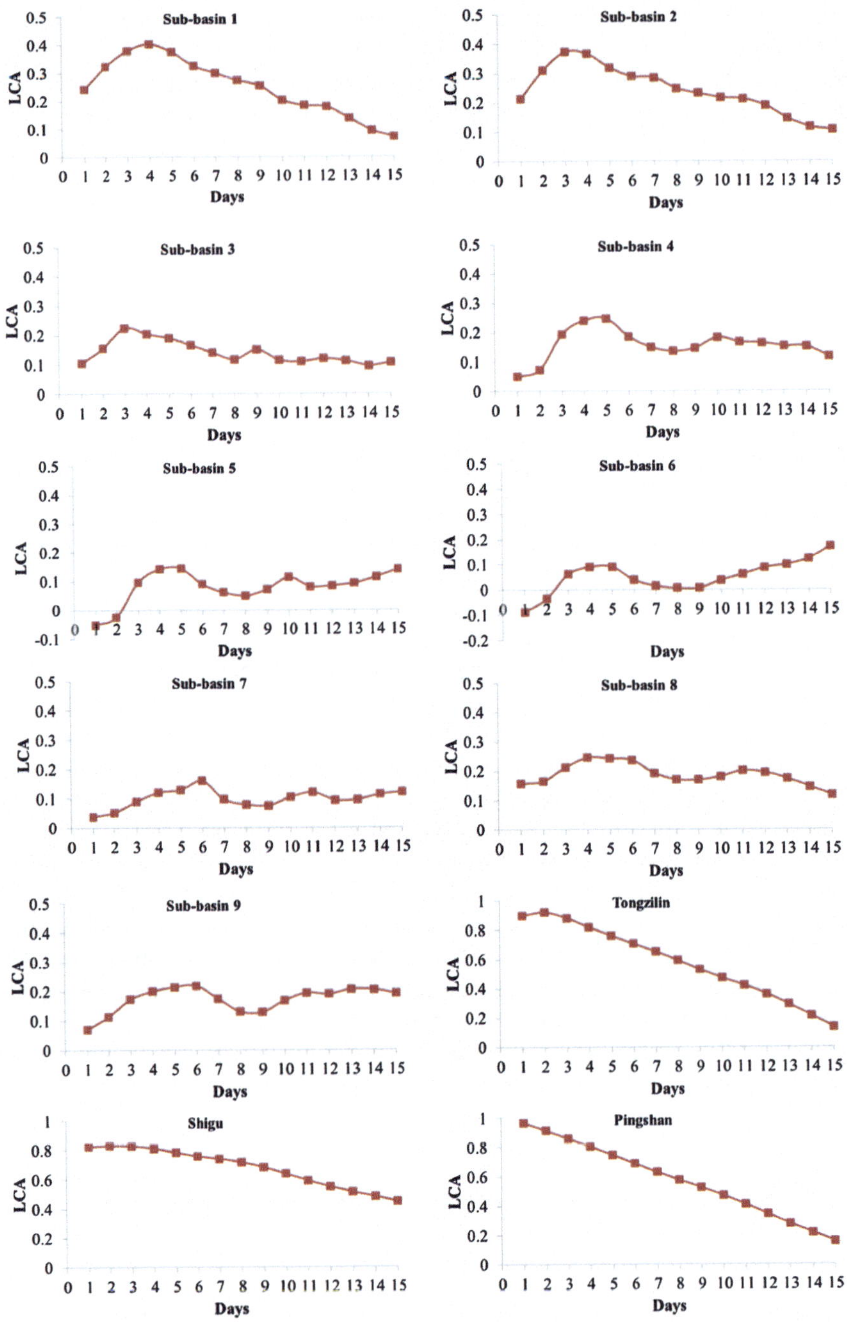

Fig. 10.7 Linear correlation coefficients between potential inputs and output of ANN model

Table 10.16 Comparison of results obtained with different ANN models

Methods	ANN	NSE		RMSE (m^3/s)		Qualified rate	
		Training	Validation	Training	Validation	Training	Validation
Copula entropy	MLF	0.9781	0.9524	548	751	0.9880	0.9786
	GRNN	0.9797	0.9427	528	825	0.9675	0.8846
	RBF	0.9395	0.9008	911	1086	0.9077	0.8426

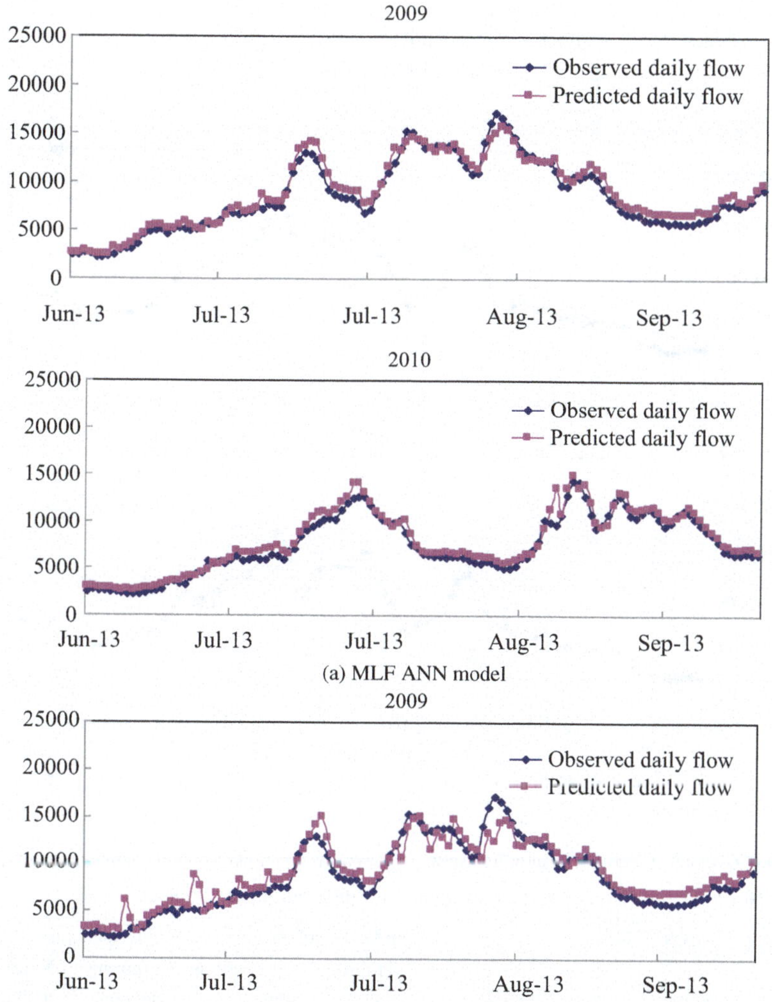

Fig. 10.8 Comparison between observed and predicted runoff values

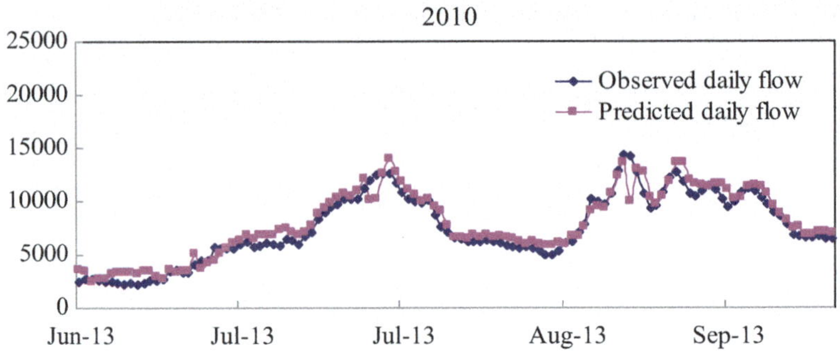

(b) RBF ANN model

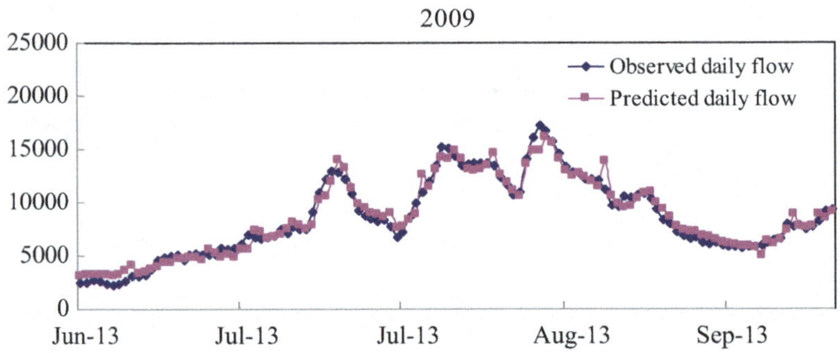

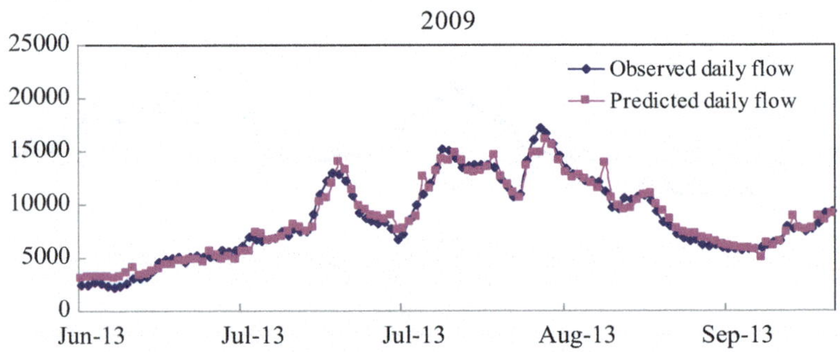

(c) GRNN ANN model

Fig. 10.8 (continued)

Table 10.17 Comparison of results obtained with different input variables

Methods	ANN	NSE		RMSE (m^3/s)		Qualified rate	
		Training	Validation	Training	Validation	Training	Validation
CE	MLF	0.9781	0.9524	548	751	0.9880	0.9786
LCC	BP	0.9674	0.9170	649	906	0.9738	0.9393

selected by the CE method are smaller than those based on the inputs selected by LCC method, and the *NSE* and Qualified rate based on the inputs selected by the CE method are higher than those based on the inputs selected by the LCC method.

References

Abrahart RJ, See L (2002) Multi-model data fusion for river flow forecasting: an evaluation of six alternative methods based on two contrasting catchments. Hydrol Earth Syst Sci 6(4):655–670

Angulo JM, Madrid AE, Ruiz-Medina MD (2011) Entropy-based correlated shrinkage of spatial random processes. Stoch Env Res Risk A 25(3):389–402

Berntson J, Espelid TO, Genz A (1991) An adaptive algorithm for the approximate calculation of multiple integrals. ACM Trans Math Softw 17:437–451

Birikundavyi S, Labib R, Trung HT, Rousselle J (2002) Performance of neural networks in daily streamflow forecasting. J Hydrol Eng 7(5):392–398

Bowden GJ, Dandy GC, Maierb HR (2005a) Input determination for neural network models in water resources applications. Part 1-background and methodology. J Hydrol 301:75–92

Bowden GJ, Maierb HR, Dandy GC (2005b) Input determination for neural network models in water resources applications. Part 2. Case study: forecasting salinity in a river. J Hydrol 301:93–107

Calsaverini RS, Vicente R (2009) An information-theoretic approach to statistical dependence: copula information. Europ Phys Lett 88(6):3–12

Castellano-Méndeza M, González-Manteigaa W, Febrero-Bande M, Prada-Sáncheza MJ, Lozano-Calderón R (2004) Modeling of the monthly and daily behavior of the runoff of the Xallas river using Box-Jenkins and neural networks methods. J Hydrol 296:38–58

Chen L, Ye L, Singh VP, Zhou J, Guo S (2014a) Determination of input for artificial neural networks for flood forecasting using the copula entropy method. J Hydrol Eng 19 (11):04014021

Chen L, Singh VP, Guo S, Zhou J, Ye L (2014b) Copula entropy coupled with artificial neural network for rainfall–runoff simulation. Stochastic environmental research and risk assessment. Stoch Env Res Risk 28(7):1755–1767

Davies L, Gather U (1993) The identification of multiple outliers. J Am Statist Assoc 88 (423):782–792

de Vos NJ, Rientjes THM (2005) Constraints of artificial neural networks for rainfall-runoff modelling: trade-offs in hydrological state representation and model evaluation. Hydrol Earth Syst Sci Discuss 2:365–415

Elman JL (1988) Finding structure in time, In: CRL Technical Report 8801, University of California at San Diego, Centre for Research in Language

Fernando TMKG, Maier HR, Dandy GC (2009) Selection of input variables for data driven models: an average shifted histogram partial mutual information estimator approach. J Hydrol 367:165–176

Hecht-Nielsen R (1987) Counterpropagation networks. Appl Opt 26:4979–4984

Hopfield JJ (1987) Learning algorithms and probability distributions in feed-forward and feed-back networks. P Natl Acad Sci USA 84:8429–8433

Hsu KL, Gupta HV, Sorooshian S (1995) Artificial neural network modeling of the rainfall–runoff process. Water Resour Res 31(10):2517–2530

Islam MN, Liong SY, Phoon KK, Liaw C-Y (2001) Forecasting of river flow data with a general regression neural network. Integrated water resources management (Proceedings of a symposium held at Davis. California. IAHS Publ. no. 272

Jain SK, Das A, Drivastava DK (1999) Application of ANN for reservoir inflow prediction and operation. J Water Resour Plann Manage 125(5):263–271

Jayawardena DA, Fernando AK, Zhou MC (1997) Comparison of multilayer perceptron and radial basis function networks as tools for flood forecasting. Destructive water: water-caused natural disasters, their abatement and control (Proceedings of the conference held at Anaheim, California, June 1996). IAHS Publ. no. 239

Jeong DI, St-Hilaire A, Ouarda TBMJ, Gachon P (2012) Comparison of transfer functions in statistical downscaling models for daily temperature and precipitation over Canada. Stoch Env Res Risk A 26(5):633–653

Karunanithi N, Grenney WJ, Whitley D, Bovee K (1994) Neural networks for river flow prediction. J Comput Civil Eng 8:201–219

Kasiviswanathan KS, Sudheer KP (2013) Quantification of the predictive uncertainty of artificial neural network based river flow forecast models. Stoch Env Res Risk A 27(1):137–146

Kohonen T (1982) Self-organized formation of topologically correct feature maps. Biol Cybern 43:59–69

Lachtermacher G, Fuller JD (1994) Backpropagation in hydrological time series forecasting. In: Hipel KW, McLeod AI, Panu US, Singh VP (eds) Stochastic and statistical methods in hydrology and environmental engineering. Kluwer Academic, Dordrecht

Lekkas DF, Onof C, Lee MJ, Baltas EA (2004) Application of Artificial Neural networks for flood forecasting. Global Nest: Int J 6(3):205–211

Li X, Guo SL, Liu P, Chen GY (2010) Dynamic control of flood limited water level for reservoir operation by considering inflow uncertainty. J Hydrol 391:124–132

Liang GC, Kachroo RK, Kang W, Yu XZ (1992) Applications of linear modeling techniques for flow routing on large basins. J Hydrol 133:99–140

Maier Holger R, Dandy Graeme C (2000) Neural networks for the prediction and forecasting of water resources variables: a review of modeling issues and applications. Environ Model Softw 15:101–124

May Robert J, Dandy Graeme C, Maier Holger R, Nixon John B (2008a) Application of partial mutual information variable selection to ANN forecasting of water quality in water distribution systems. Environ Model Softw 23:1289–1299

May Robert J, Maier Holger R, Dandy Graeme C, Fernando TMK (2008b) Non-linear variable selection for artificial neural networks using partial mutual information. Environ Modell Softw 23:1312–1326

McCulloch WS, Pitts W (1943) A logical calculus of the ideas immanent in nervous activity. B Math Biol 5:115–133

Ministry of Water Resources (MWR) (2006) Regulation for calculating design flood of water resources and hydropower projects. Chinese Shuili Shuidian Press, Beijing (in Chinese)

Mishra AK, Singh VP (2009) Analysis of drought severity-area-frequency curves using a general circulation model and scenario uncertainty. J Geophys Res 114:D06120. https://doi.org/10.1029/2008JD010986

Mishra AK, Ines AVM, Singh VP, Hansen JW (2013) Extraction of information content from stochastic disaggregation and bias corrected downscaled precipitation variables for crop simulation. Stoch Env Res Risk A 27(2):449–457

Park J, Sandberg IW (1991) Universal approximations using Radial-Basis-Function networks. Neural Comput 3(2):246–257

Powell MJD (1987) Radial basis functions for multivariable interpolation: a review. In: Mason JC, Cox MG (eds) Algorithms for approximation. Clarendon Press, Oxford, pp 143–167

Raman H, Sunilkumar N (1995) Multivariate modeling of water resources time series using artificial neural networks. Hydrol Sci J 40(2):145–163

Rumelhart DE, Hinton E, Williams J (1986) Learning internal representation by error propagation. Parallel Distrib Process 1:318–362

Shamseldin AY (1997) Application of a neural network technique to rainfall-runoff modeling. J Hydrol 199:272–294

Sharma A (2000) Seasonal to interannual rainfall probabilistic forecasts for improved water supply management: Part 1 a strategy for system predictor identification. J Hydrol 239:232–239

Specht DF (1991) A general regression neural network. IEEE Trans Neural Netw 2(6):568–576

Svozil D, KvasniEka V, Pospichal J (1997) Introduction to multi-layer feed-forward neural networks. Chemometr Intell Lab 39:43–62

Thirumalaiah K, Deo MC (2000) Hydrological forecasting using neural networks. J Hydrol Eng 5 (2):180–189

Tongal H, Demirel MC, Booij MJ (2013) Seasonality of low flows and dominant processes in the Rhine River. Stoch Env Res Risk A 27(2):489–503

Wu CL, Chau KW (2011) Rainfall–runoff modeling using artificial neural network coupled with singular spectrum analysis. J Hydrol 18:394–409

Zhao N, Linb WT (2011) A copula entropy approach to correlation measurement at the country level. Appl Math Comput 218(2):628–642

Zupan J, Gasteiger J (1993) Neural networks for chemists: an introduction. In: Zupan J, Gasteiger J (eds) Weinheim, VCH (Germany)

Chapter 11
Correlations Among Rivers Using Copula Entropy

11.1 Introduction

According to historical records, 1092 large flood events have occurred since 206 BC in China during a period of 2155 years, averaging once every two years (Technical support unit 2004). Disastrous floods can be caused by unusual combinations of hydro-meteorological factors and river basin conditions. Topography, land cover, and temporal and spatial distribution of rainfall play the dominant role in the generation of floods, which can be reflected in the contributions that major tributaries make to the mainstream flow. The coincidence of flood flows of the mainstream and its tributaries may determine peak flow. Therefore, analysis of the dependence of these tributaries and the influence of upper tributaries on the mainstream is important for hydraulic design, flood prevention and risk control.

The Yangtze River and its flood characteristic had been introduced in Chap. 9. There are also two hydrometric stations in the Yichang and Datong which are considered as the separating points for the three basins. The Three Gorges Dam (TGD), is located in the city of Yichang in Hubei province, China, which is also the world's largest capacity hydroelectric power station with a total generating capacity of 18,200 MW. An important function of this dam is to control flooding, which is a major problem for the seasonal Yangtze. Millions of people live downstream of the dam, with many large, important cities like Wuhan, Nanjing, and Shanghai situated adjacent to the river. Plenty of farmlands and China's most important industrial areas are built along the river. For flood control by TGD, the larger the information, the more accurate the decisions are made for reservoir operation and the smaller the chances are for under-design or over-design for peak discharge. Thus, it is important to investigate the characteristics of inflow to TGD, which mainly stems from five rivers in the upper Yangtze River. Most of the common practices for analysis of the associated risk for TGD, however, have relied on frequency analysis of flows of only one of the rivers, ignoring the structure dependence between flows of rivers solely for mathematical simplicity. Actually, there exists plenty of

L. Chen and S. Guo, *Copulas and Its Application in Hydrology and Water Resources*, Springer Water, https://doi.org/10.1007/978-981-13-0574-0_11

evidence for the dependence among flows of these rivers. Failure to taking into account the dependence between them may lead to overestimating or underestimating of the design peak flow and associated risk. Dependence and random characteristics of the upstream rivers are of practical value and important for flood forecasting and reservoir management.

Often there is a need to evaluate the dependence among more variables (Alfonso et al. 2010). Another way of investigating multivariate dependence is by assessing the total amount of information that is shared by all variables at the same time (Alfonso et al. 2010). In probability theory and information theory, the total correlation (Watanabe 1960) is one of several generalizations of mutual information. It is also known as the multivariate constraint (Garner 1962) or multi information (Studený and Vejnarová 1999). It quantifies the redundancy or dependency among a set of d random variables. Total correlation tells in the most general sense how cohesive or related a group of variables are. A near-zero total correlation indicates that the variables in the group are essentially statistically independent; they are completely unrelated, in the sense that knowing the value of one variable does not provide any clue as to the values of the other variables.

Since runoff is stochastic and a response of a dynamic and potentially nonlinear system, we choose to employ information theory and a copula function, nonlinear techniques, to extract relations between the mainstream and tributaries or between the gauge station upstream or downstream of a river site. The total correlation method, related to a d-dimensional copula-based joint distribution function, which is an extension of the mutual information, can be applied to measure the dependence of multivariate functions. The objective of this chapter is therefore to establish multivariate distributions associated with different dependencies based on copula functions and analyze the total correlation among rivers in the upper Yangtze River Bain by using the copula entropy theory (Chen et al. 2013).

11.2 Total Correlation

There is a need to evaluate the dependence among several variables, which involves a difficult assessment of multivariate joint probabilities. A number of pairwise approximations have been proposed for this assessment (Lewis 1959; Chow and Liu 1968; Kirshner et al. 2004). Another way of looking at multivariate dependence is by assessing the total amount of information that is shared by all variables at the same time. The total correlation can be defined as (McGill 1954; Watanabe 1960; Alfonso et al. 2010):

$$T(X_1, X_2, X_3, \ldots, X_d) = \sum_{i=1}^{d} H(X_i) - H(X_1, X_2, X_3, \ldots, X_d) \tag{11.1}$$

where T is the total correlation. It can be noted that for the case of $d = 2$, total correlation is equivalent to the well-known mutual information (or trans-information).

According to Eq. 2.61, Eq. 11.1 can be written as

$$T(X_1, X_2, X_3, \ldots, X_d) = -H_c(\mathrm{x}) \tag{11.2}$$

The total correlation is always positive since the sum of the entropy of all of the variables will always be greater than the joint entropy of all of them, and T is equal to zero if and only if all the variables being considered are independent. However, T is greater than zero if two of the variables have some dependence, even though the rest of the variables are independent (Alfonso et al. 2010).

In light of Eq. 11.2, the total correlation of random variables is equivalent to their negative copula entropy, which is invariant under arbitrary choices of marginal densities $f_i(x)$. The proposed method only needs to calculate the copula entropy instead of marginal or joint entropy, which estimates the total correlation more directly and avoids the accumulation of systematic bias inherent in terms $H(X_i)$ and $H(X_1, X_2, X_3, \ldots, X_N)$. It also may be found that the copula-entropy based total correlation only relies on the copula density function that is determined by the copula parameter. Therefore, only the copula parameter is required for the estimation of total correlation. Thus, the proposed method provides an effective way to calculate the total correlation and reduces the complexity and computational requirements, which makes it easier to use than the previous methods.

11.3 Application

11.3.1 Data

The upstream Yangtze River is taken into account in this study. Its major characteristics have been described in Chaps. 6 and 9. The annual maximum (AM) sample method is used in this study. The POME method is applied to obtain the marginal distribution whose parameters are given in Table 11.1. The marginal distributions of tributaries of the upper Yangtze River are shown in Fig. 11.1, in which the line represents the theoretical distribution and the crosses the empirical frequency distribution. The figure indicates that all theoretical distributions fit the observed data reasonably well.

Table 11.1 Estimated parameters of POME method

Rivers	λ_0	λ_1	λ_2	λ_3
Jinsha	17.7103	−2.2973	0.1064	−0.0014
Jialing	8.2541	−0.5538	0.0168	−0.0001
Min	17.0250	−2.4144	0.1233	−0.0019
Tuo	5.1232	−1.8333	0.2813	−0.0106
Wu	12.1272	−2.3221	0.1711	−0.0038

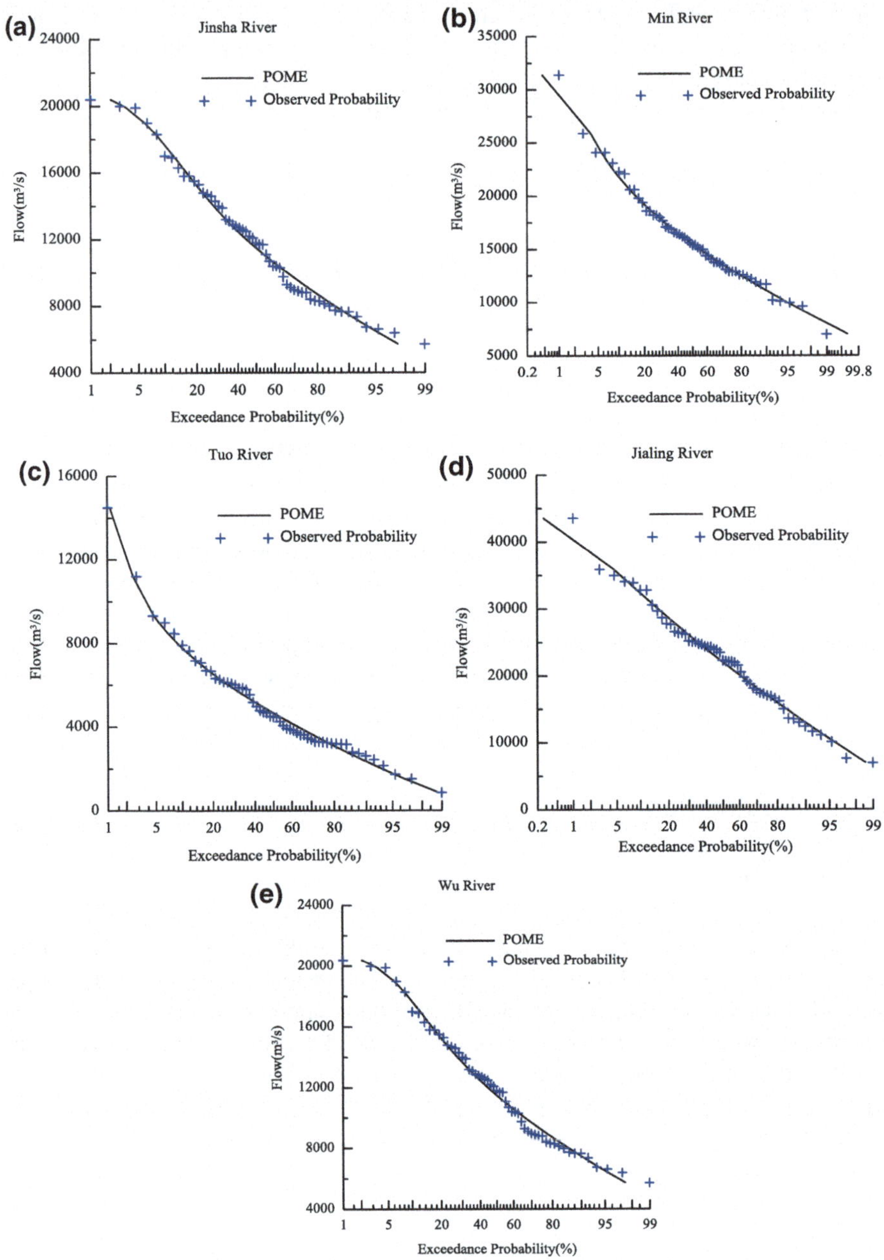

Fig. 11.1 Frequency curves of flood magnitudes based on AM samples

When quantifying dependence, it is a common practice to start by measuring linear correlation, namely Pearson correlation coefficients. The Pearson's correlations are also computed to estimate the dependence of the AM series from 1951 to 2007, and the results are given in Table 11.2. Two assumptions need to be satisfied for Pearson correlation coefficient, which are mentioned in Chap. 10 as well. One is that the variable must follow the multivariate normal distribution, and the other is that the pairwise dependency is linear. According to Fig. 10.4 of Chap. 10, the data of the Min, Tuo, and Wu Rivers show high kurtosis and skewness, and the assumption of normality is found to be inappropriate. In order to test the validity of the assumption that the pairwise dependence is linear, the time series of flow data is also divided into two segments. The t-test is used for the significance test of Pearson correlation coefficient. The *P*-values are calculated. If *P*-values are small, less than α, then the Pearson correlation coefficient is significant at the level (α is equal to 0.1 in this study). Otherwise, the correlation coefficient is not significant and equal to 0. The calculated correlation coefficient and *P*-values are given in Table 11.3, which indicates that Pearson's correlation is changing over time. In addition, the linear correlation coefficients are not valid. Kendall's coefficients for all pairs of the variables are calculated, because of their rank based characteristic. Genest and Verret (2005) indicated that Kendall's coefficients or the Spearman's correlation is known to be robust to departures from normality, while remaining powerful. The calculated results of Kendall's and Pearson's coefficients are given in Table 11.2. It is indicated that there are some dependencies between these variables.

11.3.2 Two-Variable Model

First, joint distributions of any two rivers in the upper Yangtze River are determined. According to five rivers in this area, ten bivariate joint distributions are built. Gumbel, Clayton, Frank, normal and student copulas are, respectively, used for modeling the dependence among the five stations. A pseudo-likelihood technique involving the ranks of data is used for estimating parameters of these copulas hereafter. The Cramer-von Mises functional *Sn* defined by Genest et al. (2009) is used for the goodness of fit test. The P-values are calculated. The *AIC* values of the Archimedean and meta-elliptical copulas are shown in Table 11.4. The selected copulas and their parameters are given in Table 11.5. Generally, the Archimedean

Table 11.2 Dependence measures for the upper Yangtze River based on annual maximum data

Rivers	Jinsha	Jialing	Min	Tuo	Wu
Jinsha	1.00	−0.08	0.11	0.13	0.12
Jialing	−0.12	1.00	0.03	0.18	−0.15
Min	0.16	0.13	1.00	0.36	0.01
Tuo	0.19	0.37	0.50	1.00	−0.05
Wu	0.16	−0.26	−0.02	−0.10	1.00

Note the super-diagonal elements are the Kendall correlation; the sub-diagonal elements are the Pearson correlation

Table 11.3 Calculated Pearson correlation coefficients and their corresponding P-values

Rivers	Periods	Correlation coefficients	P-values
Jinsha-Jialing	1951–1957	−0.73	0.09
	1958–2007	−0.08	0.54
Jinsha-Min	1951–1960	−0.26	0.46
	1961–2007	0.26	0.07
Jinsha-Tuo	1951–1960	−0.37	0.28
	1961–2007	0.28	0.06
Jinsha-Wu	1951–1970	−0.04	0.86
	1971–2007	0.29	0.08
Jialing-Min	1951–1970	−0.19	0.43
	1971–2007	0.29	0.09
Jialing-Tuo	1951–1970	−0.19	0.43
	1971–2007	0.53	0.001
Jialing-Wu	1951–1970	−0.19	0.42
	1971–2007	−0.30	0.08
Min-Tuo	1951–1970	0.56	0.01
	1971–2007	0.49	0.003
Min-Tuo	1951–1999	−0.16	0.28
	2000–2007	0.74	0.06
Tuo-Wu	1951–1999	−0.22	0.1
	2000–2007	0.22	0.77

copulas give a better fit than the meta-elliptical ones. Only the P-values of the selected copulas are given in the brackets of Table 11.5. Results indicate that the selected bivariate copulas cannot be rejected.

Multiple integration and Monte Carlo methods are used for calculating the copula entropy, respectively. For the first method, the multiple integration method proposed by Berntsen et al. (1991) is applied. For the second method, 10,000 pairs of $\mathbf{u}$ are generated, and the average value of the $\ln[c(\mathbf{u})]$ are calculated and given in Table 11.6, which indicates that the two methods are similar. It is also indicated from Table 11.6 that the total correlation values are not so large. This is in accordance with the meteorological conditions. The reasons are as follows: First, there are two important and large rainfall zones in the study area. The Jialing and Min River basins belong to different rainfall zones, respectively. Second, in the upper Jinsha River, as the annual average temperature is below 0 °C, the flow is mainly from snow-melting. The flow of Jinsha River downstream of the Zhimenda is formed by snow-melting and rainfall together. Third, the Wu River is the only one located on the right bank of the Yangtze River in the study area. Generally, in normal years, the flood occurrence time of the tributaries located on the right bank of Yangtze River is earlier than those on the left bank. Due to the different causes of floods in these rivers, the dependence among them is relatively small. The largest total correlation is 0.33 between Min and Tuo Rivers. This is because the distance between the two rivers is the smallest and they belong to the same rainfall zone.

Table 11.4 The log maximum likelihood and AIC values of bivariate joint distributions

Rivers	Copula	Jinsha	Jialing	Min	Tuo	Wu
Jinsha	Gumbel		–	0.85	1.11	0.56
	Frank		0.39	0.87	0.81	1.02
	Clayton	–	–	2.19	2.51	1.05
	Normal		0.34	1.22	1.73	1.05
	Student		0.32	1.55	1.68	1.02
Jialing	Gumbel	–		0.53	3.22	–
	Frank	1.22		0.08	2.12	0.94
	Clayton	–	–	0.12	2	–
	Normal	1.32		0.32	2.81	1.38
	Student	1.36		0.39	3	1.36
Min	Gumbel	0.3	0.94		7.91	–
	Frank	0.26	1.84		8.36	0.01
	Clayton	−2.38	1.76	–	11.2	–
	Normal	−0.44	1.36		9.85	0.02
	Student	−1.1	1.22		10.12	3.88
Tuo	Gumbel	−0.22	−4.44	−13.82		–
	Frank	0.38	−2.24	−14.72		0.14
	Clayton	−3.02	−2	−20.4	–	–
	Normal	−1.46	−3.62	−17.7		0.01
	Student	−1.36	−4	−18.24		0.07
Wu	Gumbel	0.88	–	–	–	
	Frank	−0.04	0.12	1.98	1.72	
	Clayton	−0.1	–	–	–	–
	Normal	−0.1	−0.76	1.96	1.98	
	Student	−0.04	−0.72	−5.76	1.86	

Note The super-diagonal elements are the maximum likelihood values; the sub-diagonal elements are the AIC values

Table 11.5 Selections of copulas and determination of the parameters

Parameters	Copula				
	Jinsha	Jialing	Min	Tuo	Wu
Jinsha	–	Frank	Clayton	Clayton	Clayton
Jialing	−0.74(0.71)	–	Gumbel	Gumbel	Normal
Min	0.41(0.60)	1.09(0.10)	–	Clayton	Student
Tuo	0.43(0.49)	1.23(0.94)	1.21(0.38)	–	Frank
Wu	0.30(0.39)	−0.25(0.91)	0.04, 2.0(0.81)	−0.45(0.94)	–

Note The super-diagonal elements are the selected copula; the sub-diagonal elements are the parameters corresponding to the selected copulas; the values in the bracket are P-values

Table 11.6 Total correlation values of two tributaries in upstream Yangtze River

Total correlation	Jinsha	Jialing	Min	Tuo	Wu
Jinsha	1	0.008	0.053	0.057	0.032
Jialing	0.009	1	0.013	0.056	0.032
Min	0.049	0.014	1	0.245	0.083
Tuo	0.055	0.054	0.235	1	0.003
Wu	0.033	0.033	0.085	0.004	1

Note the super-diagonal elements are the values based on multiple integration method; the sub diagonal elements are the values based on Monte Carlo method

There are some dependencies among Jinsha, Min and Tuo Rivers. The floods of 1931, 1954 and 1998 are caused by the large floods of the three rivers. In a normal year, rainfall on the left and right banks of Yangtze River does not happen simultaneously, but the dependence between Min and Wu Rivers cannot be neglected. The dependence between Jialing and Min or Tuo Rivers also exists. The dependence between Min and Jialing is not obvious, but we still need to pay more attention.

In order to compare the Pearson correlation coefficients (PCC) with the proposed methods, the total correlation values are calculated and given in Table 11.7 which

Table 11.7 Comparison of total correlation calculated by different methods

Comparisons	Methods	Jinsha	Jialing	Min	Tuo	Wu
Jinsha	Proposed	—	0.008	0.053	0.057	0.032
	Equation (19)		0.007	0.013	0.018	0.013
	Absolute error		0.001	0.040	0.039	0.019
Jialing	Proposed	0.009	—	0.013	0.083	0.032
	Equation (19)	0.007		0.009	0.074	0.035
	Absolute error	0.002		0.004	0.009	−0.003
Min	Proposed	0.049	0.014	—	0.245	0.083
	Equation (19)	0.013	0.009		0.144	0.000
	Absolute error	0.036	0.005	1.000	0.101	0.083
Tuo	Proposed	0.055	0.082	0.235	—	0.003
	Equation (19)	0.018	0.074	0.144		0.005
	Absolute error	0.037	0.008	0.091		−0.002
Wu	Proposed	0.033	0.033	0.085	0.004	—
	Equation (19)	0.013	0.035	0.000	0.005	
	Absolute error	0.020	−0.002	0.085	−0.001	

Note the super-diagonal elements are the values based on multiple integration method; the sub diagonal elements are the values based on Monte Carlo method. Absolute error means the values of the proposed method minus the values calculated by PCC indicates that the total correlation calculated by the proposed methods is generally larger than that calculated by Pearson correlation coefficients. In other words, the Pearson correlation coefficient underestimates the dependence of variables.

11.3.3 Three-Variable Model

Joint distributions of any three rivers in the upper Yangtze River are determined. According to five rivers in this area, ten bivariate joint distributions are derived. The three-dimensional Gumbel, Clayton and Frank copulas belonging to the Archimedean class and Normal and Student copulas belonging to the meta-elliptical class are used for modeling the dependence among the three stations. Since the asymmetric Archimedean copula can only simulate the positive correlation, for some cases only normal and Student copulas are used to build the joint distribution. A pseudo-likelihood technique involving the ranks of the data is used for estimating parameters. The estimated parameters of these copulas are given in Table 11.8. The *AIC* values of the Archimedean and meta-elliptical copulas are also shown in that Table. The dependence of the symmetric Archimedean copulas is the same for all variables. The dependencies of asymmetric Archimedean copula functions are different corresponding to different pairs, but they can only simulate the dependencies between $d-1$ variables. It is indicated that the asymmetric copulas give a better fit than the symmetric one in the Archimedean family. For the positive dependence cases, generally the asymmetric Clayton copula gives a better fit. No significant difference exists between the *AIC* values of the Normal and Student copulas. The copula corresponding to the smallest *AIC* values are selected to calculate the theoretical probabilities. The empirical joint probabilities are plotted against theoretical probabilities calculated by the joint distribution, as shown in Fig. 11.2, which shows that no significant differences between empirical and theoretical joint probabilities can be detected.

The first method, namely the multiple integration method, is used to calculate the total correlation, and results are also given in Table 11.8. It is shown that for a specific case, there are some differences in the total correlation values when selecting different copula functions. In other words, choosing an appropriate copula is important to measure the total correlation of variables. Taking the Jinsha, Min and Tuo Rivers as an example, the total correlation corresponding to the smallest *AIC* value −3.16 is 0.28. However, if the Gumbel copula, which is usually selected, is used, the calculated total correlation is only 0.10. There is a large difference between them. Generally, the copulas with maximum copula entropy value also have the smallest *AIC* values, such as in the case of numbers 4 and 5 in Table 11.8. The smallest *AIC* values −3.16 and −0.47 are corresponding to the maximum copula entropy values 0.28 and 0.08, respectively. The calculated results seem to be justified from the point of view that the three-variable correlation is larger than that of the two-variable copula. By analyzing the three-dimensional total correlation, the correlation amongst Jinsha, Min and Tuo Rivers, Jialing, Min and Tuo Rivers, and Min, Tuo and Wu Rivers are much higher.

Table 11.8 Total correlation analysis of trivariate joint distribution

Number	Rivers			Copula	θ_1 ρ_1	θ_2 ρ_1	ρ_3	υ	AIC	Total correlation
1	Jinsha	Jialing	Min	Normal	-0.12	0.23	0.12		1.58	0.05
				t	-0.13	0.23	0.1	16.97	1.53	0.05
2	Jinsha	Jialing	Tuo	Normal	-0.11	0.27	0.34		-0.46	0.13
				t	-0.13	0.26	0.33	16.4	-0.49	0.13
3	Jinsha	Jialing	Wu	Normal	-0.13	0.22	-0.25		1.12	0.06
				t	-0.13	0.21	-0.25	35	1.18	0.06
4	Jinsha	Min	Tuo	Gumbel	1.17	1.17			0.01	0.1
					1.06	1.53			-2.14	0.19
				Frank	1.41	1.41			-0.57	0.076
					0.68	3.75			-2.34	0.17
				Clay	0.52	0.52			-2.36	0.19
					0.32	1.19			-3.16	0.28
				Normal	0.24	0.28	0.59		-1.94	0.26
				t	0.22	0.23	0.57	7.7	-1.99	0.25
5	Jinsha	Min	Wu	Gumbel	1.09	1.09			2.33	0.04
					1.07	1.16			1.41	0.05
				Frank	0.74	0.74			1.90	0.02
					0.57	1.19			1.44	0.03
				Clay	0.28	0.28			-0.20	0.03
					0.22	0.42			-0.47	0.08
				Normal	0.24	0.22	0.04		1.35	0.05
				t	0.21	0.15	0.09	4.43	0.53	0.1
6	Jinsha	Tuo	Wu	Normal	0.28	0.22	-0.009		0.87	0.07
				t	0.26	0.22	-0.02	65.51	0.90	0.06
7	Min	Tuo	Wu	Normal	0.59	0.03	-0.15		-1.59	0.24
				t	0.55	0.05	-0.05	3.37	-2.02	0.28
8	Jialing	Min	Tuo	Gumbel	1.24	1.24			-1.65	0.16
					1.15	1.53			-2.49	0.22
				Frank	1.73	1.73			-1.37	0.11
					0.99	3.73			-2.44	0.18
				Clay	0.48	0.48			-1.86	0.17
					0.24	1.21			-3.01	0.27
				Normal	0.14	0.34	0.58		-2.10	0.27
				t	0.07	0.32	0.56	5.42	-2.22	0.29
9	Jialing	Tuo	Wu	Normal	0.35	-0.25	-0.03		0.10	0.1
				t	0.33	-0.24	-0.03	26.73	0.09	0.1
10	Jialing	Min	Wu	Normal	0.12	-0.25	0.03		1.86	0.04
				t	0.05	-0.2	0.05	4.23	0.66	0.09

Note Values mean maximized likelihood values for these models. The same meaning is hereafter

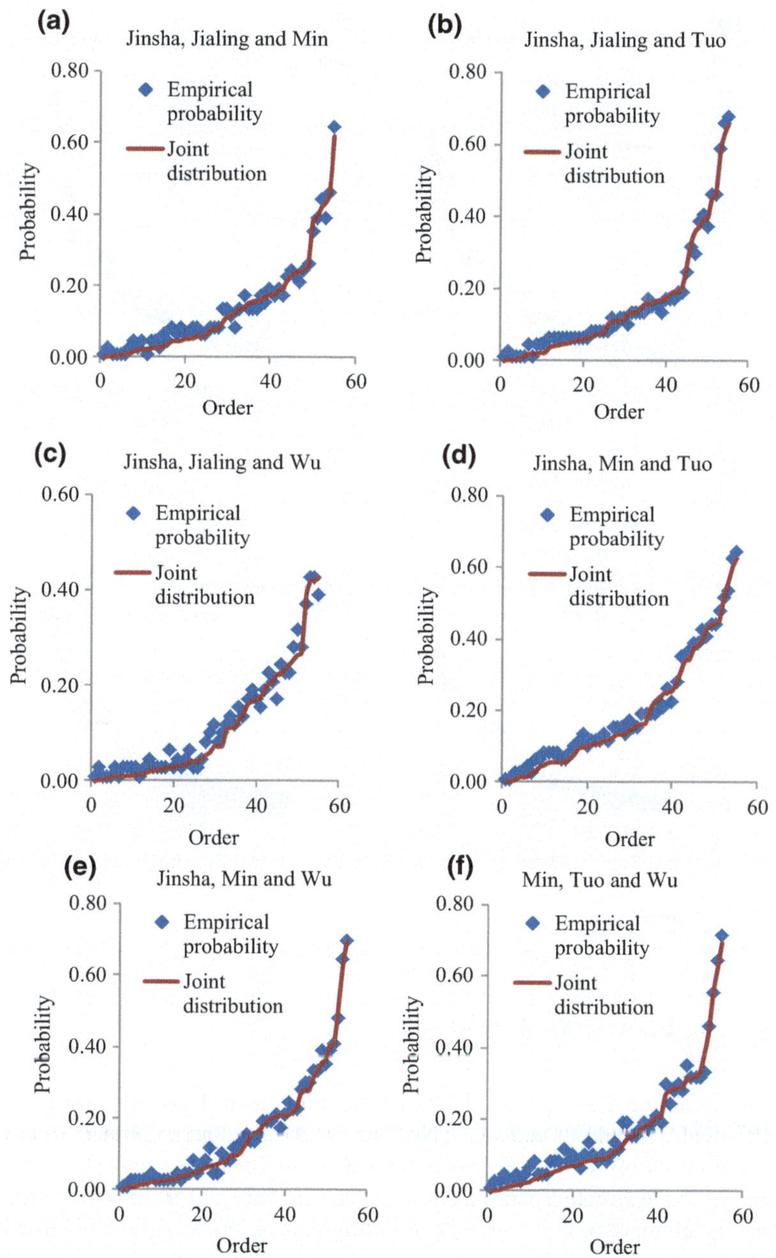

Fig. 11.2 Trivariate joint distribution and empirical probabilities of observed combinations

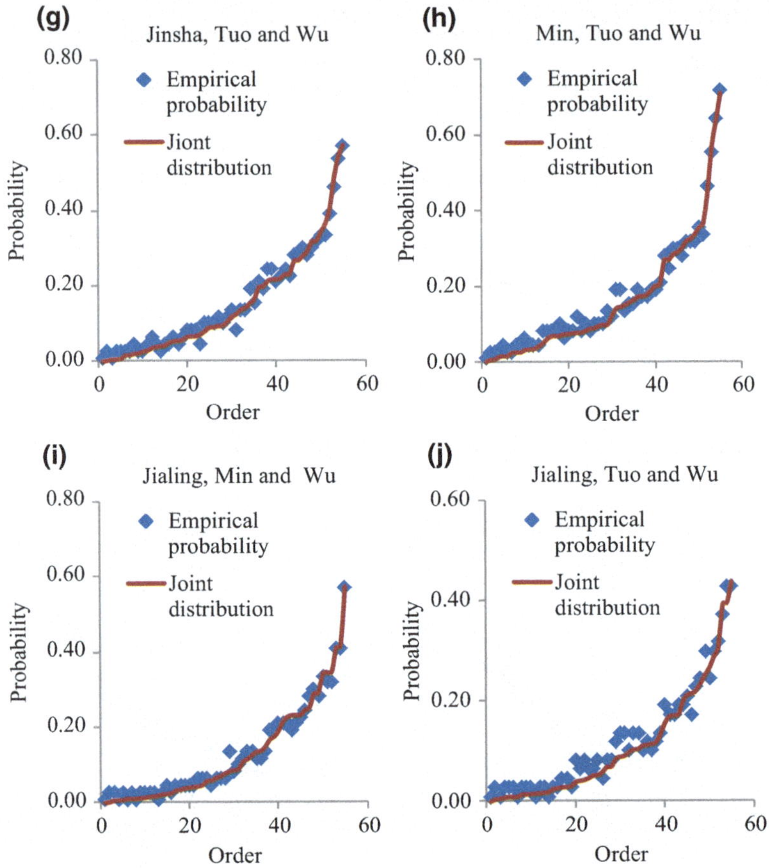

Fig. 11.2 (continued)

11.3.4 Multivariable Model

The four-variable copula is built. Due to the complicated dependence structure, the meta-elliptical copulas, namely Normal and Student copulas are used. A pseudo-likelihood technique involving the ranks of data is used for estimating parameters. The estimated parameters are given in Table 11.9, in which parameter ρ_i means the element of the Pearson correlation matrix. For example, the first line of Table 11.9 can be written as a matrix in the following:

$$\begin{bmatrix} 1 & -0.11 & 0.24 & 0.27 \\ -0.11 & 1 & 0.13 & 0.34 \\ 0.24 & 0.13 & 1 & 0.58 \\ 0.27 & 0.34 & 0.58 & 1 \end{bmatrix}$$

Similarly, the same meanings hold for others:

$$\begin{bmatrix} 1 & \rho_1 & \rho_2 & \rho_3 \\ \rho_1 & 1 & \rho_4 & \rho_5 \\ \rho_2 & \rho_4 & 1 & \rho_6 \\ \rho_3 & \rho_5 & \rho_6 & 1 \end{bmatrix}$$

The empirical and theoretical bivariate joint probabilities of four variables are shown in Fig. 11.3, which indicates that the theoretical curves fit the empirical probabilities well.

From Table 11.9, one can see the highest total correlation with a value of 0.36 exists for the combination of Jinsha, Min, Tuo and Jialing rivers. All of the four rivers are located on the left bank. Thus, some climatic factors may be similar. Due to the large dependence and average annual rainfall in Jinsha, Min, Jialing and Tuo River basins, the flows in the four rivers have an important impact on the flood occurrence in the upper Yangtze River and provide a bigger flood threat to the middle reach of the river.

The joint distribution of five rivers is built. The meta-elliptical copulas, namely Normal and Student copulas are used. The estimated parameters are listed in Table 11.10. The empirical and theoretical bivariate joint probabilities of five variables are shown in Fig. 11.4, which indicates that the theoretical curves fit the empirical probabilities well. The calculated total correlation value of the both Normal and Student copula is 0.39 which is larger than the four-variable total correlation. From this point of view, it is rational.

Table 11.9 Total correlation analysis of four-dimensional joint distribution

Number	Rivers				Copulas	ρ_1	ρ_2	ρ_3	ρ_4	ρ_5	ρ_6	υ	AIC	Total correlation
1	Jinsha	Jialing	Min	Tuo	Normal	-0.11	0.24	0.27	0.13	0.34	0.58		0.48	0.33
					t	-0.13	0.24	0.25	0.09	0.33	0.57	7.94	1.39	0.36
2	Jinsha	Jialing	Min	Wu	Normal	-0.13	0.23	0.22	0.12	-0.25	0.04		3.09	0.10
					t	-0.13	0.22	0.17	0.08	-0.24	0.05	11.24	3.91	0.10
3	Jinsha	Jialing	Tuo	Wu	Normal	-0.12	0.27	0.22	0.34	-0.25	-0.02		1.87	0.18
					t	-0.12	0.26	0.21	0.33	-0.25	-0.03	57.63	2.88	0.19
4	Jialing	Min	Tuo	Wu	Normal	0.13	0.34	-0.25	0.58	0.03	-0.03		0.68	0.31
					t	0.06	0.29	-0.22	0.56	0.04	-0.03	5.49	1.47	0.29
5	Jinsha	Min	Tuo	Wu	Normal	0.24	0.28	0.22	0.59	0.04	-0.01		0.87	0.29
					t	0.21	0.21	0.17	0.58	0.08	-0.04	5.81	1.68	0.31

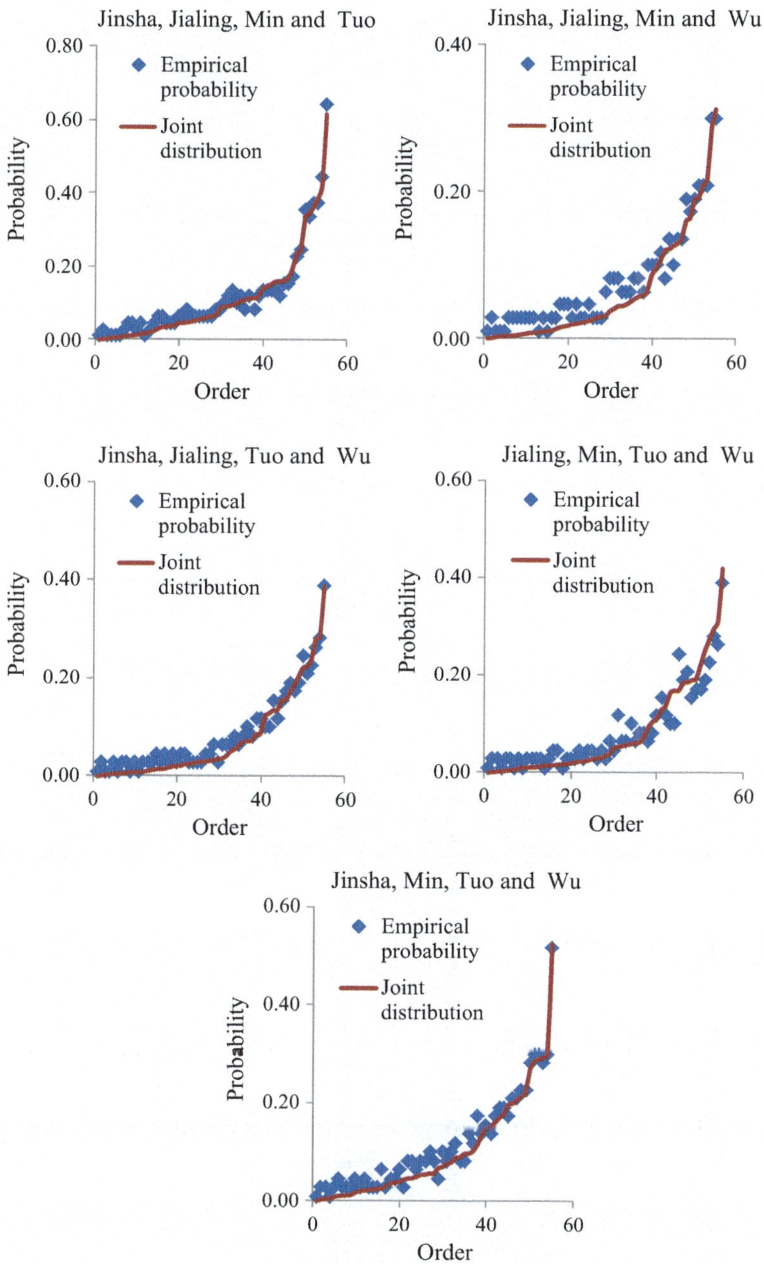

Fig. 11.3 Four-dimensional joint distribution and empirical probabilities of observed combinations

Table 11.10 Total correlation analysis of five-dimensional joint distribution

Copulas	ρ_1	ρ_2	ρ_3	ρ_4	ρ_5	ρ_6	ρ_7	ρ_8	ρ_9	ρ_{10}	υ	AIC	Total correlation
Normal	−0.11	0.24	0.27	0.22	0.13	0.33	−0.25	0.59	0.03	−0.02		4.21	0.39
Student	−0.13	0.23	0.25	0.17	0.09	0.31	−0.24	0.57	0.06	−0.03	8.86	5.11	0.39

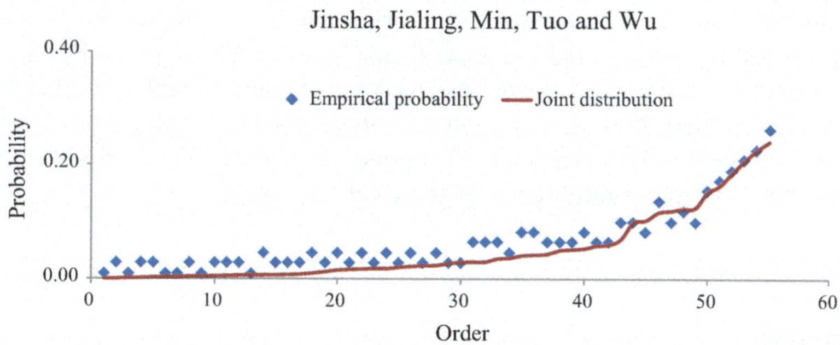

Fig. 11.4 Five-dimensional joint distribution and empirical probabilities of the observed combination

11.4 Conclusions

This chapter analyzes the dependence among the five rivers in the upper Yangtze River. Copula entropy, which is constructed by the copula and entropy theory, is first introduced in the hydrological field. Because the non-linear correlation structure and multivariate variables involved, the total correlation method is calculated based on the copula-entropy model. Using two-step algorithm, first, the copula function is built with the parameter estimation. Second, the total correlation values are obtained from the copula entropy. The conclusions are given as follows:

(1) Both the Archimedean and meta-elliptical copulas are applied to build bivariate and trivariate joint distributions. Generally, the Archimedean copula gives a better fit for the lower dimension cases. For higher dimensions, due to the complicated dependence structure, only the metaelliptical copula is used. All of the built joint distributions fit the empirical probabilities well.
(2) The copula entropy can measure the linear and non-linear dependencies based on information theory and copula function. It makes no assumptions about the marginal distributions and can be used for higher dimensions. Furthermore, the proposed method only needs to calculate the copula entropy instead of the marginal or joint entropy, which estimates the total correlation more directly and avoids the accumulation of systematic bias.
(3) Multiple integration and the Monte Carlo methods are used to obtain the total correlation values, and both methods lead to similar results. For a specific case, several kinds of copulas are employed. There is a significant difference in total correlation values when using different copula functions. Therefore, it is important to select an appropriate copula for the dependence estimation.
(4) Application results indicate that the total correlations among rivers are not so large, which is in accordance with by the climatic characteristics of the study area. The largest total correlation is 0.33 between Min and Tuo Rivers because the distance between the two rivers is the shortest and they belong to the same rainfall zone. There are some dependencies among Jinsha, Min and Tuo Rivers.

In normal years, rainfall on the left and right banks of Yangtze River does not happen simultaneously, but the dependence between Min and Wu River cannot be neglected. Due to the large dependence and average annual rainfall in Jinsha, Min, Jialing and Tuo Rivers, flows in these four rivers have an important impact on the flood occurrence in the upper Yangtze River and provide a bigger flood threat to the middle reach of the river.

References

Alfonso L, Lobbrecht A, Price R (2010) Information theory-based approach for location of monitoring water level gauges in polders. Water Resour Res 46:W03528. https://doi.org/10.1029/2009WR008101

Berntson J, Espelid TO, Genz A (1991) An adaptive algorithm for the approximate calculation of multiple integrals. ACM Trans Math Softw 17:437–451

Chen L, Singh VP, Guo S (2013) Measures of correlations for rivers flows based on copula entropy method. J Hydrol Eng 18(12):1591–1606

Chow CK, Liu CN (1968) Approximating discrete probability distributions with dependence trees. IEEE Trans Inform Technol 14(3):462–467

Garner WR (1962) Uncertainty and structure as psychological concepts. Wiley, New York

Genest C, Verret F (2005) Locally most powerful rank tests of independence for copula models. J Nonparametric Stat 17:521–539

Genest C, Rémillard B, Beaudoin D (2009) Goodness-of-fit tests for copulas: a review and a power study. Insur Math Econ 44:199–213

Kirshner S, Smyth P, Robertson AW (2004) Conditional Chow–Liu tree structures for modeling discrete-valued vector time series. In: Chickering M, Halpern J (eds) Proceedings of the twentieth conference on uncertainty in artificial intelligence UAI-04, AUAI Press, pp 317–324

Lewis PM (1959) Approximating probability distributions to reduce storage requirement. Inform Control 2:214–225

Ma J, Sun Z (2008) Mutual information is copula entropy. Tsinghua Sci Technol 16(1):51–54

McGill WJ (1954) Multivariate information transmission. Psychometrika 19:97–116

Studený M, Vejnarová J (1999) The multi-information function as a tool for measuring stochastic dependence. In: Jordan MI (ed) Learning in graphical models. MIT Press, Cambridge, MA, pp 261–296

Watanabe S (1960) Information theoretical analysis of multivariate correlation. IBM J Res Dev 4:66–82

Zhang L, Singh VP (2006) Bivariate flood frequency analysis using the copula method. J Hydrol Eng 11(2):150–164

Zhao N, Linb WT (2011) A copula entropy approach to correlation measurement at the country level. Appl Math Comput 218(2):628–642

The manufacturer's authorised representative in the EU is Springer Nature Customer Service Centre GmbH, Europaplatz 3, 69115 Heidelberg, Germany. If you have any concerns regarding our products, please contact ProductSafety@springernature.com

Printed and bound by CPI Group (UK) Ltd, Croydon, CR0 4YY
25/09/2026
02222900-0014